Artificial Intelligence in Process Engineering

Artificial Intelligence in Process Engineering

Edited by

MICHAEL L. MAVROVOUNIOTIS

Chemical Engineering and Systems Research Center
University of Maryland
College Park, Maryland

ACADEMIC PRESS, INC.

Harcourt Brace Jovanovich, Publishers
Boston San Diego New York
London Sydney Tokyo Toronto

This book is printed on acid-free paper. ♾

ACADEMIC PRESS, INC.
1250 Sixth Avenue, San Diego, CA 92101

United Kingdom Edition published by
ACADEMIC PRESS LIMITED
24–28 Oval Road, London NW1 7DX

Library of Congress Cataloging-in-Publication Data

Artificial intelligence in process engineering / edited by Michael L. Mavrovouniotis.
p. cm.
ISBN 0-12-480575-2
1. Chemical process control—Data processing. 2. Artificial intelligence. I. Mavrovouniotis, Michael L.
TP155.75.A78 1990
660'.281563—dc20 89-18269
CIP

Printed in the United States of America
90 91 92 93 9 8 7 6 5 4 3 2 1

Contents

Contributors

Numbers in parentheses refer to the pages on which the authors' contributions begin.

Bryan D. Basham (37), *Systems Development and Simulation Division Engineering Directorate—EF5, NASA Lyndon B. Johnson Space Center, Houston, Texas 77058*

D. T. Dalle Molle (1), *Amoco Chemical Co., Mailstation C2, P.O. Box 400 Naperville, Illinois 60566*

A. D'Ambrosio (189), *APEIRON, 1147 East Broadway, Suite 469, Glendale, California 91205*

James F. Davis (81), *Department of Chemical Engineering, Ohio State University, Columbus, Ohio 43210*

T. F. Edgar (1), *Department of Chemical Engineering, University of Texas at Austin, Austin, Texas 78712–1062*

L. T. Fan (311), *Department of Chemical Engineering, Kansas State University, Manhattan, Kansas 66506*

Paul E. Felix (223), *Honeywell Incorporated, Sensor and System Development Center, 1000 Boone Avenue North, Golden Valley, Minnesota 55427*

Hector García (161), *Departmento de Simulación, Instituto de Investigaciones Eléctricas, Cuernavaca, Morelos, México*

Richard A. Harris (37), *BP Microsystems, Houston, Texas 77043*

David M. Himmelblau (123), *Department of Chemical Engineering, University of Texas at Austin, Austin, Texas 78579*

Josiah C. Hoskins (123), *Schlumberger Laboratory for Computer Science, 8311 North, P.O. Box 200015, RR620, Austin, Texas 78720–0015*

Y. W. Huang (311), *Department of Chemical Engineering, Kansas State University, Manhattan, Kansas 66506*

Charles E. Hurley II (81), *Westvaco, Bleached Board Division, Covington, Virginia 24426*

A. F. Konar (223), *Honeywell Incorporated, Sensor and System Development Center, 1000 Boone Avenue North, Golden Valley, Minnesota 55427*

Ram Lavie (267), *Department of Chemical Engineering, Technion, I.I.T., Haifa 32000, Israel*

Daniel R. Lewin (267), *Department of Chemical Engineering, Technion, I.I.T., Haifa 32000, Israel*

Jane T. Malin (37), *Systems Development and Simulation Division Engineering Directorate—EF5, NASA—Lyndon B. Johnson Space Center, Houston, Texas 77058*

Eduardo Morales (161), *The Turing Institute, George House, 36 North Hanover Street, Glasgow G1 2AD, Scotland*

Douglas R. Myers (81), *Department of Chemical Engineering, Ohio State University, Columbus, Ohio 43210*

Charles Siletti (295), *Mobil Technical Center, P.O. Box 1026, Princeton, New Jersey 08543–1026*

B. M. Thuraisingham (223), *Honeywell Incorporated, Sensor and System Development Center, 1000 Boone Avenue North, Golden Valley, Minnesota 55427*

Preface

The tremendous interest in Artificial Intelligence applications that has flourished within the engineering community since the 1970s did not infiltrate chemical and biochemical process engineering until the mid 1980s. One bright exception worth mentioning is the visionary work of Powers, Siirola, and Rudd, who in the 1960s developed the AIDES system for process synthesis. Excluding this system, the early efforts for AI applications in process engineering entailed, quite predictably, the development of rule-based expert systems. While many of these applications were small and trivial in their technical accomplishments, they were nonetheless a useful vehicle for acquainting researchers and practitioners of computer-aided process engineering with AI. These applications successfully prompted the process engineering community to take a more serious interest and look deeper into the wealth of AI, beyond simple expert systems.

The flourishing of AI methods within the process engineering community has recently been augmented by an active interest from the Artificial Intelligence research community in the field of chemical and biochemical engineering, which is now viewed as an exciting domain for exploration of AI concepts for real-world applications. Thus, process engineering has the potential to become one of the driving forces for the development of AI methods.

The tasks to which AI methods are applied include design of process systems (such as reactions pathways, separation systems, and process flowsheets), planning of process operations, diagnosis of process faults, analysis of process trends, qualitative process simulation, etc.

The characteristics of chemical and biochemical processing systems make them an especially challenging domain for applications of Artificial Intelligence methods. First, the physical systems in question share a networklike structure with analog or digital circuits, but their operation is, in general, much more complicated. For example, the operation of a flash drum or a distillation column depends strongly on nonlinear thermodynamic properties of mixtures. Second, information available on the systems is often qualitative in nature. Challenges arise in the representation of available knowledge at different qualitative levels. Furthermore, one never has all the information relevant to carrying out a particular task. One must use partial information aggressively to arrive at partial results, which are valuable in determining what other information should be sought. Finally, many tasks (notably fault diagnosis) need to be carried out online, in real time.

The aim of this book is to present a diverse sample of Artificial Intelligence applications in process engineering, to help both the AI and Process Engineering communities evaluate the current state of the field and explore future directions. The editor hopes that, by reading this book, the AI researcher and practitioner can appreciate the value and challenge of process engineering problems, while the process engineer can realize the value of AI methods. Both communities can benefit by receiving guidance, through the examples provided here, in creating successful combinations of problems and methods.

Perhaps the most important criteria in the selection of the contributions for this book were educational value and diversity of AI methods and process engineering application domains. To achieve inclusion of a variety of AI techniques and concepts, the coverage of many important and mature streams of research had to be limited. For example, qualitative reasoning has been applied extensively and in depth by several research groups, and only a small fraction of that work could be chosen for this volume. Diversity was also pursued with respect to research groups: Each of the contributed chapters originates from a different research group.

To date, most AI applications in process engineering carry out fault diagnosis, a task that is well suited to AI approaches. Our understanding of process fault diagnosis is in an intermediate state, as we possess some of the necessary knowledge and rough methods but seldom have well defined algorithmic techniques. In addition, the relevance of symbolic and qualitative information has been established. For these reasons, the use of AI methodologies for carrying out process fault diagnosis is well justified. Other process engineering tasks tend to be either too easy (routine design, such as equipment sizing), too difficult (creative design of a new process), or too numerical in nature (optimizing control of continuous processes). The dominance of process fault diagnosis as an application field is evident in this book: Chapters 1 through 6 entail diagnosis applications; Chapters 7 and 8 involve process control; and Chapters 9 and 10 process design applications. It is interesting to note in the fault diagnosis chapters that the AI techniques employed are quite diverse. A brief overview of the contents of this book follows.

In Chapter 1 *(Qualitative Modeling of Chemical Reaction Systems)*, Dalle Molle and Edgar discuss the use of qualitative reasoning for modeling and simulation of chemical systems. Qualitative reasoning entails the use of inequality relationships and the signs of parameters as qualitative values. In dynamic systems, the direction of change of each parameter is also represented by a qualitative value. Qualitative values are related through qualitative constraints, derived empirically or from algebraic quantitative models. The qualitative simulation technique is broadly applicable and has been extensively used within process engineering, mostly for fault diagnosis tasks. The detailed examples in this chapter illustrate very effectively the power (and limitations) of qualitative reasoning. Malin, Basham, and Harris expand on the qualitative simulation theme in Chapter 2 *(Use of Qualitative Models in Discrete Event Simulation to Analyze Malfunctions in Processing Systems)* with a discussion of CONFIG, a fault diagnosis tool that can handle systems with time delays and features graphic construction of system models.

The fault diagnosis theme carries on in Chapter 3 *(An Expert System for Diagnosis of a Sequential, PLC-Controlled Operation)*, where Myers, Davis, and Hurley discuss the diagnosis of faults in processes that are controlled by Programmable Logic Controllers. Unlike qualitative simulation, which gives a flavor of logic to parameters and constraints that are fundamentally quantitative, this chapter deals with inherently logical (Boolean) relationships and values.

In Chapter 4 *(Fault Detection and Diagnosis Using Artificial Neural Networks)*, Hoskins and Himmelblau carry out fault diagnosis using the technique of neural networks, which are parallel computational structures consisting of large numbers of interconnected processing elements. Neural networks are usually trained through examples. Interest in neural networks is on a sharp rise, after almost two decades of dormancy. One of the advantages of neural networks in fault diagnosis is that they offer an easy avenue for using quantitative information to reach qualitative or logical conclusions.

The issue of quantitative versus qualitative information is debated in Chapter 5 *(A Modular Approach to Multiple Faults Diagnosis)* by Morales and García. This chapter provides an interesting discussion of a system that performs constraint propagation through qualitative and quantitative constraints, with a divide-and-conquer strategy, aimed at the detection of multiple faults. The debate of quantitative versus qualitative knowledge is continued in Chapter 6 *(Modeling Real-World Processes: Deep and Shallow Knowledge Integrated with Approximate Reasoning in a Diagnostic Expert System)*, by D'Ambrosio. This chapter also focuses on the comparison and integration of deep knowledge, such as mathematical models, and shallow knowledge, such as experiential heuristics.

Chapter 7 *(XIMKON—An Expert Simulation and Control Program)*, by Konar, Thuraisingham, and Felix, introduces us to the incorporation of AI systems as coordinators and front ends for existing, traditional software modules—in this case, FORTRAN programs. This kind of application can have an immediate impact, because it builds on an extensive body of previous work—which may have limitations but cannot be replaced quickly. The enhancement of the applicability and ease of use of such large information bases is one avenue for AI to infiltrate and compete with traditional problem-solving and computing styles. XIMKON focuses on the control of continuous processes. Lewin and Lavie balance the process control picture with a discussion of the control of batch processes in Chapter 8 *(Exothermic Batch Chemical Reactor Automation via Expert System)*. The system described in this chapter is based on the gradual abstration of process features and their utilization in an adaptive fashion.

Moving on to process design, Siletti discusses in Chapter 9 *(Design of Protein Purification Processes by Heuristic Search)* a system that synthesizes bioseparation processes based on product properties and design objectives. This rule-based system initially constructs abstract processes that describe the basic purification steps that must be used; it then refines this process description and completes the design by selecting specific process equipment. Chapter 10 *(An Adaptive Heuristic-Based System for Synthesis of Complex Separation Sequences)*, by Huang and Fan, also involves process design, but in the domain of chemical (rather than biochemical) systems. The synthesis is carried out in an evolutionary manner by rule bases, which can adapt by adjusting the priorities of the heuristics employed.

This volume represents the first collection of AI applications in process engineering to be published in the form of a book. As additional reading material, the editor recommends a special issue of the journal *Computers and Chemical Engineering* (Volume 12, Issue 9/10, 1988), which contains a set of technical papers with very little overlap with the contributions in this book.

Michael L. Mavrovouniotis
College Park, Maryland

1 Qualitative Modeling of Chemical Reaction Systems

D. T. DALLE MOLLE[1] AND
T. F. EDGAR[2]

Department of Chemical Engineering
University of Texas at Austin
Austin, Texas

Abstract

Developing accurate models of chemical reactors is important for process design and model-based monitoring and control. While numerical methods for analyzing kinetic mechanisms and reactor models exist, these methods are unsuitable when parameter values and functional relationships are unknown or poorly known. Qualitative simulation is a promising technique for reasoning about dynamic systems with incomplete knowledge because it generates all possible behaviors of a system from qualitative descriptions of the physical system. In this work, qualitative models for several different kinetic mechanisms are developed. These models have the ability to generate qualitative descriptions of the dynamic behavior of the components in the reaction mechanism. Two different types of chemical reactors are also modeled qualitatively. A parallel-plate radial flow reactor is developed that predicts the radial variations at steady state of the components in a simplified plasma etcher model. An adiabatic CSTR is modeled illustrating some of the many behaviors possible for this type of reactor. In all cases, the models presented in this work capture all the possible qualitative behaviors that can be exhibited by a corresponding quantitative model. When quantitative information is included in some of the models, the number of possibe behaviors is reduced and the resulting behaviors are more descriptive, providing numerical ranges on the new steady states.

[1] Present address, Amoco Chemical Company, Naperville, Illinois.
[2] Author to whom all correspondence should be addressed.

ISBN 0-12-480575-2

1. Introduction

Recently, there has been considerable interest in methods for analyzing physical processes qualitatively. The goal of these qualitative reasoning methods is to provide alternative ways for understanding, representing, and analyzing the function and behavior of physical systems when faced with an incomplete description of the real system. A variety of techniques for qualitative reasoning has appeared in both the artificial intelligence and engineering literature. Much of the recent work in qualitative reasoning has been based on the previous work of de Kleer and Brown (1984), Forbus (1984), and Kuipers (1984). For example, Oyeleye and Kramer (1988) have combined the use of signed directed graphs with confluences (de Kleer and Brown, 1984) to analyze qualitatively the steady-state effects of faults in chemical process plants. Dalle Molle *et al.* (1988) used qualitative simulation (Kuipers, 1984, 1986) to predict the dynamic behavior of several systems from chemical engineering.

Other recent research efforts have developed new approaches to qualitative reasoning not aligned with any other technique. Sacks (1988) has developed a program for analyzing the phase-plane behavior of nonlinear systems described by ordinary differential equations using piecewise linear abstraction. Mavrovouniotis and Stephanopoulos (1988) present a formal analysis of order-of-magnitude reasoning and have applied their approach to reason qualitatively about chemical and biochemical engineering systems. Williams (1988) has developed a qualitative symbolic algebra for reasoning with algebraic equations. This approach overcomes many of the weaknesses exhibited by methods that reason with confluence equations. Iwasaki and Bhandari (1988) presented an approach for abstracting systems with a large number of variables into subsystems that can be treated independently. The behavior of the overall system is described in terms of the subsystems. However, the method is based on a linear system representation, which may not always be appropriate for a given physical system.

Chemical reactor design and control require a reasonable understanding of the kinetic mechanisms through which the reactants are converted to products. While a complete fundamental mechanism is desirable, approximate models are often adequate for control and design purposes. Filippi *et al.* (1986) and Filippi-Bosey *et al.* (1989) describe a method called *tendency modeling* for determining approximate reaction networks that can be used for reactor design and analysis. The method requires a variety of concentration and temperature data that are analyzed to determine whether to accept or refute a postulated reaction step. This *gray-box* approach uses a mixture of analytical tools and heuristics. Recently, Bonvin and Rippin (1988) presented an approach for deriving or testing stoichiometric equations for complex reaction systems based on various methods of linear analysis. The method includes ways to incorporate previously known reaction steps but was found to be sensitive to noisy data. Mavrovouniotis and Stephanopoulos (1988) used order-of-magnitude reasoning to simplify proposed biochemical reaction mechanisms.

This work is concerned with providing qualitative dynamic descriptions of chemical reaction systems. In particular, the dynamic behavior of several kinetic mechanisms and reactor models is analyzed using qualitative simulation as described by Kuipers (1986). While other methods of qualitative reasoning may also be applicable to this type of analysis, qualitative simulation is used because it provides dynamic descriptions of systems of any complexity. The methods proposed by Oyeleye and Kramer (1988) and Williams (1988) are limited to steady-state, algebraic analyses. The work of Sacks (1988) requires complete knowledge of parameter values and functional relationships, and is restricted to second-order systems. Qualitative simulation allows for representing systems with incomplete knowledge in terms of Qualitative Differential Equations (QDEs) and is guaranteed to capture all possible dynamic descriptions of the system.

2. The QSIM Algorithm for Qualitative Simulation

2.1. The Basic QSIM Algorithm

The QSIM algorithm by Kuipers (1986) provides a framework for developing qualitative models of physical mechanisms. The algorithm is an abstraction of Ordinary Differential Equations (ODEs) and allows for representing incomplete knowledge of functional relationships and parameter values in the form of Qualitative Differential Equations (QDEs). The QSIM algorithm has been used to model the dynamic behavior of a variety of physical processes including mass-and-spring systems (Lee and Kuipers, 1988), certain physiological mechanisms (Kuipers 1987), fluid flow through tanks (Chiu, 1988), and mixing tanks (Dalle Molle *et al.*, 1988). The generality of the algorithm makes it a suitable choice for modeling qualitatively complex chemical reaction processes.

The QSIM algorithm is described in detail by Kuipers (1986); only a summary will be presented here. Within the QSIM framework, a qualitative model consists of a set of variables (or parameters) of the system and constraints that relate the variables to one another. Variables have quantity spaces with landmark (qualitatively significant) values that represent the range of values a variable may have. The qualitative state of a variable is a list of the qualitative value (at or between landmark values) and the qualitative derivative (increasing, decreasing, or steady) of the variable. The constraints represent qualitative versions of common mathematical operations such as addition, multiplication, and differentiation. Given an incomplete set of variable states and a set of constraints, QSIM determines all possible completions of the state that are consistent with the constraints.

In determining the dynamic behavior of a system, QSIM uses a strategy of proposing and filtering sets of qualitative transitions to predict all possible subsequent states for the following time point or open interval. Inconsistent states are detected by application of continuity and model constraints. Global filters identify states as inconsistent, quiescent, divergent, or identical to a previous state (indicating oscillatory behavior); the remaining states are again subjected to the propose/filter procedure. A branching tree of states results, and each path through the tree constitutes a possible behavior of the system.

Kuipers (1986) has proven that the set of possible behaviors generated by the QSIM algorithm is guaranteed to include all the behaviors exhibited by the original ODE from which the QDE was derived. While this approach is sound, it is not complete. The algorithm cannot be guaranteed to recognize and eliminate all spurious behaviors (predictions that represent no real physical system behavior). Recent developments in qualitative reasoning have been aimed at reducing the number of spurious predictions in qualitative simulation. Some of these results follow.

2.2. Recent Extensions of the QSIM Algorithm

While the QSIM algorithm as presented by Kuipers (1986) provides a complete language for developing and simulating qualitative models, many situations require additional qualitative reasoning techniques to reduce or elimiate spurious behaviors. The most common source of spurious behaviors is ambiguity in the derivative of complex expressions or variables such as the highest-order derivative in a model. For example, if $z = xy$, and x is increasing at the same time y is decreasing, then the direction of charge of z can be increasing, steady, or decreasing if constrained only by this constraint. While derivatives of variables are constrained by continuity and model constraints, they do not have the more complete representation of quantity spaces and landmark values. Kuipers and Chiu (1987) presented two methods for taming intractable branching caused by under-constrained derivatives. One method changes the level of description to aggregate a number of behaviors whose distinctions are real but are not considered important. This approach amounts to ignoring the direction of change of a given variable. Their second approach, discussed in detail by Chiu (1988), places additional constraints on the derivatives of variables that may otherwise be constrained only by continuity. These second-derivative, or *curvature constraints*, are applied when the derivative of a variable is zero to verify or refute curvatures proposed by the QSIM algorithm. In many cases, curvature constraints can be derived automatically from the original set of constraints. In other cases, manually deriving curvature constraints is necessary and may also lead to constraint expressions with a minimal amount of potential ambiguity. An example of this situation appears later in this chapter. Recent work by Dalle Molle (1989a) extended this approach to evaluate automatically the value of the third-order derivative in cases where the curvature constraint vanishes. Both curvature and third-order constraints are required by many of the models in this work. These various methods for taming otherwise intractable branching were summarized by Kuipers *et al.* (1989).

Another manifestation of the loss of knowledge that occurs when abstracting to qualitative models is the absence of phase relationships between variables in systems that exhibit oscillatory behavior. Recent work by Lee and Kuipers (1988) and Struss (1988) identified system independent global filters for qualitative simulation derived from constraints that exist on the trajectories of variables in the phase plane. Lee and Kuipers (1988) showed how the Non-Intersection of phase-plane Constraint (NIC) can eliminate large classes of spurious behaviors in mass-and-spring systems. Dalle Molle and Edgar (1989) used the NIC filter to achieve a tractable qualitative simulation of a PI controller on a first-order system.

While purely qualitative models are useful for identifying the possible solutions of a given QDE from a particular initialization, the resulting set of behaviors may be too large to be manageable or the descriptions not detailed enough for the purpose of the model. In many situations, functional relationships are known to some degree, and parameter values may be known with some uncertainty. In such cases, the modeler usually wants to incorporate this partial quantitative knowledge into the qualitative model. Kuipers and Berleant (1988) have presented a method for representing incomplete quantitative knowledge into QSIM-type qualitative models. Unknown functions are represented with upper and lower envelopes, and parameters can have ranges or precise numerical values. With this additional knowledge, qualitative states with inconsistent quantitative values can be eliminated, behaviors with quantitatively inconsistent states can be pruned, and models that produce only inconsistent behaviors can be refuted. The elimination of inconsistent models by using incomplete quantitative knowledge is discussed further by Dvorak and Kuipers (1989). Examples of qualitative models with partial quantitative knowledge are discussed later in this chapter.

3. Building Qualitative Models of Reaction Systems

The process of designing chemical reaction systems consists of two fundamental steps. In the first part of the design, the chemist or engineer develops a model that describes at a molecular level the pathway or mechanism by which the reactants are converted through various intermediates into the desired (and undesired) products. Rate equations for elementary reaction steps can be derived directly from the physical mechanisms involved (i.e., two-body collisions). For more complicated mechanisms involving adsorption or diffusion, rate equations are assumed that approximate the actual physical processes and that are consistent with experimental data. When a kinetic model has been established, the engineer develops a model for the reactor based on the kinetic model that takes into account the effects of the physical system (size and shape of the reactor, temperature and pressure conditions, etc.) and the effects of fluid flow and heat and mass transfer. Reactor models often characterize temperature and pressure variations in addition to changes in concentrations. In both reactor and kinetic model development, functional relationships between process variables and the parameters for these relationships are assumed to be known in order to solve a quantitative model. In the next two sections, qualitative kinetic and reactor models are developed and stimulated without requiring detailed knowledge of process parameters or functional relationships. While qualitative descriptions are somewhat weaker than quantitative model predictions, qualitative models capture all possible behaviors of a model for all possible values of the unknown parameters.

3.1. *Qualitative Analysis of Kinetic Models*

Developing useful reactor models relies on identifying the proper kinetic mechanism of the reaction system. A common experimental technique for identifying kinetic mechanisms is to perform studies on batch reactions where the reactants are mixed together and held at constant conditions (i.e., temperature and pressure). The qualitative trends of concentration versus time plots can provide insight into the series of individual reactions that take place in converting the reactants to the final products. Detailed regression analyses can also be performed on the experimental data to verify the proposed reaction pathway and to determine the reaction order of each step, as well as to provide estimates of the coefficients of the rate expressions. However, these analyses often require large amounts of data from experiments under controlled conditions. In this work, the qualitative trends of concentration-versus-time profiles for several common reaction mechanisms are predicted without requiring detailed knowledge of process parameters.

3.1.1. Reactions in Series

In the production of many chemicals, several reactions often occur in series. In the simplest case, a reactant is converted in one step to some product. In more complicated schemes, a reactant may go through one or more intermediates before reaching its final state. A generic representation of this type of reaction pathway is

$$\mathrm{A} \xrightarrow{k_1} \mathrm{B} \xrightarrow{k_2} \mathrm{C} \xrightarrow{k_3} \mathrm{D} \xrightarrow{k_4} \mathrm{E}.$$

This pathway consists of four irreversible first-order reaction in series. If the reaction conditions are assumed to be isothermal with constant density, then the following equations can be derived from component mass balances:

$$\frac{dC_\mathrm{A}}{dt} = -k_1 C_\mathrm{A} \tag{1}$$

$$\frac{dC_\mathrm{B}}{dt} = k_1 C_\mathrm{A} - k_2 C_\mathrm{B} \tag{2}$$

$$\frac{dC_\mathrm{C}}{dt} = k_2 C_\mathrm{B} - k_3 C_\mathrm{C} \tag{3}$$

$$\frac{dC_\mathrm{D}}{dt} = k_3 C_\mathrm{C} - k_4 C_\mathrm{D} \tag{4}$$

$$\frac{dC_\mathrm{E}}{dt} = k_4 C_\mathrm{D}, \tag{5}$$

where k_i ($i = 1, 2, 3, 4$) are the reaction rate constants. In the examples that follow, only A is present in the concentration $C_{\mathrm{A}0}$ at $t = 0$.

Two Reactions in Series. Qualitative models can be developed for the cases of one, two, three, or four reactions occurring in series by directly translating the ODE to QDEs. When only one reaction occurs, the qualitative simulation results are similar to those of other first-order systems studied previously (Dalle Molle *et al.*, 1988). They will not be discussed here. For the case of two reactions occuring (A → B → C), a quantitative model could consist of Equations 1, 2, and 3 ($k_3 = 0$). Mathematically, this two-reaction system could also be represented by equations 1 and 2, and an overall mass balance of the form $C_\mathrm{A} + C_\mathrm{B} = C_{\mathrm{A}0}$ (or by two other sets of equations). However, in the qualitative model, it is necessary to use equations 1, 2, and 3 (to constrain properly the derivatives of the concentrations) *and* the overall mass balance (to constrain properly the values of the concentrations). The use of redundant constraints in qualitative simulation has been proven useful in other systems (Dalle Molle *et al.*, 1988). Oyeleye and Kramer (1988) refer to redundant constraints as latent constraints and have found them essential in determining the qualitative steady-state behavior of chemical processes.

The qualitative model based on Equations 1, 2, and 3 and the overall mass balance generates one possible behavior for the initialization described earlier. Figure 1 shows the qualitative plots for the concentrations and their derivatives. Figure 1 captures several qualitatively important features for this system. First, C_A decreases monotonically from C_{A0} to 0. Second, C_B increases from 0 to a maximum at T1, then decreases to 0. Finally, C_C increases monotonically from 0, with 0 initial derivative (positive second derivative), to a final concentration of C_{A0}. In addition to Equations 1, 2, and 3, this simulation also required the use of curvature constraints, which were derived automatically from the constraints of the model, to eliminate otherwise intractable branching in dC_B/dt and dC_C/dt.

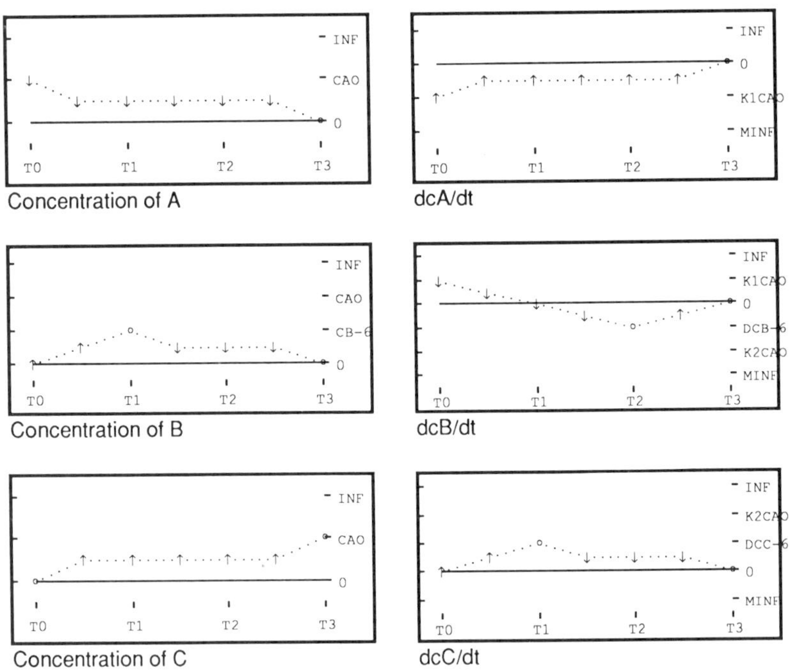

Figure 1. Unique qualitative behavior of two reactions in series (A → B → C).

Three Reactions in Series. For the case where three reactions occur in series (A → B → C → D), the qualitative model must also be developed from the four component mass balances and the overall mass balance. This system also requires the use of curvature constraints to eliminate a major source of spurious behaviors. When the qualitative model for this system is simulated from the initial condition of only A present at $t = 0$, five behaviors are generated. Inspection of the analytical solution to Equations 1, 2, and 3 (tractable in this case) reveals that two of the five behaviors predicted are spurious. One of these spurious behaviors is shown in Figure 2. The spurious nature of this behavior occurs at time

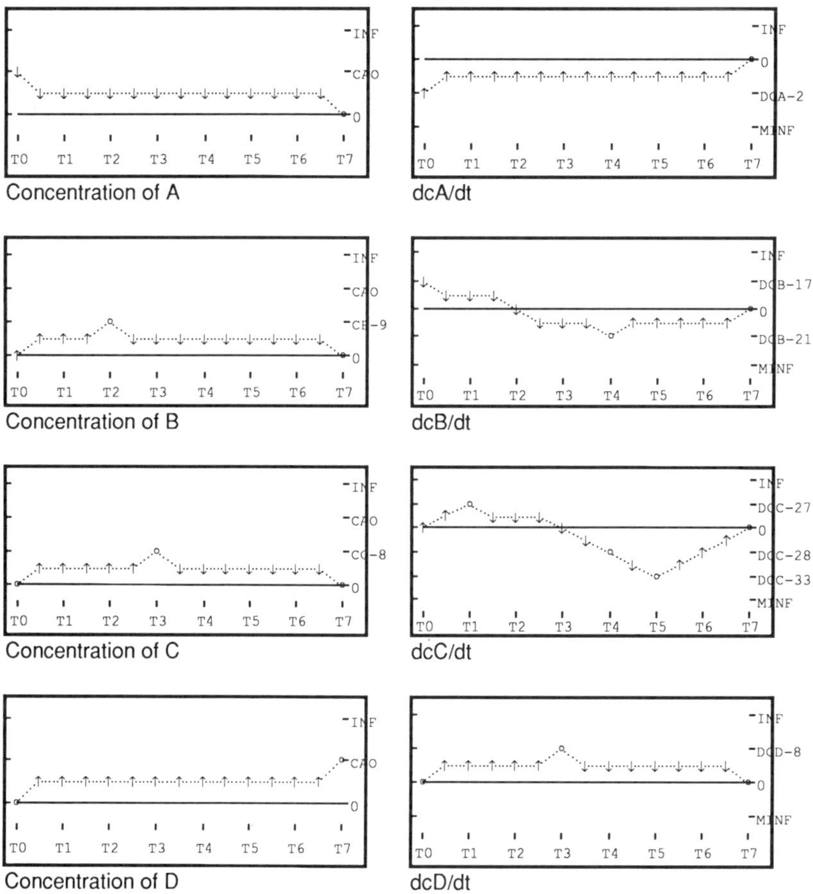

Figure 2. A spurious prediction for three reactions in series (A → B → C → D). Without third-order constraint evaluation, dC_C/dt has a spurious inflection.

point T4, where dC_C/dt is steady, then continues in a negative direction. This behavior is allowed by the qualitative simulation because the curvature (second derivative) of dC_C/dt vanishes and must be determined by the next-highest nonzero derivative, which is currently unknown. Dalle Molle (1989a) showed how the value of the third-order derivative can be determined directly from the constraints and variable values without any additional modeling effort. When the automatic third-order constraint evaluation technique is used in the simulation, only the three real behaviors are predicted.

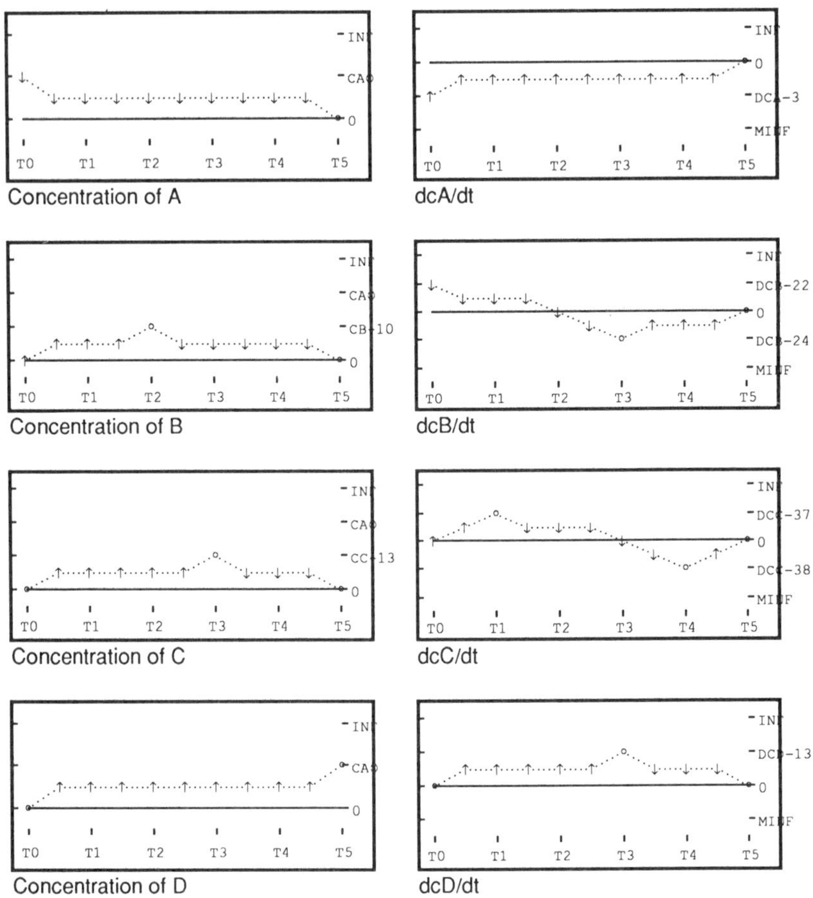

Figure 3. One of three behaviors for three reactions in series (A → B → C → D) with third-order constraint evaluation. In this behavior, the critical points $dC_C/dt = 0$ and $d^2C_B/dt^2 = 0$ occur simultaneously. Two other behaviors (not shown) have these critical points occurring one before the other.

Figure 3 shows one of the three real behaviors predicted by the qualitative model for this system. This result also captures the same important qualitative features of the system behavior as discussed earlier for the two-reaction case. In addition, the correct trajectory for C_D (first- and second-order derivatives equal to zero initially, its value increasing monotonically from 0 to C_{A0}) is also captured in the simulation. The other two real behaviors for this system are similar to those in Figure 3, except that the critical points $d^2C_B/dt^2 = 0$ and $dC_C/dt = 0$ do not occur at the same time. Either one of these two critical points can occur before the other, giving a total of three behaviors. This phenomenon is referred to as *occurrence branching.* An analysis (not shown here) of the system equations reveals that the two critical points $d^2C_B/dt^2 = 0$ and $dC_C/dt = 0$ occur simultaneously whenever $(k_1 + k_2)/2k_3 = 1$. Figure 4 shows a numerical simulation of this system for the case where $k_1 = k_2 = k_3 = 1$, depicting the case where these two critical points occur at the same time. When $(k_1 + k_2)/2k_3 > 1$, $d^2C_B/dt^2 = 0$ occurs first; when $(k_1 + k_2)/2k_3 < 1$, $dC_C/dt = 0$ occurs first.

Four Reactions in Series. When all four reactions in series are considered ($A \rightarrow B \rightarrow C \rightarrow D \rightarrow E$), the simulation also requires the use of curvature constraints and the automatic evaluation of third-order constraints in addition to the model constraints, Equations 1–5, and the overall mass balance. The qualitative model for this system yields 39 genuine behaviors. As in the case of three reactions in series, the distinctions occur because certain critical points can be achieved before, at the same time, or after other critical points, depending on the values of k_1, k_2, k_3, and k_4. Table 1 lists the five pairs of critical points that exhibit occurrence branching. Note that while five pairs of variables, with three distinctions per pair, could result in 243 distinct possibilities, only 39 possible behaviors are generated. The reason why not every combination is generated by the QSIM algorithm is that some of the pairs of variables are not independent of each other. For example, if the pair $(dC_B/dt\ d^2C_{D+}/dt^2)$[1] occurs together, then d^2C_{D+}/dt^2 cannot also occur at the same time as d^2C_{B-}/dt^2 since d^2C_{B-}/dt^2 always occurs after dC_B/dt. Because this system has four parameters (the rate constants) that lead to five pairs of critical points that may occur in any of three ways, analyzing the model equations to identify the inequalities between the relative rate constants that determine the conditions that a given pair of critical points will occur at the same time is nontrivial and has not been done for this system. Nevertheless, the

[1] See footnote in Table 1 for an explanation of "+" and "−" critical points.

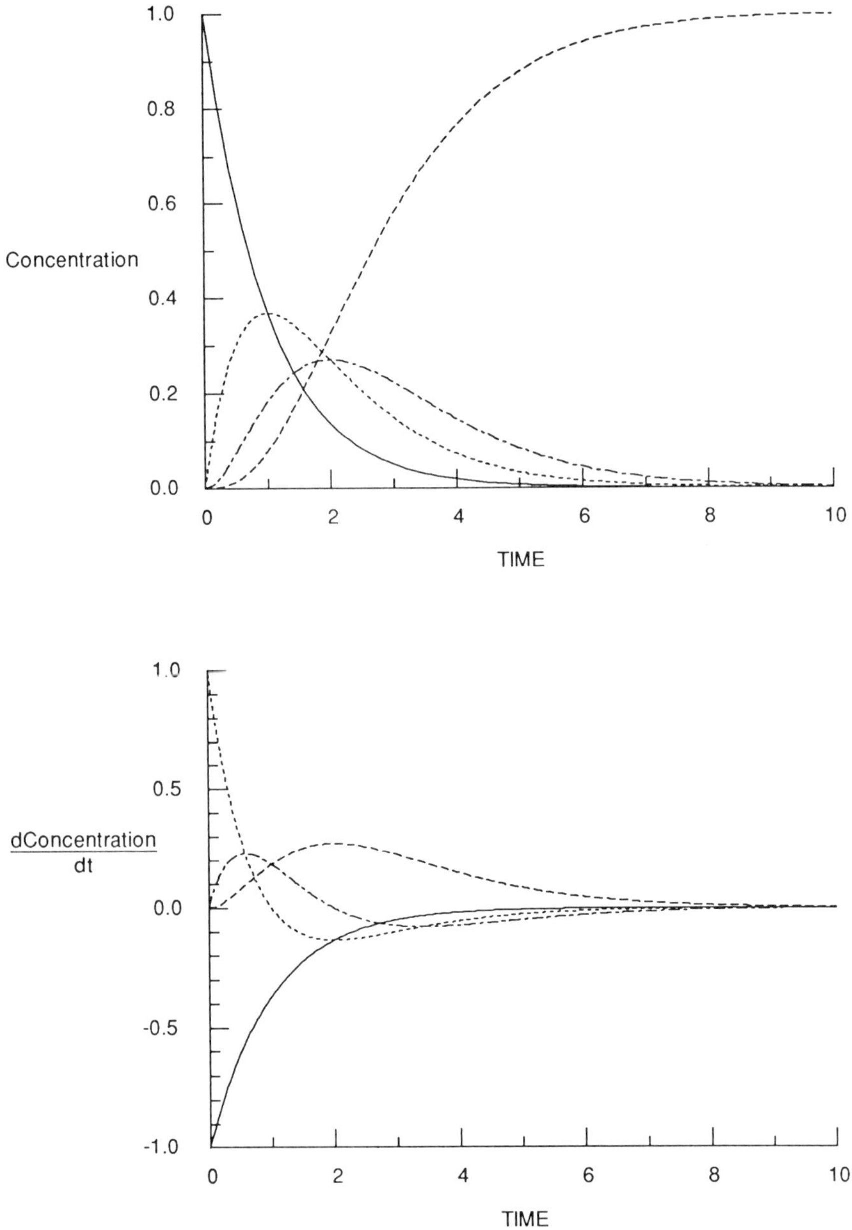

Figure 4. Numerical simulation of three reactions in series with $k_1 = k_2 = k_3 = 1$. This quantitative behavior corresponds to the qualitative behavior in Figure 3. (C_A: ————, C_B: --------, C_C: —-——-—, C_D: — — — — .)

Table 1. Critical points with occurrence branching for four reactions in series.*

Pair 1	Pair 2	Pair 3	Pair 4	Pair 5
$\frac{dC_B}{dt}$	$\frac{d^2C_{B-}}{dt^2}$	$\frac{d^2C_{B-}}{dt^2}$	$\frac{d^2C_{B-}}{dt^2}$	$\frac{d^2C_{C-}}{dt^2}$
$\frac{d^2C_{D+}}{dt^2}$	$\frac{d^2C_{D+}}{dt^2}$	$\frac{dC_C}{dt}$	$\frac{dC_D}{dt}$	$\frac{dC_D}{dt}$

* The "+" and "−" signs indicate that the critical point occurred when the value variable that achieved the critical point was positive or negative. Some variables have "+" and "−" critical points.

qualitative model produces a tractable solution set for this system of five differential equations.

For reactions in series, the qualitative models developed here capture the qualitative behavior of these systems; the depletion of the initial reactant, the generation and subsequent consumption of a number of intermediate species, and the production of a finite amount of the final reaction product. Further, the models predict correct descriptions of the derivatives of each species. The qualitative simulation also identifies the possible critical points of each trajectory and properly identifies all the possible occurrences of these critical points. Table 2 summarizes these results for two, three, and four reactions in series. Extrapolating these trends to even more reactions in series suggests that more critical points will exist, leading to a greater number of possible behaviors. When the number of behaviors generated in qualitative simulation increases to unmanageable levels, methods of aggregation (Weld, 1986) could help identify the distinctions, which are real, and classify behaviors into groups with similar characteristics.

Table 2. Comparison of two, three, and four reactions in series.

Reaction System	Curvature Constraints Required?	Third-Order Constraints Required?	Number of Behaviors	Pairs of Critical Points with Occurrence Branching
A → B → C	Yes	No	1	0
A → B → C → D	Yes	Yes	3	1
A → B → C → D → E	Yes	Yes	39	5

3.1.2. Equilibrium Reactions

While some reaction pathways may consist of irreversible reactions in series, reversible reactions in series are also quite common. These types of reactions pose challenging engineering problems when the desired products are not favored by the equilibrium conditions. Changes in temperatures and pressures, and the addition or removal of reactants or products are often necessary to drive the reaction in the desired direction. A generic representation of such a system is given by the equilibrium reactions

$$\mathrm{A} \underset{k_2}{\overset{k_1}{\rightleftarrows}} \mathrm{B} \underset{k_4}{\overset{k_3}{\rightleftarrows}} \mathrm{C}.$$

Given that the reaction conditions are isothermal with constant density, then mass balances on each component yield the following set of equations:

$$\frac{dC_{\mathrm{A}}}{dt} = k_2 C_{\mathrm{B}} - k_1 C_{\mathrm{A}} \tag{6}$$

$$\frac{dC_{\mathrm{B}}}{dt} = k_1 C_{\mathrm{A}} - k_2 C_{\mathrm{B}} - k_3 C_{\mathrm{B}} + k_4 C_{\mathrm{C}} \tag{7}$$

$$\frac{dC_{\mathrm{C}}}{dt} = k_3 C_{\mathrm{B}} - k_4 C_{\mathrm{C}}. \tag{8}$$

When only A or only C is present initially, then the system is described as two reversible reactions in series. When only B is present initially, the system is described as two reversible reactions in parallel. Levenspiel (1972) discussed the possible behaviors for both types of systems as a function of the rate constants, k_i ($i = 1, 2, 3, 4$). The qualitative model developed below determines the possible behaviors without the need for numerical values.

A qualitative model for this system of reactions is again developed directly from the component mass balance equations and an overall mass balance. When applying the automatic generation of curvature constraints, this model generates an intractable set of behaviors from the initialization that has only A present in the amount C_{A0} at $t = 0$. The application of third-order constraints is inappropriate here since the problem lies in an ambiguous curvature constraint. Specifically, the curvature constraint for the variable dC_{B}/dt (derived by twice differentiating Equation 7) is

$$\mathrm{sd2}\left(\frac{dC_{\mathrm{B}}}{dt}\right) = k_1\,\mathrm{sd1}\left(\frac{dC_{\mathrm{A}}}{dt}\right) + k_4\,\mathrm{sd1}\left(\frac{dC_{\mathrm{C}}}{dt}\right) \qquad \left(\mathrm{at\ sd1}\left(\frac{dC_{\mathrm{B}}}{dt}\right) = 0\right), \tag{9}$$

where sd1 means the sign of the first derivative and sd2 means the sign of the second derivative. In a typical behavior, k_1 sd1(dC_C/dt) has the value "+" and k_4 sd1(dC_A/dt) has the value "−". The qualitative result for the addition of "+" and "−" is ambiguous, and the qualitative simulation branches. Third-order constraints are only useful when a curvature constraint vanishes. However, taking the second derivative of the overall mass balance, $C_A + C_B + C_C = C_{A0}$, yields

$$\frac{d^2C_A}{dt^2} + \frac{d^2C_B}{dt^2} + \frac{d^2C_C}{dt^2} = 0, \tag{10}$$

and, using the fact that sd1(dC_B/dt) $= 0$, yields $d^2C_A/dt^2 = -d^2C_C/dt^2$. Substituting this result into equation 9 gives the resulting curvature constraint

$$\mathrm{sd2}\left(\frac{dC_B}{dt}\right) = (k_4 - k_1)\,\mathrm{sd1}\left(\frac{dC_C}{dt}\right) \qquad \left(\text{at sd1}\left(\frac{dC_B}{dt}\right) = 0\right). \tag{11}$$

With this manually derived curvature constraint, three initial states are generated depending on whether $(k_4 - k_1)$ is less than, equal to, or greater than 0. Note that when $(k_4 - k_1) = 0$, the curvature constraint vanishes, and the automatic evaluation of third-order constraints cannot derive a nonzero result when $(k_4 - k_1) = 0$. A similar manual derivation of a third-order constraint yields

$$\mathrm{sd3}\left(\frac{dC_B}{dt}\right) = ({k_1}^2 - {k_4}^2)\,\mathrm{sd1}\left(\frac{dC_C}{dt}\right) \qquad \left(\text{at sd1}\left(\frac{dC_B}{dt}\right) = 0\right), \tag{12}$$

which also has the value 0 when $(k_4 - k_1) = 0$. Chiu (1989) shows, for systems of this type, that if sd1, sd2, and sd3 of a variable vanish simultaneously, then all derivatives must be zero, and the system must be quiescent.

Given the curvature constraint analysis described above, the simulation for the two reversible reactions in series generates four behaviors from the three initial states. In general, these behaviors agree with the results presented by Levenspiel (1972) to the degree that trajectories of all the concentrations and their derivatives are identical. The initial state where $(k_4 - k_1) < 0$ generates the two behaviors shown in Figure 5. The second of these behaviors is identical to the two behaviors generated from the other two initial states. The first of the two behaviors in Figure 5 suggests that C_B has a maximum in its trajectory. Figure 6 shows a numerical simulation for the case where $k_1 = 3$, $k_2 = 1$, $k_3 = 2$, $k_4 = 1$, justifying the maximum in the trajectory of C_B. In fact, this maximum is present in all simulations

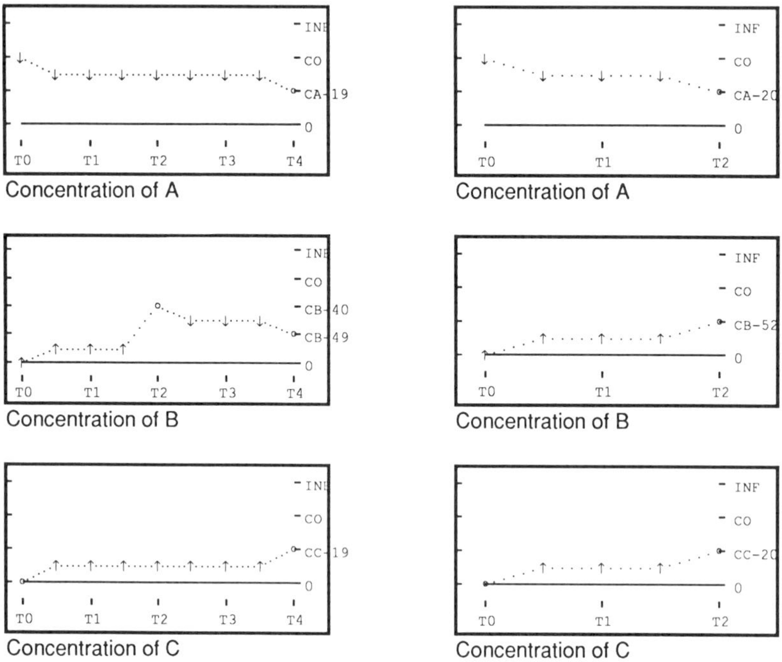

(a) Behavior 1 (b) Behavior 2

Figure 5. Behaviors 1 and 2 for two reversible reactions (A $\rightleftarrows$ B $\rightleftarrows$ C).

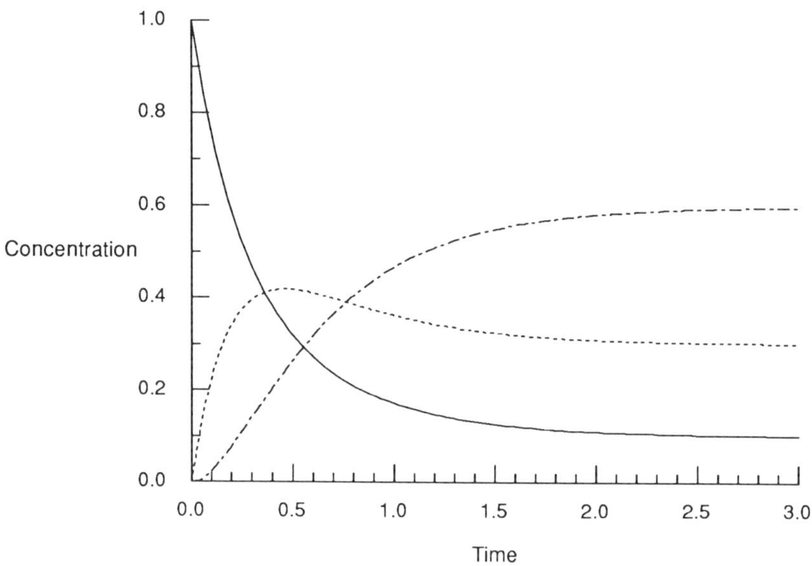

Figure 6. Numerical simulation of two-equilibrium reactions with $k_1 = 3$, $k_2 = 1$, $k_3 = 2$, $k_4 = 1$. (C_A:————, C_B:-------, C_C:—-—-—.)

where $(k_4 - k_1) > 0$, indicating that the second behavior in Figure 5 is spurious. These results illustrate the fact that, while qualitative simulation is guaranteed to capture all real behaviours, spurious behaviors cannot always be eliminated.

While the qualitative model for the two-equilibrium reaction system generates accurate descriptions of the trajectories of the concentrations, these results provide no insight as to the relative concentrations of A, B, and C when the system reaches steady-state. Since each concentration has its own quantity space, the relative magnitudes of each concentration cannot be determined from this model. However, additional constraints could be added to express the ratios C_A/C_B, C_B/C_C, and C_C/C_A. In developing expressions for those ratios, the terms K_{AB} $(= k_1/k_2)$ and K_{BC} $(= k_3/k_4)$ would have to be defined. Since both K_{AB} and K_{BC} could be less than, equal to, or greater than 1, the number of initial states would increase to 27, and the number of behaviors would be 36. While this approach can be used to describe fully this system in a qualitative sense, a more reasonable approach is to consider ranges or values for each rate constant and apply the quantitative/qualitative method of Kuipers and Berleant (1988) to this system. This approach is applied to this system of equilibrium reactions in Section 4.

3.2. *Qualitative Analysis of Reactor Models*

While understanding the kinetic mechanism of a reaction system is important to the design of a reactor, other factors such as flow and temperature conditions also play a crucial role in the design and control of a chemical reactor. Reactor designs vary widely depending on a number of factors such as the kinetics, thermodynamic properties, and the phases of the reactants (solids, liquids, or gases). The two most common types of reactors are *plug-flow* reactors and *CSTRs* (Continuously Stirred Tank Reactors). In plug-flow reactors, the concentrations vary along the length of the (typically tubular) reactor at steady-state. Since the equations that model these types of reactors are similar to those for batch reactions discussed in the previous section, these types of reactors will not be discussed here. Instead, this work concentrates on developing qualitative models of a CSTR and a parallel-plate reactor with radial plug-type flow. Radial-flow reactors are used in a process called plasma etching, an important step in the manufacturing of semiconductor devices.

3.2.1. A Continuously Stirred Tank Reactor (CSTR)

In general, CSTRs can exhibit a variety of behaviors. Under certain conditions, a CSTR can operate at several steady-states for the same set of process parameters and input settings. Dynamically, CSTRs can exhibit unusual behaviors such as inverse response, bifurcation, and limit cycles. One of the first comprehensive theoretical analyses of CSTRs was done by Aris and Amundson (1958). Another major study was later done by Uppal *et al.* (1974). While the variety of CSTR configurations and the large parameter space make identifying all the possible behaviors an enormous task, certain CSTR configurations with simple kinetic mechanisms can be analyzed without unreasonable difficulty. In particular, a constant volume, constant flow rate, adiabatic CSTR with an irreversible first-order reaction is consider in this chapter. The results presented here do not demonstrate all behaviors for all CSTRs, just all behaviors for the particular configuration described above with the initializations discussed below. For a more detailed analysis, the interested reader should refer to Uppal *et al.*, (1974).

Figure 7 shows the CSTR configuration and nomenclature considered in this work. Following the mass and energy balance derivation of Stephanopoulos (1984), the adiabatic CSTR in Figure 7 with the irreversible exothermic first-order reaction A → B can be modeled by the equations.

$$\frac{dC_{\mathrm{A}}}{dt} = \frac{C_{\mathrm{Ai}} - C_{\mathrm{A}}}{\tau} - R_{\mathrm{A}} \tag{13}$$

$$\frac{dT}{dt} = \frac{T_{\mathrm{i}} - T}{\tau} - \Delta H R_{\mathrm{A}}, \tag{14}$$

where

$$R_{\mathrm{A}} = K_{\mathrm{A}} C_{\mathrm{A}} \tag{15}$$

$$K_{\mathrm{A}} = k_0 e^{-E/T} \tag{16}$$

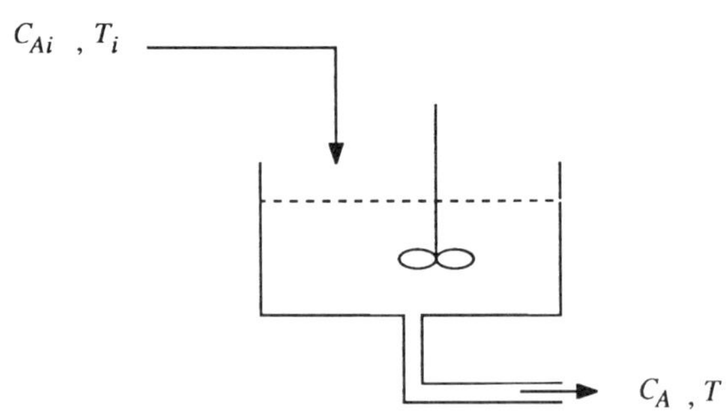

Figure 7. Configuration and nomenclature for a Continuous Stirred Tank Reactor.

In Equations 13 through 16, C_A is the concentration of A in the tank (and exit stream), C_{Ai} is the concentration of A in the inlet stream, T is the temperature in the tank (and exit stream), T_i is the temperature of the inlet stream, τ is the residence time or space velocity (tank volume/inlet flow rate), R_A is the rate of reaction of A, K_A is the reaction rate constant, ΔH is the heat of reaction (divided by the product of the solution density and heat capacity), k_0 is the rate constant pre-exponential, and E is the activation energy (divided by the universal gas constant). Models for a CSTR, or *mixed flow* reactor, assume that the fluid in the tank is uniformly mixed so that the properties of the tank are the same everywhere. In addition, the physical parameters are assumed to be constant. The concentration of B at any time can be determined from an overall mass balance.

A qualitative model for the CSTR in Figure 7 can be developed by translating Equations 13 through 16 into qualitative constraints. However, the qualitative constraint language of the QSIM algorithm can only describe the exponential relationship between the rate constant K_A and the temperature T as $K_A = M^+(T)$ or $K_A = M^-(E/T)$, where M^+ and M^- mean a monotonically increasing or decreasing function, respectively. Since the M^+ and M^- relationships capture a large class of functions, the results generated by the QSIM algorithm may not be only those of exponential relationships. While the results for exponential functions are guaranteed to be included, additional behaviors representing other M^+ and M^- relationships may also be predicted.

In addition to the mass balance and reaction rate constraints, curvature constraints are essential for a tractable solution. Because of the nonlinear character of the model equations, curvature constraints cannot be derived automatically as described by Chiu (1988) due to limitations in the algebraic manipulation portion of the algorithm for automatically deriving curvature constraints. Furthermore, an automatic derivation of curvature constraints for models with M^+ or M^- constraints necessarily requires the use of a *smoothness* assumption, which replace with "is similar to" making linear approximations at certain times in the simulation. Kuipers *et al.* (1989) discuss the use and limitations of the smoothness assumption in more detail. Nevertheless, curvature constraints can be derived manually for this system without the need for any assumptions on the curvature of M^+ or M^- constraints since the true function is known in this case to be exponential and its curvature can be derived directly.

Contrary to suggestions by Kuipers and Chiu (1987), curvature constraints can be derived for any variable in the constraint model, not just the highest-order derivatives. By inspection, the variables dC_A/dt, C_A, dT/dt, T, and R_A are all candidates for curvature constraints since they are involved in constraints with potential ambiguity: R_A is the product of two variables while dC_A/dt and dT/dt are sums of two variables, etc. Since deriving curvature constraints involves differentiation, curvature expressions often result in expressions with as much potential ambiguity as the original constraints, providing no useful information to the simulation. For the CSTR model above, curvature constraints can be derived without much difficulty for C_A, T, and R_A. For dC_A/dt and dT/dt, the method of ignoring irrelevant distinctions (Kuipers and Chiu, 1987) is used instead of potentially ambiguous curvature constraints. The derivation of the curvature constraints for C_A, T, and R_A can be found in the Appendix.

The qualitative model of the CSTR based on Equations 13 through 16 and the curvature constraints for C_A, T, and R_A (Equations A2, A3, and A15 in the Appendix) can be initialized by setting the system parameters C_{Ai}, T_i, τ, k_0, and E to positive values and ΔH to a negative value. While QSIM generates one initial state for these parameters, three initial states are mathematically possible for some sets of the system parameters. Since the three initial states are qualitatively identical (all variables have the same qualitative values), the qualitative model cannot make a distinction among the three initial steady-states that are physically possible. However, the behaviors generated from the one qualitative initial state will capture all the behaviors possible from all of the three mathematically possible initial states. In Section 4, quantitative information is used in the initialization and simulation of the CSTR system to eliminate potential initial states and behaviors.

To demonstrate the number of possible behaviors for this system, a step decrease in the initial concentration, C_{Ai}, is considered. With this initialization, three initial states are possible. The new value of C_{Ai}, C_{Ai-}, could be greater than, equal to, or less than the steady-state value of C_A. Since some of the possible values for C_A with multiple steady-states may be close to C_{Ai}, these three initial states are all feasible, not just pathologically distinct. From the three initial states, eight behaviors are predicted by the qualitative model. During the simulation, automatic evaluation of third-order constraints was required in addition to the use of curvature constraints. In all eight behaviors, the temperature in the tank decreases monotonically to a new value. The distinctions in the dynamic behaviors can be found in the behavior of C_A. Figure 8 shows the four behaviors of C_A generated by the initial state where $C_{Ai-} > C_A$. The first and third of these behaviors are also generated by the other two initial states (giving the other four behaviors). The important conclusion from Figure 8 is that, with a decrease in C_{Ai}, C_A may (1) decrease monotonically to a new value; (2) decrease to a minimum and return to its initial value (no net change); (3) decrease to a new value via a minimum; or (4) decrease initially to a minimum, then increase to a new value greater than its initial value

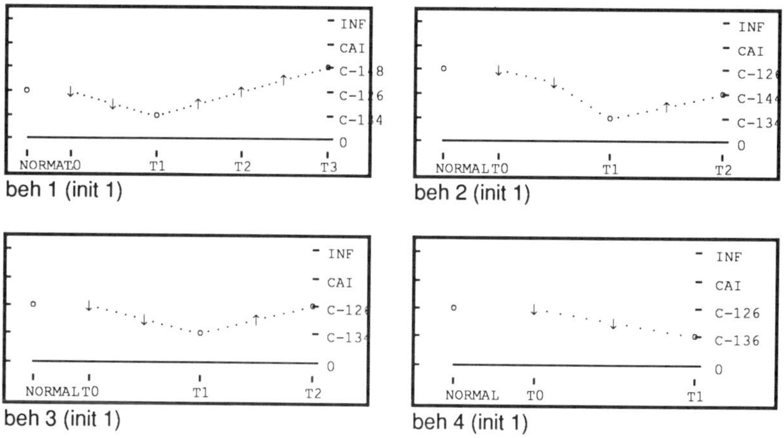

Figure 8. Four possible qualitative behaviors of C_A for the CSTR in response to a step decrease in C_{Ai}.

(inverse response). The real system behavior depends on the actual parameters of the system and the particular initial steady-state (if more than one steady-state can exist).

The eight behaviors for the response of the CSTR to decreases in C_{Ai} have been verified by Dalle Molle (1989b). For changes in T_i, multiple initial states are also possible, but the behaviors are less complicated. Dalle Molle (1989b) has discussed the behaviors from other initializations in more detail. In all cases, the qualitative model captures all the behaviors possible for any parameter set of the problem description. The number of possible behaviors is greatly reduced when quantitative knowledge is incorporated into the qualitative model. This issue is discussed later in Section 4.

3.2.2. A Radial-Flow Parallel-Plate Reactor

In the manufacturing of semiconductor devices, plasma etching has become the preferred method for the selected removal of material from a silicon wafer. In plasma etching, low-pressure gases containing a variety of species partially ionized by a high-frequency potential are passed over a substrate masked with a desired pattern. The etchant species react with the substrate to create gaseous products that are carried away with the bulk gas flow. Several researchers have investigated radial-flow reactors from the etching of silicon by CF_4 where the flow of reactants is from the center of the reactor outward (Dalvie *et al.*, 1986; Venkatesan *et al.*, 1987). Figure 9

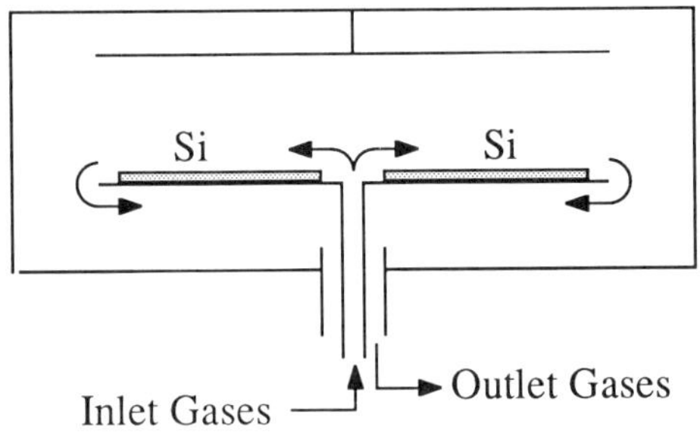

Figure 9. Parallel-plate radial-flow reactor configuration and flow pattern.

shows a typical plasma reactor configuration with radial outward flow. As the dynamics of these processes are often negligible, these models of plasma etching concentrate on the radial variations of the reactants.

Since plasma etching processes have complex chemistry and a large, poorly known parameter space, many assumptions are required to formulate a tractable model. In addition to the assumptions used by Venkatesan *et al.*, (1987), a simple qualitative model can be developed by further assuming that the gas transport is a fully developed plug-type flow and that the electron density is constant over the radius of the reactor. While many species are present in the reactor, only the important ones will be considered in this model. The following kinetic mechanism is used:

Reaction	*Rate*
$e + CF_4 \rightarrow CF_3 + F + e$	$R_1 = k_1 n_e C_{CF4}$
$F + Si \rightarrow SiF(ads)$	$R_2 = k_2 C_F$
$3F + SiF(ads) \rightarrow SiF_4(gas)$	$R_3 = k_3 C_F^n$,

where C indicates the concentration, n_e is the electron density, and k represents a reaction rate constant. The first reaction represents the dissociation of CF_4 into the reactive species CF_3 and F. The free fluorine then reacts with the silicon substrate in the rate-limiting step (Venkatesan *et al.*, 1987). The third-step merely lumps all subsequent reactions into one step, since it does not play a part in modeling the system when Step 2 is rate limiting. The CF_3 species reacts with other species such as F (recombination) and other CF radicals (C_2F_6 formation, polymer formation), but these reactions are not considered here.

Given the assumptions above, the system can be completely described by the conservation equations. For radial symmetry and the reaction steps R_1 and R_2 above, the following equations are obtained for the species CF_4,

F, and SiF_4:

$$\frac{d}{dr}(ruC_{CF_4}) + k_1 rn_e C_{CF_4} = 0 \tag{17}$$

$$\frac{d}{dr}(ruC_F) - k_1 rC_{CF_4} n_e + k_2 rC_F = 0 \tag{18}$$

$$\frac{d}{dr}(ruC_{SiF_4}) - k_2 rC_F = 0, \tag{19}$$

where u is the local velocity of the gases (and $ru = r_o u_o$ where the subscript o indicates the outer edge of the reactor). Equations 17 through 19 differ from the equations of the other models in this study because they 1) describe variations in position rather than time and 2) are nonautonomous (the independent variable appears explicitly in the equation). Hence, the independent variable is the radial position r rather than time. To generate the independent variable, an additional constraint is added such that $dr/dt\ (r) > 0$. In addition to Equations 17 through 19, an overall mass balance and curvature constraints for dC_{CF4}/dr, dC_F/dr, dC_{SiF4}/dr, and R_F $(=R_1 - R_2)$ are included in the qualitative model. Because Equations 17 through 19 are nonautonomous, the curvature constraints must be derived manually (see the Appendix for details).

For the system described above, the initial conditions are, at $r = 0$, $C_{CF_4} = C_{CF_{40}}$, $C_F = 0$, and $C_{SiF_4} = 0$. With these initial conditions and using curvature and third-order constraints, the qualitative model generates 11 possible behaviors (concentration-versus-radius profiles) when $r \to \infty$. The 11 behaviors differ by the relative order of three pairs of critical points that exhibit occurrence branching. Table 3 lists the pairs of critical points that exhibit occurrence branching. As in the case of four reactions in series, not every combination of occurrence branching is possible, as some occurrences are dependent on others. Figure 10 shows the behavior of the system where critical points that can exhibit occurrence branching occur at the same time. This arrangement generates the behavior with the fewest distinguished time (radial position) points.

If it were not necessary that the model maintain a detailed description of the variable R_F defined above, then the method of ignoring irrelevant distinctions could be used on that variable. With this change, the reactor model produces only three behaviors differing by the occurrence branching of Pair 1 in Table 3. Since the method of ignoring irrelevant distinction sacrifices the ability to create new landmarks (achieve a critical point), then Pairs 2 and 3 in Table 3 cannot exist. With this smaller set of critical points, it can be determined numerically that the critical points of Pair 1 occur

Table 3. Critical points with occurrence branching for a parallel-plate radial-flow reactor.*

Pair 1	Pair 2	Pair 3
$\frac{d^2C_{CF_4-}}{dr^2}$	$\frac{d^2C_{SiF_4+}}{dr^2}$	$\frac{d^2C_{CF_4-}}{dr^2}$
$\frac{dC_F}{dr}$	$\frac{dR_{F-}}{dr}$	$\frac{dR_{F-}}{dr}$

*The "+" and "−" signs indicate that the critical point occurred when the value variable that achieved the critical point was positive or negative. Some variables have "+" and "−" critical points.

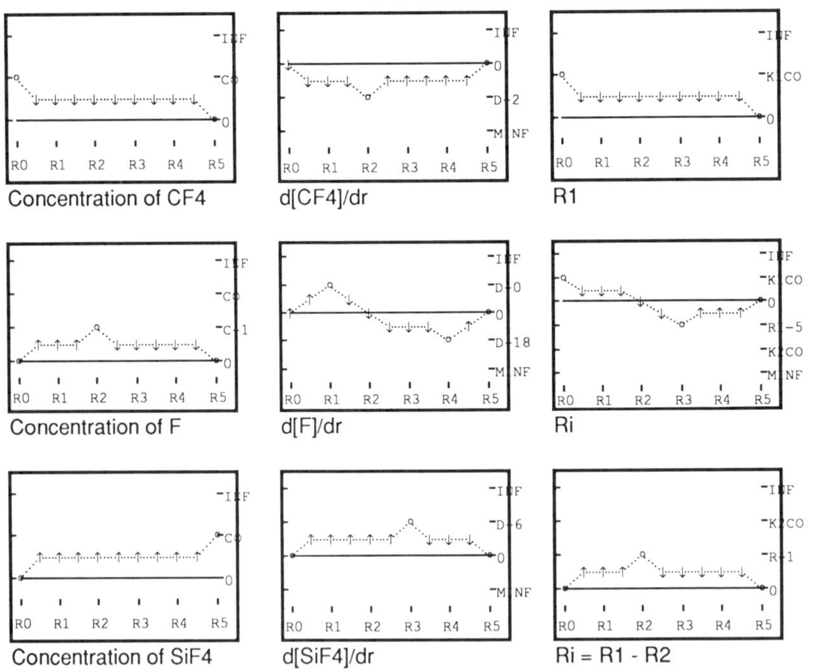

Figure 10. Most concise (most simultaneously occurring critical points) behavior for two reactions in series in a radial-flow reactor.

simultaneously when $3k_1 = k_2$. This analysis is nontrivial when all three pairs of critical points are considered.

While the kinetic mechanism for this model is essentially two reactions in series, the nonautonomous nature of the equations creates several qualitative distinctions. First, Figure 10 shows that all of the initial derivatives of the components are 0 since $r = 0$ initially. In the batch kinetics example of two reactions in series, only the final product had a zero initial derivative. Besides requiring manual derivations of curvature constraints, the radial-flow reactor occasionally required the use of third-order constraints, which were not needed in the batch system. Also, the radial-flow reactor had three pairs of variables that exhibited occurrence branching; there were no such pairs in the batch case. These points illustrate the differences in behavior between a model of a kinetic mechanism and the model of a particular reactor with the same kinetic mechanism.

3.3. Qualitative Models of Reaction Systems—Discussion

For simple reaction systems such as two reactions in series, the qualitative model predicts a single behavior that represents the only real behavior of the system for any possible (positive) values of the rate constants. For more complex reaction systems, the number of qualitative predictions typically increases, indicating possible qualitative distinctions in the behavior of the system. In some cases, these distinctions are as subtle as one critical point occurring before, at the same time, or after another critical point (occurrence branching). In other cases, different behaviors are distinguished by variables of a model having entirely different trajectories, such as a monotonic decrease or inverse response. In all cases, these distinctions represent potentially real features of the physical system. With the exception of the second behavior in Figure 5 for two equilibrium reactions, all of the behaviors generated in this work are genuine and can be verified by numerical simulation. In general, however, it is difficult to verify or refute qualitative predictions without resorting to analytical or numerical techniques. Even with a tractable solution, considering all possible values for the parameters of the model is tedious, and for large systems, impractical. Hence, in qualitative simulation, there is a trade-off between identifying all the possible behaviors of the system (which is guaranteed) and identifying only the "real" behaviors of the system (ones that can be predicted from an equivalent quantitative model).[2] One way to minimize

[2] Strictly speaking, real behaviors are those exhibited by the actual physical system. Mathematical models are often used as the standard by which to analyze qualitative predictions (Kuipers, 1986). Both qualitative and quantitative models, however, can miss real system behaviors if invalid assumptions or inaccurate parameters are used in the model.

this problem is to include quantitative information in the model, reducing the number of predictions generated. This approach is considered in the next section.

4. Partial Quantitative Knowledge in Qualitative Models

As qualitative models are used to describe processes of greater complexity and higher order, the number of possible behaviors generated by a purely qualitative simulation may become quite large. Kuipers and Berleant (1988) presented a method for incorporating partial knowledge of quantitative information into a qualitative model. In their approach, called Q2, unknown monotonic functions can be constrained to lie between upper and lower curves or "envelopes." Landmarks may have known numerical ranges rather than just ordinal relations. Since a considerable amount of quantitative information is known for many systems, the number of behaviors generated by a qualitative simulation can be greatly reduced by including knowledge about functional relationships and parameter ranges. Furthermore, the results from a semiqualitative/semiquantitative simulation are more descriptive than purely qualitative results. In this section, partial quantitative information is used to provide a more descriptive result to the system of two-equilibrium reactions and to eliminate behaviors in the CSTR system.

4.1. Two-Equilibrium Reactions

One of the limitations of a purely qualitative simulation is that the results are often too qualitative to be useful. For the system of two-equilibrium reactions discussed in Section 3.1.2., the qualitative model was able to generate correct qualitative descriptions of the dynamic behavior of the system. However, the model (without additional constraints) was unable to predict the relative amounts of each species at equilibrium. However, if the unknown rate constants could take on numerical values or ranges, then this information could be used by Q2 to reduce the uncertainty in the equilibrium values of the concentrations. Table 4 shows the equilibrium values determined through the range propagation scheme of Q2 for several cases of ranges on the rate constants. For the nominal values of $k_1 = 3$, $k_2 = 1$, $k_3 = 2$, and $k_4 = 1$, Q2 can reduce the ranges of the equilibrium concentrations to agree with the analytical solution up to four decimal places.[3] As

[3] Actually, a tolerance parameter can be adjusted by the modeler to loosen or tighten the relative accuracy of the numerical calculations.

Table 4. Ranges on equilibrium values for two-equilibrium reactions.*
(Nominal Values: $k_1 = 3$, $k_2 = 1$, $k_3 = 2$, $k_4 = 1$)

% Error in Nominal Values	C_{Alb}	C_{Aub}	C_{Blb}	C_{Bub}	C_{Clb}	C_{Cub}
Analytical	0.1000	0.1000	0.3000	0.3000	0.6000	0.6000
0	0.1000	0.1000	0.3000	0.3000	0.6000	0.6000
1	0.0912	0.1092	0.2792	0.3212	0.5696	0.6296
5	0.0601	0.1509	0.1998	0.4096	0.4394	0.7412
10	0.0283	0.2116	0.1037	0.5305	0.2534	0.8680

* *lb* and *ub* designate lower-bound and upper-bound, respectively.

the uncertainty in the ranges increases, the ranges on the equilibrium concentrations increase. Clearly, the uncertainty in the rate constants is not linearly related to the uncertainty in the ranges of the equilibrium concentrations. As expected, the propagation of parameter uncertainty broadens the variable ranges rather than decreasing them.

4.2. An Adiabatic CSTR

In Section 3.2.1., a qualitative model was developed for a constant-volume adiabatic CSTR with an irreversible reaction ($A \rightarrow B$) taking place. In general, a CSTR can exhibit a variety of behaviors, depending on its configuration, kinetics, specific parameter values, and initializations. For the specific model developed in Section 3.2.1., four qualitative behaviors were predicted for a step decrease in the inlet concentration of C_A, as shown in Figure 8. While all four of these behaviors are real for some set of parameter values and at some of the possible initial states (Dalle Molle, 1989b), not all of the four behaviors are necessarily possible for a specific set of parameters from a particular initial state (if more than one initial state is possible). When the parameters of the model are known, this information can be used by the quantitative range reasoner, Q2, to refute some of the predictions generated from a purely qualitative simulation.

Table 5 lists a set of parameters for the CSTR that will be provided to Q2 (the parameters are defined in Section 3.2.1.). Given these parameters, the system of Equations 13 through 16 gives rise to only one possible initial state with the values $T = 471.07$ °K and $C_A = 0.28930$ (moles/liter). While Q2 does not explicitly solve systems of nonlinear equations, its range reasoning capabilities can determine whether an initial state is numerically consistent. For the CSTR problem, Q2 can verify from an initial guess of

Table 5. Parameter values for the quantitatives CSTR model.*

Parameter	Value	Units
C_{Ai}	1	Moles/liter
T_i	400	°K
τ	10	Minutes
k_0	10,000	Minutes^{-1}
E	5,000	°K^{-1}
ΔH	−100	Liter-°K/mole

* Section 3.2.1. defines these parameters.

ranges on the temperature whether these ranges bound the actual solution. For example, if the initial guess for T were (460.0 465.0), Q2 would refute the initial state since the ranges specified were not consistent with the model constraints and the parameters values. However, if Q2 is given the range for T of (470.0 472.0), then the initial state is not refuted. Furthermore, Q2 determines that C_A must be in the range (0.280 0.300). Smaller initial ranges on T lead to smaller ranges on C_A, so any desired accuracy in the initial state can be achieved.

Figure 11 shows the only behavior predicted by Q2 for the qualitative/quantitative model of the CSTR for a step decrease in the inlet concentration of C_A from 1.0 to 0.9. From an initial state where $T = (470.5\ 471.5)$

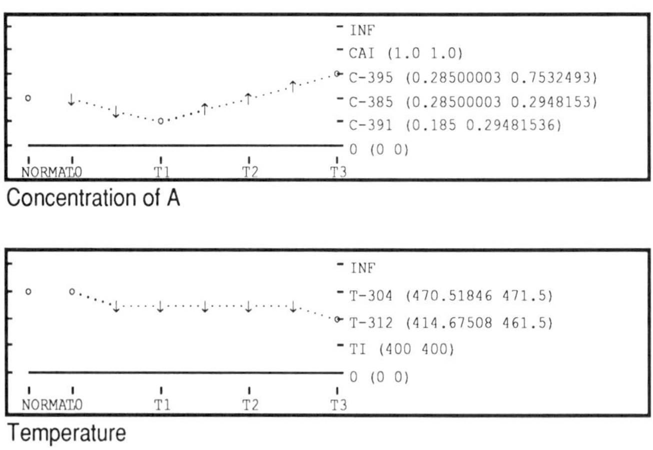

Figure 11. Unique prediction of the CSTR model with quantitative information for a step decrease in C_{Ai}.

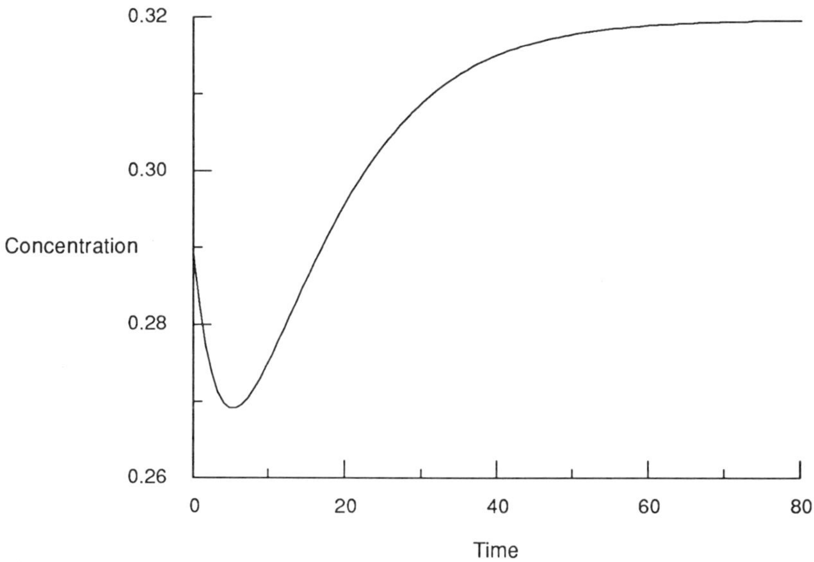

Figure 12. Numerical simulation showing inverse responsible for C_A resulting from a step decrease in C_{Ai} for an adiabatic CSTR.

and $C_A = (0.285\ 0.295)$, the model predicts that only the inverse response behavior of Figure 8 is consistent with the parameters of the model listed in Table 5. Figure 12 shows the trajectory of C_A from a numerical simulation of the CSTR for the same set of parameters and initialization, verifying the qualitative/quantitative model predictions. While the qualitative/quantitative predictions are not able to determine small ranges on the new steady-state values of T and C_A, the quantitative information has allowed the simulation to eliminate quantitatively inconsistent states and behaviors.[4]

5. Discussion

Developing accurate models of chemical reactors is important for process design and model-based monitoring and control. Understanding the reaction pathway and determining an appropriate kinetic mechanism is essential for developing reactor models. While numerical methods for analyzing kinetic mechanisms exist, these methods are unsuitable when parameter values and functional relationships are unknown or poorly known. Qualitative simulation is a promising technique for reasoning about dynamic

[4] Recently, Dalle Molle (1989b) has shown how *internal subdivision* (Berleant, 1990) can significantly reduce broad ranges that occasionally result from the standard Q2 analysis.

systems with incomplete knowledge. While traditional methods of analysis require a complete description of the process and generate a unique prediction, qualitative simulation produces all possible behaviors of a system from qualitative descriptions of the physical system. In addition, quantitative information can be included in qualitative simulation to reduce the number of possible behaviors and to provide a more descriptive prediction of the behavior of the system.

In this work, qualitative models for several different kinetic mechanisms were developed. These models have the ability to generate qualitative descriptions of the dynamic behavior of the components in the reaction mechanism. Two different types of chemical reactors were also modeled qualitatively. A parallel-plate radial flow reactor was developed that predicted the radial variations at steady-state of the components in a simplified plasma etcher model. An adiabatic CSTR was modeled illustrating some of the many behaviors possible for this type of reactor. All of the models required the use of curvature constraints, and most of them needed third-order constraint evaluation to achieve tractable solutions. In all cases, the models presented in this work captured all of the possible qualitative behaviors that can be exhibited by a corresponding quantitative model. When quantitative information was included in some of the models, the number of possible behaviors was reduced, and the resulting behaviors were more descriptive, providing numerical ranges on the new steady-states.

This work brought together several recent developments in AI in the area of qualitative reasoning. By combining the QSIM algorithm (Kuipers, 1986) with other recent works in qualitative simulation, tractable solutions to complex reaction systems were obtained. However, other issues in qualitative simulation still need to be addressed. Qualitative model building is still something of an art. While qualitative models can be developed easily from systems of differential equations, models for poorly understood processes are not as straightforward to develop. Curvature constraints were essential to the results presented in this work; however, it is not always known *a priori* which variables in a qualitative model will require curvature constraints for a tractable solution. Furthermore, curvature constraints for many systems must be developed manually by the modeler. Automating the curvature constraint generation for a larger class of systems would reduce the bottleneck of qualitative simulation, model building.

As qualitative model building becomes more refined, larger, more complex qualitative models will be possible. More complicated models naturally lead to more possible behaviors. Additional reasoning techniques may be required to reduce or eliminate a potential increase in the number of spurious behaviors. Furthermore, analyzing large behavior trees resulting from the simulation of a complex model may be a difficult task. Since qualitative simulation distinguishes different qualitative behaviors by distinctions such as behavior patterns (trajectories) and occurrence branching, the process of identifying distinctions and classifying similar behaviors from large behavior trees will also need to be automated. Methods of aggregation used in other qualitative reasoning methods may be useful in qualitative simulation.

Acknowledgment

This research was funded by the Texas Advanced Technology Program (TATP).

Appendix: Curvature Constraint Derivations

A.1. Curvature Constraints for CSTR Model

Curvature constraints are expressions of the second derivative of a given variable. For the CSTR model in Section 3.2.1., curvature constraints are needed for C_A, T, and R_A. Curvature constraints could also be derived for dC_A/dt and dT/dt, but the method of ignoring irrelevant distinctions is used instead for these variables. For C_A, differentiating Equation 13 gives

$$\ddot{C}_A = -\frac{\dot{C}_A}{\tau} - \dot{R}_A. \tag{A1}$$

Since the curvature constraint is applied when the first derivative of C_A is steady ($\dot{C}_A = 0$), Equation A1 simplifies to

$$\ddot{C}_A = -\dot{R}_A. \tag{A2}$$

For temperature, differentiation of Equation 14 with the condition $\dot{T} = 0$ gives the curvature constraint

$$\ddot{T} = -\Delta H \dot{R}_A. \tag{A3}$$

For the reaction rate R_A, differentiation of Equation 15 once and twice gives

$$\dot{R}_A = \dot{K}_A C_A + K_A \dot{C}_A \tag{A4}$$

$$\ddot{R}_A = \ddot{K}_A C_A + 2\dot{K}_A \dot{C}_A + K_A \ddot{C}_A. \tag{A5}$$

While Equation A5 is a valid constraint, it cannot be used to evaluate $\ddot{R}_A$ since $\ddot{K}_A$ and $\ddot{C}_A$ are not known. This problem occurs because QSIM only maintains the value of a variable and its first derivative, and only a value for variables when derivatives are ignored. Hence, expressions for $\ddot{K}_A$ and

$\ddot{C}_A$ must be derived, and the result must be simplified. If the expression cannot be simplified to a few terms, then the chance of an ambiguous value resulting is high. Since this curvature constraint is only applied when $\dot{R}_A = 0$, the following expressions, derived from Equations 13, 14, and A4, respectively, can be used in developing an expression for $\ddot{R}_A$:

$$\ddot{C}_A = -\frac{\dot{C}_A}{\tau} \tag{A6}$$

$$\ddot{T} = -\frac{\dot{T}}{\tau} \tag{A7}$$

$$\dot{K}_A C_A = -K_A \dot{C}_A . \tag{A8}$$

An expression for $\ddot{K}_A$ can be obtained by twice differentiating Equation 16:

$$\dot{K}_A = \frac{E}{T^2} K_A \dot{T} \tag{A9}$$

$$\ddot{K}_A = \frac{E}{T^2}\left(\dot{K}_A \dot{T} + K_A \ddot{T} - \frac{2}{T} K_A \dot{T}^2\right). \tag{A10}$$

Using Equations A7 and A9, Equation A10 simplifies to

$$\ddot{K}_A = \dot{K}_A\left[\frac{\dot{T}}{T}\left(\frac{E}{T} - 2\right) - \frac{1}{\tau}\right]. \tag{A11}$$

The second and third terms on the right-hand side of Equation A5 can be rewritten using Equations A9 and A6 as

$$2\dot{K}_A \dot{C}_A = 2\frac{E}{T^2} K_A \dot{T} \dot{C}_A \tag{A12}$$

$$K_A \ddot{C}_A = -K_A \frac{\dot{C}_A}{\tau}. \tag{A13}$$

Substituting Equations A11, A12, and A13 into Equation A5 gives

$$\ddot{R}_A = \dot{K}_A\left[\frac{\dot{T}}{T}\left(\frac{E}{T} - 2\right) - \frac{1}{\tau}\right] C_A + 2\frac{E}{T^2} K_A \dot{T} \dot{C}_A - K_A \frac{\dot{C}_A}{\tau}. \tag{A14}$$

Using Equation A8 and simplifying, the following curvature constraint results:

$$\ddot{R}_A = \frac{K_A \dot{T} \dot{C}_A}{T}\left(\frac{E}{T} + 2\right). \tag{A15}$$

Since the term in parentheses is always positive, no ambiguity is possible with this expression.

A.2. Curvature Constraints for the Radial-Flow Reactor Model

Because the quantity $ru + r_o u_o$ (where the subscript o indicates the outer edge of the reactor) is constant, Equations 17 through 19 for the parallel-plate radial-flow reactor can be written as

$$\frac{dC_{CF_4}}{dr} = -k_1 r n_e C_{CF_4} \tag{A16}$$

$$\frac{dC_F}{dr} = k_1 r n_e C_{CF_4} - k_2 r C_F \tag{A17}$$

$$\frac{dC_{SiF_4}}{dr} = k_2 r C_F. \tag{A18}$$

For the derivative of the concentration CF_4, a curvature constraint can be developed by twice differentiating Equation A16 and simplifying the result with the fact that $\ddot{C}_{CF_4} = 0$ when evaluating the curvature constraint for $\dot{C}_{CF_4}$:

$$\dddot{C}_{CF_4} = -2k_1 r n_e \dot{C}_{CF_4}. \tag{A19}$$

The curvature constraint for $\dot{C}_F$ can be derived in the same way. However, the resulting expression for $\dddot{C}_F$ contains several terms and is therefore likely to be ambiguous when evaluated:

$$\dddot{C}_F = k_1 n_e(2\dot{C}_{CF_4} + r\ddot{C}_{CF_4}) - k_2(2\dot{C}_F + r\ddot{C}_F). \tag{A20}$$

Note that the $\ddot{C}_F$ term will vanish when the curvature constraint is evaluated, so Equation A20 actually has just three nonzero terms. Simplifying and rearranging Equation A20 gives

$$\dddot{C}_F = 2(k_1 n_e \dot{C}_{CF_4} - k_2 \dot{C}_F) + k_1 n_e r \ddot{C}_{CF_4}. \tag{A21}$$

Using the variable R_F defined in Section 3.2.2. (see also Equations A26 and A27), the constraint can be further simplified to

$$\dddot{C}_F = 2\dot{R}_F + k_1 n_e r \ddot{C}_{CF_4}. \tag{A22}$$

Equation A22 describes the curvature of $\dot{C}_F$ with only two terms, but it is still potentially ambiguous. Furthermore, Equation A22 cannot be used in conjunction with the method of ignoring the direction of change in R_F. However, at $\ddot{C}_F = 0$, $\dot{R}_F = -R_F/r$ (see Equation A28), and Equation A22 can be rewritten as

$$\dddot{C}_{\mathrm{F}} = r\left(k_1 n_e \ddot{C}_{\mathrm{CF}_4} - 2\frac{R_{\mathrm{F}}}{r^2}\right). \tag{A23}$$

Equation A23 is valid regardless of the description level chosen for R_{F}. However, it can still become ambiguous when $\ddot{C}_{\mathrm{CF}_4}$ and R_{F} have the same sign. Simulation of the reactor model with the curvature constraints in Equations A19, A23, and A29 (developed below) reveal that $\ddot{C}_{\mathrm{CF}_4}$ and R_{F} have the same sign in the later stages of the simulation (after five distinguished position points) and intractable branching occurs. Since the branching due to ambiguity occurs at later stages (large r), Equation A23 can be approximated by only the $\ddot{C}_{\mathrm{CF}_4}$ term since R_{F} has a $1/r^2$ dependence (and is assumed to be negligible). At small r, $\ddot{C}_{\mathrm{CF}_4}$ and R_{F} have the different signs, so that the sign of either one determines the sign of Equation A23. Therefore, for all r, Equation A23 can be represented by

$$\dddot{C}_{\mathrm{F}} \approx r k_1 n_e \ddot{C}_{\mathrm{CF}_4}. \tag{A24}$$

Equation A24 can also be derived by a twice differentiating (and simplifying) Equation A17 linearized at a given radius. The curvature constraint for $\dot{C}_{\mathrm{SiF}_4}$ can be found simply by twice differentiating Equation A18:

$$\dddot{C}_{\mathrm{SiF}_4} = k_2(2\dot{C}_{\mathrm{F}} + r\ddot{C}_{\mathrm{F}}). \tag{A25}$$

Even though Equation A25 has the potential for ambiguity, no intractable branching occurs during the simulation due to this constraint.

When the curvature of R_{F} is maintained by the model (not ignored), a curvature constraint is required to keep it from causing spurious behaviors. Since R_{F} is defined by

$$R_{\mathrm{F}} = R_1 - R_2 = k_1 n_e C_{\mathrm{CF}_4} - k_2 C_{\mathrm{F}}, \tag{A26}$$

the curvature expression results in twice differentiating Equation A26.

$$\dot{R}_{\mathrm{F}} = \dot{R}_1 - \dot{R}_2 = k_1 n_e \dot{C}_{\mathrm{CF}_4} - k_2 \dot{C}_{\mathrm{F}} = -rk_1 R_1 + rk_2 R_{\mathrm{F}} \tag{A27}$$

$$\ddot{R}_{\mathrm{F}} = \ddot{R}_1 - \ddot{R}_2 = -k_1 n_e \ddot{C}_{\mathrm{CF}_4} - k_2 \ddot{C}_{\mathrm{F}} = k_1(R_1 + r\dot{R}_1) + k_2(R_{\mathrm{F}} + r\dot{R}_{\mathrm{F}}). \tag{A28}$$

Using Equations A27 and A28 and the fact that $\dot{R}_{\mathrm{F}} = 0$ in the curvature constraint for R_{F}, the following constraint can be derived:

$$\ddot{R}_{\mathrm{F}} = -k_1 n_e k_2 r R_{\mathrm{F}}. \tag{A29}$$

References

Aris, R., and Amundson, N. R. Analysis of Chemical Reactor Stability and Control. *Chemical Engineering Science*, 7, pp. 121–155 (1958).

Berleant, D., The Use of Partial Quantitative Knowledge with Qualitative Reasoning, Ph. D. dissertation University of Texas at Austin (1990).

Chiu, C. Higher Order Derivative Constraints and a QSIM-Based Total Qualitative Simulation Scheme. Technical Report AITR88-65, University of Texas, Austin, Texas (1988).

Chiu, C. System Property and Third-Order Constraints in Qualitative Simulation. University of Texas, Austin, Texas (1989).

Dalle Molle, D. T. Automatic Evaluation of Third-Order Constraints in Qualitative Simulation. Technical Report AITR89-98 in preparation, Department of Computer Sciences, University of Texas at Austin (1989a).

Dalle Molle, D. T. Qualitative Simulation of Dynamic Chemical Processes. Ph.D. Dissertation, University of Texas at Austin (1989b).

Dalle Molle, D. T., and Edgar, T. F. Qualitative Simulation for Process Modeling and Control. In *Proceedings of the Amerian Control Conference*, Pittsburgh (1989).

Dalle Molle, D. T., Kuipers, B. J., and Edgar, T. F. Qualitative Modeling and Simulation of Dynamic Systems. *Computers and Chemical Engineering*, 12 (9/10), pp. 853–866 (1988).

Dalvie, M., Jensen, K. F., and Graves, D. B. Modelling of reactors for plasma processing. I. Silicon etching by CF_4 in a radial flow reactor. *Chemical Engineering Science*, 41 (4), p. 653 (1986).

Dvorak, D. L., and Kuipers, B. J. Model-based Monitoring of Dynamic Systems. In *Proceedings of the 11th International Joint Conference on Artificial Intelligence (IJCAI-89)*, Morgan Kaufman, Los Altos, CA (1989).

de Kleer, J., and Brown, J. S. A Qualitative Physics Based on Confluences. *Al Journal*, 24, p. 7 (1984).

Filippi, C., Bordet, J., Villermaux, J., Marchal-Brassely, S., and Georgakis, C. Batch Reactor Optimization by Use of Tendency Models. *Computers and Chemical Engineering*, 13 (1/2), pp. 35–48 (1986).

Filippi-Bosey, C., Greffe, J. L., Bordet, J., Villermaux, J., Barnay, J. L., Bonte, P., and Georgakis, C. Tendency Modeling of Semibatch Reactors for Optimization and Control. *Computers and Chemical Engineering*, (41) (4), pp. 913–920 (1989).

Forbus, K. Qualitative Process Theory. *Al Journal*, 24 p. 85 (1984).

Iwasaki, Y., and Bhandari, I. Formal Basis for Abstraction of Dynamic Systems. In *Proceedings of the National Conference on Artificial Intelligence (AAAI-88)*, Morgan Kaufman, Los Altos, CA (1988).

Kuipers, B. "Commonsense Reasoning about Causality: Deriving Structure from Behavior. *Al Journal*, 24, p. 169 (1984).

Kuipers, B. Qualitative Simulation. *Al Journal*, 29, p. 289 (1986).

Kuipers, B. Abstraction by Time-Scale. In *Proceedings of the National Conference on Artificial Intelligence (AAAI-87)*, Morgan Kaufman, Los Altos, CA (1987).

Kuipers, B., and Berleant, D. Using Incomplete Quantitative Knowledge in Qualitative Reasoning. In *Proceedings of the National Conference on Articificial Intelligence (AAAI-88)*, Morgan Kaufman, Los Altos, CA (1988).

Kuipers, B., and Chiu, C. Taming Intractable in Qualitative Simulation. In *Proceedings of the 10th International Joint Conference in Artificial Intelligence* (*IJCAI-87*), Morgan Kaufman, Los Altos, CA, (1987).

Kuipers, B. J., Chiu, C., Dalle Molle, D. T., and Throop, D. R. Methods for Constraining Intractable Branching in Qualitative Simulation. Technical Report AITR89-117, Department of Computer Sciences, University of Texas at Austin, submitted for publication (1989).

Lee, W. W., and Kuipers, B J. Non-Intersection of Trajectories in Qualitative Phase Space: A Global Constraint for Qualitative Simulation. *In Proceedings of the National Conference on Artificial Intelligence (AAAI-88)*, Morgan Kaufman, Los Altos, CA (1988).

Levenspiel, O. S. *Chemical Reaction Engineering*, John Wiley and Sons (1972).

Mavrovouniotis, M. L., and Stephanopoulos, G. Formal Order-of-Magnitude Reasoning in Process Engineering. *Computers and Chemical Engineering*, 12 (9/10), pp. 867–880 (1988).

Oyeleye, O. O., and Kramer, M. A. Qualitative Simulation of Chemical Process Systems: Steady-State Analysis. *AIChE Journal*, 34 (9), pp. 1441–1454 (1988).

Sacks, E. Qualitative Analysis by Piecewise Linear Approximation. In *Proceedings of the National Conference on Artificial Intelligence (AAAI-88)*, Morgan Kaufman, Los Altos, CA (1988).

Stephanopoulos, G. *Chemical Process Control*, Prentice-Hall, Inc., Englewood Cliffs, New Jersey (1984).

Struss, P. Global Filters in Qualitative Behaviors. In *Proceedings of the National Conference on Artificial Intelligence (AAAI-88)*, Morgan Kaufman, Los Altos, CA (1988).

Uppal, A., Ray, W. H., and Poore, A. B. On the Dynamic Behavior of Continuous Stirred Tank Reactors. *Chemical Engineering Science*, 29, pp. 967–985 (1974).

Venkatesan, S. P., Trachtenberg, I., and Edgar, T. F. Effect of Flow Direction on Etch Uniformity in Parallel-Plate (Radial Flow) Isothermal Plasma Reactors. *Journal of the Electrochemical Society*, 134 (12), pp. 3194–3197 (1987).

Weld, D. S. The Use of Aggregation in Causal Simulation. *Al Journal*, 30, p. 1 (1986).

Williams, B. MINIMA: A Symbolic Approach to Qualitative Algebraic Reasoning. In *Proceedings of the National Conference on Artificial Intelligence (AAAI-88)*, Morgan Kaufman, Los Altos, CA (1988).

2 Use of Qualitative Models in Discrete Event Simulation to Analyze Malfunctions in Processing Systems

JANE T. MALIN AND
BRYAN D. BASHAM

NASA—Lyndon B. Johnson Space Center
Houston, Texas

RICHARD A. HARRIS

MITRE Corporation
Houston, Texas

Abstract

CONFIG is a modeling and simulation tool prototype for analyzing the normal and faulty qualitative behaviors of systems. Qualitative modeling and discrete event simulation methods have been integrated to address the challenges of analyzing the effects of faults and malfunctions in complex continuous processing systems. CONFIG is designed to support development of software and procedures for management of failures early during system design, and to support model-based development of diagnostic expert systems. The library design module supports building libraries of hierarchically organized, reusable information about elements of the domain to be analyzed, including qualitative models of types of component functions and continuous behaviors, and types of relations between components and their qualitative variables. Component models are defined in terms of modes and processes, which are defined by invocation statements

ISBN 0-12-480575-2

and effect statements with time delays. Statements are defined in terms of qualitative variables, value classes, operators, and operations, in a language library. System models are constructed graphically by connecting instances of qualitative component models and relations from the library. System models are analyzed using a modified form of discrete event simulation. CONFIG is currently in limited field test, being used to investigate the effects of failures in reconfigurable computer networks with various communication protocols and to model a portion of the Space Shuttle Remote Manipulator System.

1. Introduction

Qualitative modeling and discrete event modeling have been independently developed to analyze the behavior of complex systems. Both can be characterized as supporting simulations of system configurations made up of discrete models of the behavior of individual system components and processes. The purpose of the CONFIG project described here is to develop a practical combination of these lines of work, to address the challenges of analyzing the effects of faults and malfunctions in complex continuous processing systems. Each of these lines of work has strengths and weaknesses, and this approach is intended to use the strengths of each line of work, and to compensate for some weaknesses of each with strengths from the other.

In process engineering, development of software and procedures for management of system failures and degradations should occur early during system design. Designing, testing, and operating engineered devices requires analysis of the effects of faults and malfunctions as they propagate through configurations of components as time passes in an operational system. Such analyses are done in the development of failure management software, in the design of system sensors to support fault management, and in failure modes and effects analysis in the areas of safety and reliability engineering. An earlier study has shown how engineers do such analyses by using simplified conceptual models of both normal and faulty modes of components or subsystems, with their local consequences. By mentally simulating the effects of failures and control actions, they trace effects as they propagate through the system structure (Malin and Lance, 1987).

The purpose of the CONFIG project is to develop a generic system modeling tool that provides a software version of human qualitative analyses of system behavior. This tool should be usable by design engineers and operators, and should support development of fault management software and procedures, including model-based diagnostic expert systems. The objective is to provide computer modeling and simulation methods that correspond to the commonsense methods of human experts. Such a tool should support more consistent and rapid analyses, and should permit more complex and extensive analyses than could be performed mentally.

This goal has led to efforts to combine qualitative modeling and discrete event simulation, and to extend their capabilities. Qualitative modeling provides approaches to quantizing continuous variables into understandable discrete ranges and to modeling changes in component states and processes in terms of relationships among these qualitative variables. Discrete event simulation provides an array of capabilities for manipulating and analyzing system models of connected components, including statistical methods, in which changes in the behavior of components are represented in terms of discrete events with explicit time delays.

To combine these technologies, our approach defines the primary types of discrete events as changes in the qualitative value of a variable or in the operating mode of a component. We have developed a discrete event simulation control structure to accommodate these types of discrete events. In addition, we have developed a general capability for defining qualitative variables, values, operators, and operations, and we have developed an approach to qualitative fault representation.

In this chapter, we present a progress report. We describe the approach we have taken to combining qualitative modeling and discrete event simulation. We describe the current capabilities of the CONFIG tool prototype, and outline plans for future work in the context of the current implementation.

2. The Problem

To illustrate the types of analysis of failure effects that the tool is intended to support, an example is provided of the types of information and analyses a designer needs to develop software that performs automated differential diagnosis of malfunctions in a continuous processing system. Note that the objective is to develop diagnostic software using system design information, rather than waiting for diagnostic expertise to develop from working with malfunctions as they occur in the operational system.

In a differential diagnosis system, sensor data and results of tests and procedures are used to differentiate among a set of candidate diagnostic hypotheses (Pople, 1982). Automated diagnostic software should narrow down the possible causes by using device sensors and actuators, with minimal direct physical inspection. A method of designing such a diagnostic system is briefly summarized to illustrate how analysis of device behavior is used during design (for more details, see Malin and Lance, 1987).

First, for a system of components, the anaylst defines the sets of target failures and degradations to differentiate, and the date and tests to be used for diagnosis. Next, the sequence of effects of each failure, in terms of these data and tests, is identified by simulation. Each failure is inserted into a model of the system, and the time course of propagation of effects is determined. The mental models used by the analyst are frequently discrete and qualitative, even though the components of such systems may exhibit continuous behavior. System components are conceived of as having various normal and faulty modes, with different behaviors in each mode. Change in a component mode or input variable can result in qualitative changes in the component or flows through it. This can cause further mode changes or can result in propagation of qualitative changes in variables through the system, affecting the modes and behavior of other components. The analyst traces these changes and notes changes in sensed values at points in the system designed to detect failure, or further failures and degradations caused by the initial failure. If a change would result in the automatic running of a test or reconfiguration procedure, the effects of that procedure are also determined in the context of the failure.

The results are analyzed to evaluate the ability of the proposed data and tests to safely differentiate the target failures. If needed, new measurements and tests are then developed to differentiate more of the failures. These may not require new instrumentation, but may use composite data such as sensor comparisons or trend or rate information. Then, a new analysis of failure effects is performed using the augmented set of data and tests, and an evaluation is performed again of how well target failures can be differentiated. This process is repeated until the required level of diagnosis is achieved. Thus, model-based analysis of failure effects can directly support the development and validation of diagnostic rules and procedures, especially for sets of failures and degradations that are difficult to diagnose.

Support of this process implies the need for modular qualitative models of both faulty and normal behavior of components. Many faults can be described as local modes of components with local consequences. Once a good local model of component behavior modes is generated, the modeling tool should support its storage as library data, its repeated use in simulations in various system configurations, and its modification. Simulations should permit tracing the effects of these faults through structural or functional paths to points of interest.

3. Modeling and Simulation Background

3.1. Conventional Methods

Conventional quantitative modeling approaches are appropriate and powerful in many applications in process engineering. However, there are a number of problems with using these conventional approaches to achieve the goals described here for analysis of malfunctions in continuous processing systems. Numerical parameters are not known exactly, especially early in design, and, in any case, the designer does not usually perform the types of analysis we have described in quantitative terms. When the first malfunction occurs or an operating limit is exceeded, these models are often no longer relevant. The models are not general purpose, but usually designed and compiled for a specific analysis of normal behavior. It is difficult to use and understand the models in any other way. They are not usually expressed in terms that can be related easily to the components and organization of the system.

A tool that combines qualitative and discrete event modeling approaches cannot be expected to achieve the precision of quantitative methods, but it can provide capabilities with the fidelity of the designer's mental models, while overcoming some of the problems of conventional approaches.

3.2. Qualitative Modeling

In qualitative modeling, the dynamic behavior of continuous physical systems is described in discrete terms, using quantized state variables. Qualitative algebras, which use special forms of equations, relations, and operations, have been developed to represent interactions and compute solutions using these quantized variables. When a continuous-state variable is divided into a set of ranges of interest (often positive, zero, and negative), it is a qualitative-state variable. Thus, the qualitative value of a continuous quantity is the range it is in. In some types of qualitative modeling, the qualitative value of a variable is defined relative to "landmark" values (Kuipers, 1986), which can be thought of as defining the boundaries of the set of ranges. Forbus (1988) provides a comprehensive view of qualitative physics.

Two general types of approaches have been developed to qualitatively model systems, device-centered and process-centered approaches. Device-centered models use a set of local models of the behavior of each type of component in a system ("device"), and connections through which information is communicated between the components. For example, a device such as a buzzer might be composed of a clapper, a coil, and a battery, and the wire and fields that connect them, so that outputs from one component can become inputs of other components. The local component models may have a number of distinct behavior regions or states, and may specify not only the behavior of the component in each region (using qualitative equations), but also the conditions that cause a transition from the current behavior region to another. The device model is a network of components and connections representing the structure of the system. Reasoning is accomplished by changing an input variable and analyzing the effects in the network, as a sequence of change events involving component variables and states. The analysis method commonly used is qualitative constraint propagation. See, for example, de Kleer and Brown (1983).

In constraint propagation, constraints define dependencies among variables in the models. Known values are used to assign additional values using the network of constraints. This approach has the advantage of starting the solution process anywhere in the network, using whatever values are available. The network of constraints is used to assign values satisfying constraints, working through the network, although the system may ultimately fail to assign unambiguously all the desired values. The network is also used to recognize inconsistencies between constraints and values, a capability that serves as the basis for failure diagnosis. Spreadsheets perform a quantitative form of constraint propagation. Principles and algebras for qualitative constraint propagation have been formulated.

Process-centered models use a set of models of physical processes (e.g., flow or boiling), which need not be local, but may include a system of related components in which the process is occurring. A particular situation can be modeled as having a set of components and their relationships, such as liquid in a container on a heat source, and a set of active processes, such as heat flow and boiling. Processes are active while their conditions are met, usually defined both by component relationships (preconditions) and qualitative values of variables (quantity conditions: relationships between variable values and landmark values). When a process is active, influencing variables and relations determine the direction of change in variables. This changing variable can eventually reach a new qualitative value (relationship to landmark values), thus causing process events (activations and deactivations). Predictive analysis is accomplished by two types of deduction: resolving influences to determine the direction of change of a variable, and identifying the next possible qualitative value of a changing variable that would trigger any process event. See, for example, Forbus (1984).

Because qualitative representations are less precise than quantitative ones, ambiguities or uncertainties as to what could be possible arise as qualitative algebras and qualitative deductive methods are applied. Much of the qualitative reasoning work has focused on developing representations and reasoning methods to generate envisionments (de Kleer, 1979) that enumerate all the possible different qualitative states and transitions of a system, and on histories (Hayes, 1979) that are particular possible sequences of behavior of components of a system. Because of the large numbers of possibilities that can be produced in these branching simulations, considerable effort has been expended in developing methods to constrain and prune the results. For example, in work on generating an explanation of the function of a system based on a qualitative description of the behavior of its components and its structure, all interpretations that were not consistent with the purpose of the system were ruled out on teleological grounds (de Kleer, 1979).

Qualitative modeling approaches typically lack an explicit representation of time, duration, and delays, which are often used by a mental modeler, especially to analyze interacting dynamic processes. Williams (1986) discusses some of the limitations of these approaches, and presents a general approach to qualitatively representing time and reasoning about qualitative episodes for use in temporal constraint propagation.

Qualitative models typically have not represented the analyst's knowledge of specific faulty behaviors of components. Although the concept of components with states or operating modes is common, faulty modes are not modeled, and their effects on systems are not simulated. Inconsistencies with modeled normal system behavior are used to diagnose observed abnormal system behaviors. See Davis and Hamscher (1988) for a survey of diagnostic model-based reasoning.

Some work has used fault models for diagnosis. In fault diagnosis, components that are reachable either physically or functionally from a hypothesized fault are identified and considered (Abbott, 1988), or simulation takes the form of firing rules that explicitly code the state transitions that a change in one component causes in a connected one (Pearce, 1988). Pan (1984) has qualitatively modeled faulty and normal modes of component behavior, and has used constraint propagation to predict types of behavior events from the qualitative model, including changes in trends and mode transitions. This modeling approach includes a time-scale concept to specify delay of a mode transition and duration of trend-type events.

3.3. Discrete Event Simulation

In discrete event modeling and simulation, state changes in a system's entities, *events*, occur discretely rather than continuously (as they would in differential equation models), and occur at nonuniform intervals of time.

In the most common form of discrete event simulation, events are scheduled to occur at some interval from the present time. Throughout the simulation, new events are scheduled and added to an event list that contains records of events and the times they are scheduled to occur. Simulation processing jumps from event to next event, rather than occurring at a regular time interval, and events are executed and removed from the event list. Computation that results in creation of new events is localized in components, which are connected in a network. Any simulation run produces a particular history of the states of the system, and random variables are used in repeated runs to produce distributions of system output variables.

Discrete event simulation has been used almost exclusively to solve queueing-oriented service scheduling problems. The analyses concern distribution and selection of workload, resources, and tasks in a system of *server* entities providing services to a set of customers or *job* entities. The basic modeling objects and simulation control programs of discrete event simulation tools and languages have been tailored to these problems. These tools and languages are widely and productively used in systems engineering, operations research, and management science.

There are three main approaches to modeling for discrete event simulation for queueing-oriented problems. The predominant approach, *event scheduling*, uses a scheduled event list. As scheduled events (including events of beginning or ending a service activity) come up, they are executed unconditionally. Another approach, *activity scanning*, does not use an event list. All activities are conditional, and whenever time is advanced, all are considered for beginning or ending by a method that is analogous to the scanning of processes in process-centered qualitative reasoning. The third approach, *process interaction*, focuses on a sequence of activities and events that describe the total history of the progress of a servicing job, and interaction between processes related to several jobs. To support this approach, both a scheduled event list and a scanned list of conditional events may be used, combining the other two approaches.

In discrete event simulation, time specification permits modeling of systems with memory and feedback. There is a concept of *lumped* model simplification strategies, including quantization of variables. Since the time or delay of an event is commonly a random variable, some ambiguities can be modeled and resolved statistically. Recently, there has been work on application of discrete event simulation to continuous processing system problems, in an area called *discrete-continuous simulation*, which calls on numerical simulation when interacting continuous processes require event rescheduling. See Fishman (1978) for a review of discrete event simulation concepts and methods.

Zeigler (1984) has formalized the general fundamentals of discrete event simulation in the discrete event system specification (DEVS) formalism for specifying a range of models expressible within discrete event simulation languages, including the three main modeling approaches used for queueing-oriented problems. This formalism includes the concept of components of a system, connected together so that they interact via outputs of one component propagated to inputs of another. There is a distinction between active and passive components, and the concept of phases of component behavior. Within these phases, state transition and output functions determine how inputs are processed, how phases change, and how outputs are produced. The type of connections between components can be defined, and the variables permitted to propagate on those connections can be constrained.

It is in the context of this general definition of discrete event simulation that the CONFIG modeling and simulation approach has been developed to accomodate qualitative models and support solving of problems that require an understanding of the effects of failures and control actions as they propagate through component configurations.

CONFIG uses both event scheduling and activity scanning approaches. Event scheduling is used for component updates and delayed effects on component variables and modes. Conditional activity (or process) scanning is used within each local component update event. The concept of modes is used to localize further the scanning to processes that could be active when a component is in a particular normal or faulty behavior region.

4. CONFIG Implementation and Examples

Discrete event simulation systems are reasonable candidates for analyzing qualitative models of dynamic processing systems. There is considerable overlap in definition of the objects and events to model. However, current queueing-oriented approaches are not designed to handle qualitative models of continuous systems. A new discrete event simulation approach is needed, tailored to handle types of qualitative variables, relations, algebras, and deductive strategies. Since considerable effort in the qualitative modeling community has been expended on identification and resolution of ambiguities, means of identifying alternative local outcomes and accomodating their elimination and selection need to be provided. These methods are particularly relevant to situations in which trends in variables in the direction of qualitative landmarks might be interrupted during the delay between their beginning and their scheduled end. Since the discrete event simulation mechanism takes the place of constraint propagation, qualitative models must incorporate time and directed causality.

To develop this new discrete event simulation approach, a number of new concepts and methods were developed. These concepts and methods include a definition of a component model, a definition of the types of links connecting components (*relations* and *variable clusters*), new state transition structures (*processes*), methods for representing qualitative and quantitative functions (*process language*), and a new simulation control approach. Descriptions of these concepts and methods are presented in the sections that follow.

In addition to providing new modeling approaches, CONFIG has been developed with the goal of allowing a designer to describe the system easily, using diagrams and his or her own qualitative vocabulary. It is unreasonable to hope to achieve the fidelity attained by quantitative models using differential equations and transform functions, but the analysis capabilities of the design engineer should be enhanced. CONFIG encourages the incorporation of design knowledge into the model. Furthermore, the results can immediately be communicated to the user in the same words and diagrams that were used to describe the behavior of the system.

Since CONFIG does all processing symbolically at the level of detail chosen by the modeler, the output is already in the same language used by the modeler. This is an important feature of CONFIG, which provides modelers the tools for defining their own domain-specific language and does not restrict them to some particular modeling language. The goal is for the design engineer to become the modeler, eliminating possible misinterpretation between the engineer and the simulation expert.

CONFIG consists of four major modules as seen in Figure 1: (1) the Library Design Module, (2) the Model Building Module, (3) the Simulation Module, and (4) the Experimentation and Analysis Module.

The library designer creates his or her own qualitative language of variables, value classes, and operators to describe the component level behavior. Component functionality is defined in terms of mode dependent, mode independent, and mode transition processes that are conditionally invoked and executed with appropriate delays. A class hierarchy of components (i.e., object-oriented definition of components) with their associated mode transition diagrams[1] (Pan, 1984), and their possible types of relations are stored in a library knowledge base.

[1] Mode diagrams are very similar to deterministic finite-state machines. However, multiple events may occur on any transition.

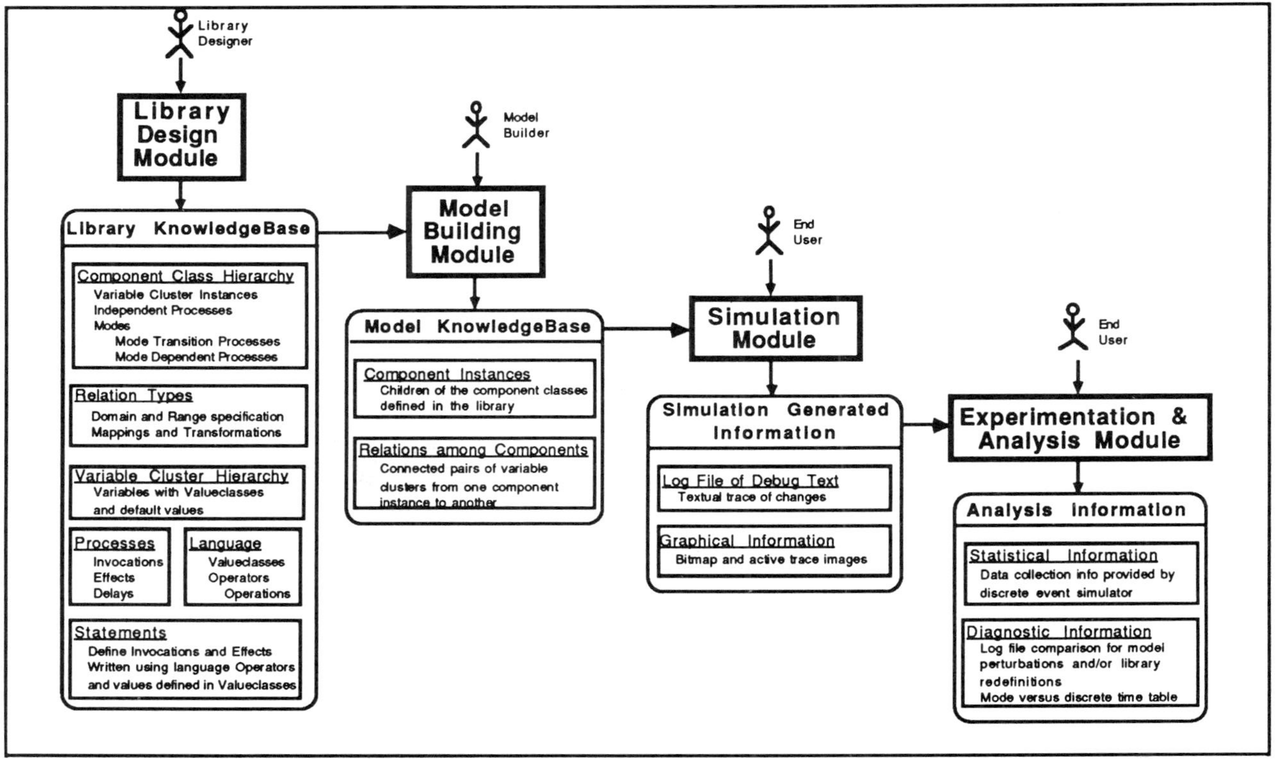

Figure 1. CONFIG modules.

To create various configurations of system models, the model builder graphically and interactively makes instances of the components from a particular library and connects them via instances of relation types. (See the large window in the lower right of Figure 2.) The model builder uses the detailed component-level functionality provided in the library. After a desired configuration is obtained, the end user may perturb the model by changing one or more variable values and then see the results by running the simulator. Several forms of diagnostic information are provided, ranging from multiple levels of debugging text to animated graphics.

It may appear at first that the library designer, model builder, and end user are different individuals using CONFIG at the different stages of design and analysis. This may be the case; in fact, it is one of CONFIG's goal to allow modular use of the environment. However, this does not preclude the user of CONFIG from performing all three roles. This main distinction to be noted about the model builder and library designer is that the library designer should be a component expert and the model builder should be a system configuration investigator/expert. In many cases, but not all, these people are one and the same.

In the sections that follow, the concepts are explained in terms of an application in the domain of thermal and fluid systems. The complete thermal library and model from which these examples are drawn is based on a two-phase thermal control system for a space system.

4.1. Libraries and Component Class Hierarchies

In the current implementation of CONFIG, IntelliCorp's KEE is used extensively as a foundation for the object-oriented aspects of the library design and model building modules. IntelliCorp's SimKit provides CONFIG with a graphical model editor and a discrete event simulator.

To manage large collections (and hierarchies) of models of components, as well as the other elements of the CONFIG system, SimKit's Library structure was used. This structure, which is based upon KEE's Knowledge Base (KB) data structure, allows the programmer to group collections of CONFIG elements into a KB, but it also provides a simple mechanism for defining KB *inheritance*. Thus, the library builder may construct one library on top of another, accessing the elements of the superlibrary and creating new ones as a subclasses of the superlibrary's elements. Currently, library inheritance is limited to a tree-like hierarchy of libraries with single inheritance. For example, if one had THERMAL and ELECTRICAL libraries, one could not create a model that includes components from both of these libraries. In Figure 3, the SIMKIT library is the base of the library

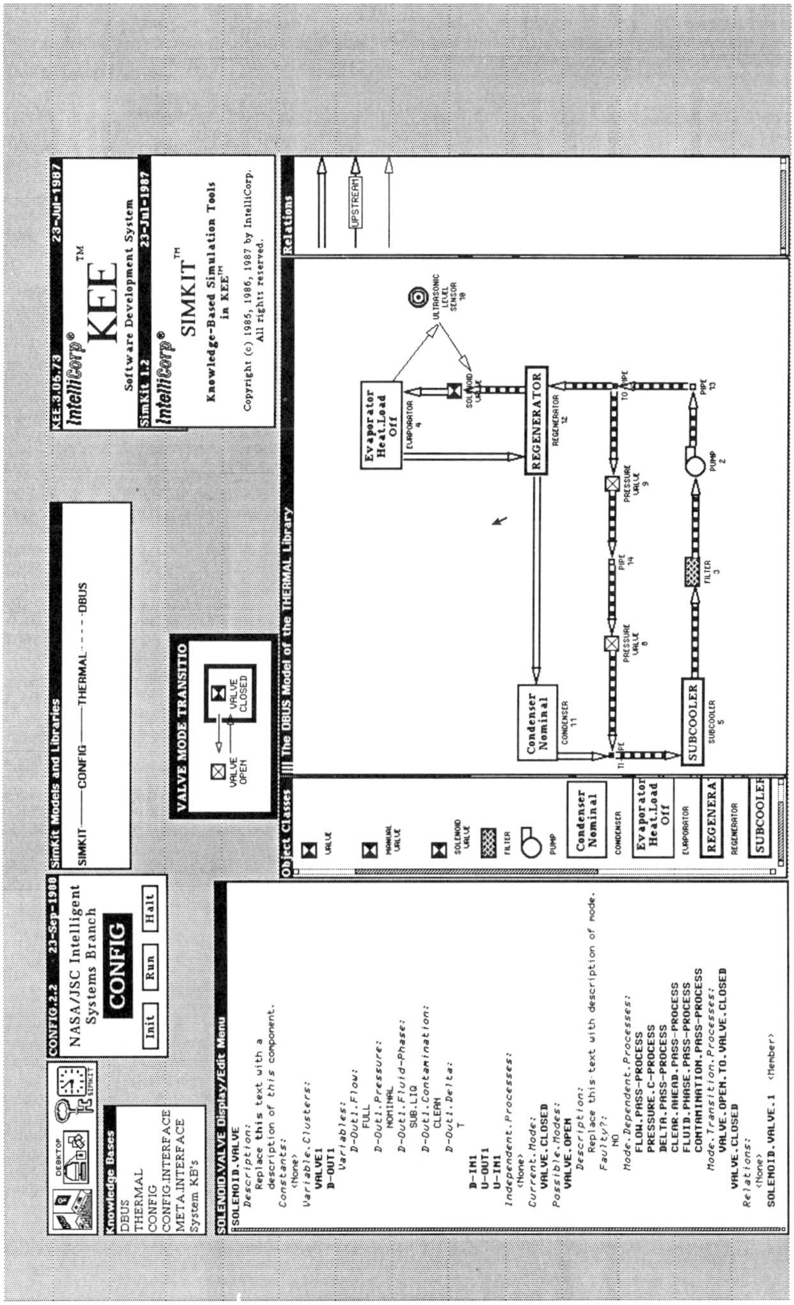

Figure 2. Thermal model in design canvas.

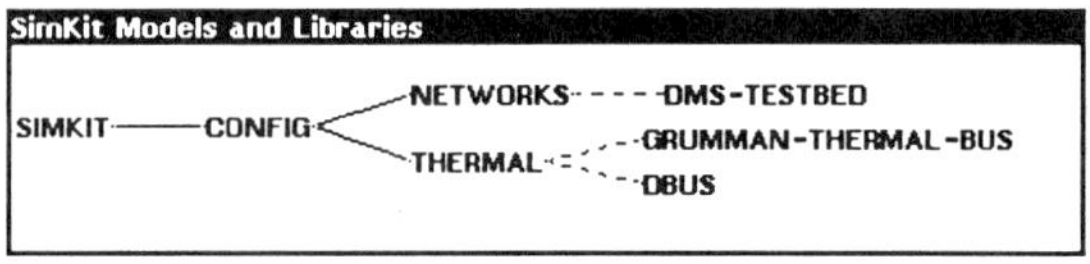

Figure 3. Library hierarchy.

tree, CONFIG is a sublibrary of SIMKIT, and THERMAL and NETWORKS are sublibraries of CONFIG. The dashed lines indicate the models that are members of the libraries. Thus, DBUS and GRUMMAN-THERMAL-BUS are models under the THERMAL library and DMS-TESTBED is a model under the NETWORKS library.

The CONFIG Library holds the elements that are basic to developing components and their functionality, and it contains the data structures and code to implement the CONFIG discrete event control structure. All other libraries are built up on top of the CONFIG library and build upon CONFIG's generic elements. Components are defined according to classification hierarchies and inherit attributes and functionality from parent component classes. The classes of components with their corresponding ancestry are stored in libraries, (e.g., see Figure 4) that may be used by multiple system designers to construct a variety of system level models. Models are collections of component instances and the relationships between them.

A component class is a type of component that is a parent to other component classes or component instances, thus passing down any attributes and/or functionality defined at its level to all children defined under it. A component class may be considered a generic component, fully functional component, or both.

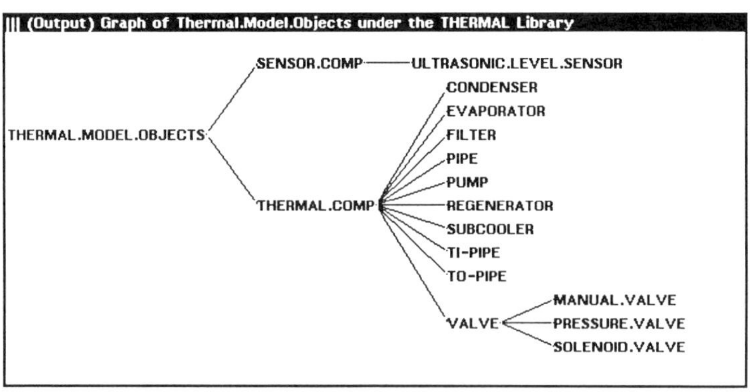

Figure 4. Thermal component hierarchy.

A component instance is a component that has been created from a component class and is not a subclass; this type of creation is commonly referred to as *instantiation*. An instance must be a child of a particular component class of a library, and in the current implementation, all component instances must reside in a model. Component instances can be viewed as the lowest level (or leaves) in the class inheritance tree.

Fully functional components are component classes that possess all the functionality required in order to create working component instances in a model. They are components that are sufficiently described to a degree that will facilitate a reasonable simulation.

Generic components are component classes from which the user is able to create subclasses. The concept of generic components is important to the overall design efficiency of the library. Generic components allow the user to make changes to groups of classes of components without having to hunt down all the components and make the same change to each. Generic components also provide the designer with levels of classification within individual libraries. Generic components may not fully functional. For example, the THERMAL. COMP class, in Figure 4, is a generic component and the CONDENSER, PIPE, VALVE, etc. are functional subclasses of THERMAL.COMP.

A component class may be used as both a generic component class and a fully functional component class. For example, the VALVE class (see Figure 5) was defined as a fully functional component that can be instantiated as a simple valve that may be closed or open, but it can also be considered the generic component class parent for other specific classes of valves such as pressure valves (which automatically open and close based on their input pressure). The subclasses inherit all the functionality of their parent and also gain any additional functionality defined at their own level.

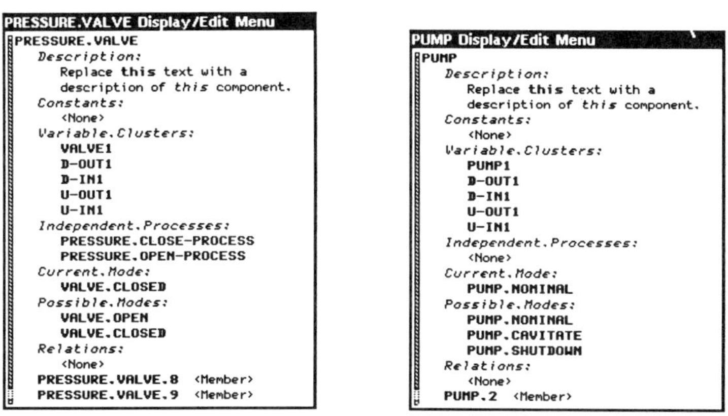

(a) (b)

Figure 5. Component class displays. (a) Pressure valve; (b) Pump.

4.2. Component Models, Modes, and Processes

Components (i.e., fundamental element models) are the basic building blocks in CONFIG used to represent the objects of a system level model. A component does not necessarily have to represent an actual physical entity; it may represent any abstract object. For example, a component may be a physical device such as a pump or a transistor, or it may be something less concrete such as a generic computational device that performs some arbitrary function. CONFIG components are models of functionality, although the physical properties of a particular device will naturally influence the behavior of the device.

All components inherit a fixed set of attributes from the CONFIG top-level component class. The following attributes are basic to the modeling representation, and allow the library designer to define how a component is to behave within the simulation (compare this with the examples in Figure 5):

- *Constants* holds the names of the constant attributes associated with a component. This allows the user to create parameters that are component-specific (and even local to the component within a model); these attributes take on a variable ValueClass and are assigned a unique value from that ValueClass.
- *Variable Clusters* (VCs) holds the names of the variable cluster instances and variables associated with a component (Discussed in Section 4.5.)
- *Independent Processes* holds the names of the mode-independent processes associated with a component. These processes are always activated for the component and are used to describe all behaviors that transcend mode distinctions. For example, a possible independent process for pipes may be the inclusion of erosion into the flow within the pipe. (Processes are defined below.)
- *Current Mode* holds the mode under which the component instance is currently operating. This attribute is important to the simulation procedure in that an event is scheduled to update the component instance when the value of this attribute changes. (Discussed in more detail in Section 4.6.)
- *Possible Modes* is a list of valid modes for the component. For example, the possible modes of a PUMP are PUMP.NOMINAL, PUMP.CAVITATE, PUMP.SHUTDOWN. (See Figure 6.)
- *Mode Transition Diagram* holds the data structure that describes the relationships between the component's Possible Modes. It shows all possible modes of the component plus the transitions between those modes. (Figure 6 shows the mode transition diagram for the PUMP class.)

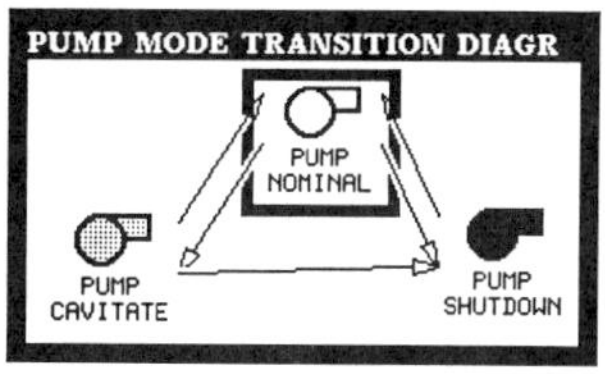

Figure 6. Pump mode transition diagram.

- The *Relations* attribute shows all the relations connected to variable clusters (VCs) of this component instances. Relations facilitate the propagation of information through the model. (Discussed in Section 4.5.)
- *Component Changes* holds the names of recent variable changes in a simulation. It is used only by the simulator, to record what has changed within the component instance during simulation. It also determines when an update for the instance is to be scheduled (i.e., when the first item is placed on this list). (Discussed in Section 4.6.)

The core of the component model is its mode transition diagram, which specifies the modes of operational behavior (both normal and faulty). It is this diagram that contains the possible operating modes a component may exhibit and the directed transitions among the modes. For an example, see Figure 6. When creating a mode diagram, modes are created with their appropriate attributes. When a transition is defined among modes, skeletal mode transition processes are automatically created. These mode transition processes may later be fleshed out (i.e., completely defined) by specifying the conditions required to make such a transition. The mode diagram is also used to designate which mode is the default mode in which a component is to be initialized.

The concept of operating modes and mode transition processes provides a capability for representing process-oriented qualitative model information within the device-oriented discrete event system. CONFIG allows the user to define component behavior in terms of processes that manipulate the values of the variables affected by the components.

Conceivably, CONFIG could be used to model an entire system by abstracting the system to a single component that represents the whole system and then defining all processes necessary to represent the system dynamics. This would defeat one of the major purposes of CONFIG, however. CONFIG allows local definition of component models, and then aids in discovering what could occur on a system level, based on a particular configuration of the components.

Computation and specification requirements are significantly reduced by focusing the level of component description on modes of operation and by specifying qualitative ranges of component variables relative to mode transition boundaries. Rather than using constraint propagation techniques that cannot specify when data should be propagated, the system uses appropriately scheduled discrete events to determine the consequences of component changes. Discrete events can be defined at the level of changes in operating modes, and simulation processing need occur only when modes or variables cross qualitative boundaries.

A mode is a particular operating region that a component may have in which there are unique operating characteristics. These unique operating characteristics are defined by the library designer by specifying the processes associated with each mode:

- *Mode Dependent Processes* contain the processes that are common to the associated mode. This type of process is evaluated only when a component is in the associated mode. (Processes are defined below.)
- *Mode Transition Processes* contain the processes that conditionally change the component from the current mode to another mode.

For example, in the mode transition diagram for the PUMP component class (in Figure 6), three modes are defined: *nominal*, where the pump is operating normally; *cavitate*, where the pump is cavitating due to input fluid consisting of both vapor and liquid; and *shutdown*, where the pumping mechanism is no longer operable (i.e, failed and has interrupted the flow of the fluid). Note that a fault or degradation can simply be modeled as a mode. In each mode, a set of processes define the mode dependent behavior. Any processes that are not mode dependent are similarly defined as Independent Processes.

Processes contain three basic pieces of information: invocation statements, effect statements, and a delay associated with each effect. When a process is active, effect statements can be executed only when all of the invocation statements are satisfied. If there are no invocation statements, the effects are unconditionally executed.[2] For example, in the PUMP.BEING.DAMAGED-PROCESS mode dependent process of the PUMP.CAVITATE mode (see Figure 7), the effect TRY. DAMAGING-PROCESS, "(TRY DAMAGING-PROCESS)," will be per-

[2] Processes are conditional forms that are similar to production rules, but the representation makes execution more efficient for the discrete event simulation than with an inference engine. When they are present, invocation statements act like the IF antecedents of production rules, determining when to "fire." Effect statements differ from the THEN consequents in most production rules, since effects schedule actions rather than assert new facts into a rule base. In addition, the invocation and effect statements are reusable over groups of component classes.

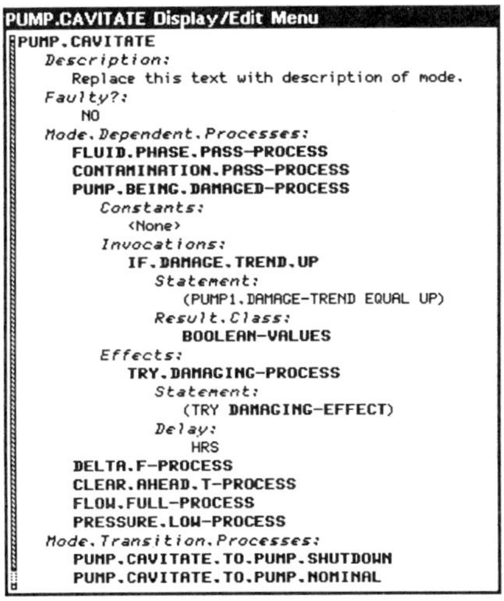

Figure 7. Pump cavitate mode display.

formed only if the IF.DAMAGED.TREND.UP, "(PUMP1.DAMAGE-TREND EQUAL UP)," is true.

Effect statements describe the actions that a process takes after a delay. There are three cases:

(1) setting a component variable (usually a qualitative function of variable values),
(2) setting the Current.Mode of the component (only in a mode transition process), or
(3) activating another process.

Each effect has an associated delay. The delay defines when (relative to the current time of the simulation) the associated effect is to be executed. The delay attribute automatically defaults to zero, which informs the simulator to perform the effect immediately. However, the delay value may be set to an integer, a qualitative constant which translates into an integer value, or a component (or process) constant attribute whose value is qualitative constant or integer.

A comparison to the process definitions in qualitative process theory (Forbus, 1984) yields the following similarities. A change in a component variable or a mode is equivalent to passing a landmark value and reaching a new qualitative range. Thus, invocation statements and the functions in effect statements can be used to determine both the direction of change and the new qualitative value that will be reached, possibly after a delay. Since a trend might be interrupted during the delay, a rescheduling mechanism is provided (discussed in the next section).

In summary, components instances are the basic constituents of CONFIG models that are connected together using relations. A component model is described by its mode independent processes and, more importantly, its modes, which contain mode dependent and mode transition processes. Modes are connected together in a mode transition diagram which delineates the transition dependencies among the individual modes. Finally, processes are the lowest level of described functionality. A process, when activated, will execute its effect statements if and only if all of its invocation statements evaluate to "true." An effect is immediately executed if the associated delay is zero; otherwise, the effect will be scheduled for later execution. This can result in mode changes and changes in the values of qualitative variables.

4.3. Qualitative Representation of Continuous Processes

The most notable feature of CONFIG is its ability to simulate typically continuous processing systems (such as heat exchange systems) with a discrete event simulator. Capabilities are provided by CONFIG to support computation only when a variable reaches a new qualitative value. Thus qualitative changes are considered to be the discrete events of a continuous process. All other changes that do not significantly alter the system's behavior are not modeled. For example, the pressure in a system can fluctuate a few pounds per square inch up or down as long as it remains within nominal boundaries. To support this approach facilities are provided for defining qualitative variables and their algebraic manipulation.

Variables exhibiting continuous behavior are partitioned into qualitative values that represent regions or ranges between landmark values. For example, pressure is continuously variable (sometimes minutely), but it does not warrant attention until it exceeds a particularly high value or goes below a particularly low value. Thus, it might be reasonable to define the magnitude of pressure as having three qualitative values representing the ranges defined by landmark values such as low, nominal, and high. If there are other interesting landmark values, the designer may add additional qualitative values to represent the new ranges, which will typically result in a higher-fidelity correspondence to the actual quantitative value. The operations in the Process Language (described in the next section) can handle any number of values for a qualitative variable without greatly affecting the speed of the simulation.[3] The drawback to simply adding

[3] Since the qualitative ranges are used as symbolic constants in a table-lookup function for most operators, the speed of this operation is linear with respect to the number of arguments in the table operator.

more ranges is that it places more demands on defining the qualitative algebra needed to manipulate these additional values. The library builder must define qualitative operations that correspond to the set of qualitative values.

The library designer can use the Process Language to define a number of types of qualitative variables, including qualitative trends, higher-order derivatives, and complex qualitative values such as *oscillating*. This language definition capability should prove useful for implementing various qualitative-value-based approaches to controlling qualitative ambiguity, such as orders of mangitude and higher-order derivatives.

Trend variables in CONFIG may be conceived as qualitative versions of first-order derivatives of the function that produces the values in the trend's associated variable. Qualitative values such as *up*, *down*, and *inactive* (or +, −, 0) can be used to indicate the direction or sign of change in a particular variable; and values such as *fast* and *slow* may be used to indicate the rate at which the particular variable is changing.

Figure 8 illustrates how the magnitude values and derivatives of a continuous function might be described using qualitative values. The qualitative values of pressure magnitude are on the vertical (y) axis and arbitrary points in time are on the horizontal (x) axis. The curved line represents the actual quantitative values of the pressure over time, $f(t)$. At time, $t0$, the pressure can be described as being low, with a fast upward trend. At time $t1$, the pressure becomes nominal with a slow upward trend. At time $t2$, it is still nominal, but with a fast downward trend. At $t4$, the pressure is low

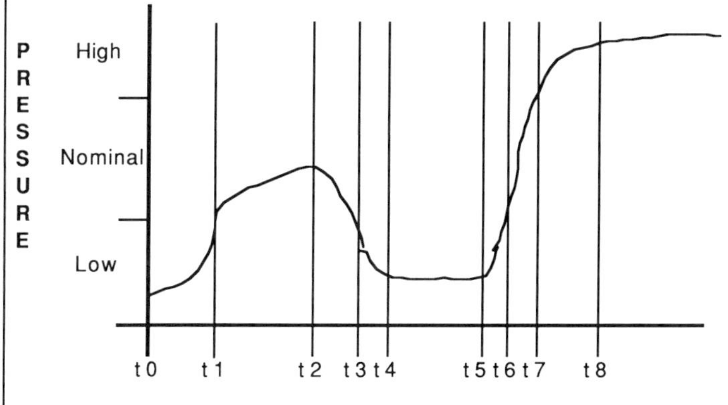

Figure 8. Qualitative description example.

with an inactive trend. At *t*5 it begins to move to nominal, then high with a fast upward trend and then begins to level out in the high pressure with a slow upward trend moving toward and inactive trend at *t*8.

The library designer can also use second-order derivatives when describing this function qualitatively. For example, it is plausible to say that at time *t*5 the pressure is not only low and increasing at a fast rate but also accelerating, and that at time *t*6 it begins to decelerate. The definition of sets of qualitative values, and thus the qualitative modeling approach, is under the control of the library designer, using the Process Language.

The other major part of defining a qualitative algebra is the definition of the operators and operations. When a component process is active, computations are performed to predict and schedule the changes in qualitative values or modes that will occur as a result of the latest changes in qualitative values of component variables. Thus, operations must be defined to determine how to combine the influences of variables. These operations must take account of the ambiguities in prediction introduced by the lack of precision in the qualitative variables. They must also be tailored to the discrete event simulation approach, which performs particular simulations with all ambiguities resolved, such as qualitative histories rather than envisionments.

The Process Language provides a flexible capability for defining many types of operators and operations. Multidimensional tables have been used to implement combination operations for qualitative magnitudes and trends, but other implementation styles are also available. Outcome of combinations can be uncertain, either because of qualitative ambiguity, or because of interacting processes that can interrupt trends before new qualitative values are reached. The former type of uncertainty is handled within combination operations. Approaches to handling ambiguous combinations are discussed in Section 4.4. The latter type of uncertainty is handled by dynamic event rescheduling.

In CONFIG, the current form of dynamic event rescheduling uses a special type of effect that is a process to check for trend interruption. For example, in the thermal model, the pump changes from a normal operating mode to a mode in which it is cavitating when the input fluid phase changes to one with some vapor in it. As the pump cavitates, it degrades until the level of damage causes the pump to change to shutdown mode. To represent this delayed shutdown effect after a period of cavitation, a try-shutdown process event is scheduled. To provide for trend interruption, an automatic mechanism checks the duration of the damage trend before actually executing the effects of the process.

A specific implementation of the PUMP.CAVITATE mode is given in Figure 7. The mode dependent process called PUMP.BEING.DAM-

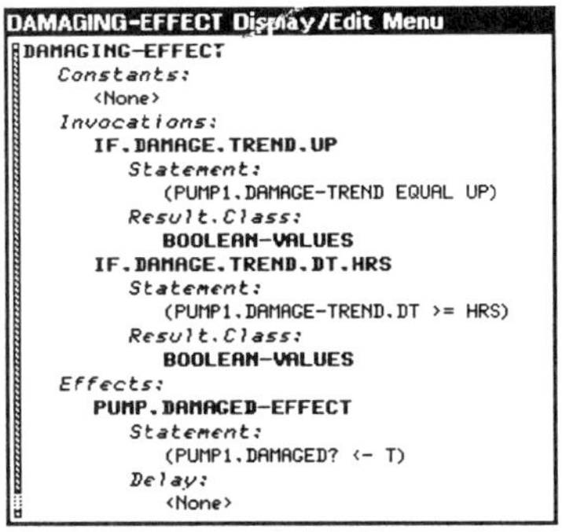

Figure 9. Damaging effect display.

AGED-PROCESS has a single invocation that checks to see if the pump damage trend is UP. If so, the TRY.DAMAGING-EFFECT process is scheduled on the event queue at HRS later. When the discrete event simulator reaches this event, it evaluates the DAMAGING-EFFECT process. As seen in Figure 9, this process checks to make sure the trend's delta time (i.e., the amount of time the current trend has lasted) is greater than or equal to HRS before it concludes that the pump is actually damaged. To summarize, this series of processes guarantees that the pump has remained in the cavitation mode with damage trend UP for at least HRS amount of time before it changes the pump to a failed shutdown mode. If something occurs during the interim that changes the Pump1.Damage.Trend variable, then the delta time will automatically get changed internally by the simulation control mechanism. Thus, the DAMAGING-EFFECT process will not be satisfied and the mode will not change until it has remained in a damaged state for the appropriate period of time to cause a shutdown.

Other approaches to rescheduling are possible. The current approach could be enhanced by numerical simulation capabilities such as those used in the discrete-continuous simulation. In addition, capabilities to inspect potentially interacting events on the discrete event queue, and even to use qualitative process limit analysis methods (Forbus, 1984) to determine rescheduling have yet to be explored.

4.4. Process Language

Early in the development of the CONFIG tool, it was a requirement that an engineer (not familiar with artificial intelligence or Lisp) should be able to construct a domain-specific qualitative language (e.g., the THERMAL language in Figure 10), of qualitative ValueClasses and Operators on those

```
THERMAL.LANGUAGE Display/Edit Menu
THERMAL.LANGUAGE
  Operators:
    GREATER.THAN.EQUAL.TO-OPERATOR
    CLOG-OPERATOR
    EVAP.FLOW-OPERATOR
    FLUID.PHASE-OPERATOR
      Symbol:
        FP
      Precedence: <Unknown>
      Operations.List:
        FLUID.PHASE-OPERATIONS
    GREATER.THAN-OPERATOR
    LEVEL.TREND-OPERATOR
    LEVEL.TREND.SPEED-OPERATOR
    PRESSURE.C-OPERATOR
    REGEN.FLUID.PHASE-OPERATOR
    TI.CONTAM-OPERATOR
    TI.D-OPERATOR
    TI.F-OPERATOR
    TI.FP-OPERATOR
    TI.P-OPERATOR
    TO.F-OPERATOR
    TO.P-OPERATOR
    AND-OPERATOR
    EQUALS-OPERATOR
      Symbol:
        EQUAL = EQ EQUALS
      Precedence: 0.75
      Operations.List:
        EQUALS-OPERATION
    OR-OPERATOR
    NOT-OPERATOR
    MEMBER-OF-OPERATOR
  Valueclasses:
    CONTAMINATION-CONSTANTS
    FILTER-CONSTANTS
    FLOW-CONSTANTS
    FLUID.PHASE-CONSTANTS
    HEAT.GRAD-CONSTANTS
    HEAT.LOAD-CONSTANTS
    HEAT.SINK-CONSTANTS
    PRESSURE-CONSTANTS
      Qualitative.Symbols:
        LOW NOMINAL HIGH
    TIME-CONSTANTS
    TREND-CONSTANTS
    TREND.SPEED-CONSTANTS
    NUMBERS
    BOOLEAN-VALUES
      Boolean.Map:
        (F)
      Qualitative.Symbols:
        T F
```

Figure 10. Thermal language display.

ValueClasses. ValueClasses and Operators are the language elements for describing qualitative ranges and their algebraic manipulations. These elements are objects and may be inherited from a superlibrary, thus allowing the library designer to create additional language elements within the appropriate library and thereby incrementally increasing the vocabulary of the language. This expandable language is called the CONFIG Process Language (see Figure 11). The Process Language is not bound to a particular qualitative scheme, but may have several independent representations. The language describes the lexical, syntactic, and semantic structure of the Statements used in the invocations and effects of the component processes.

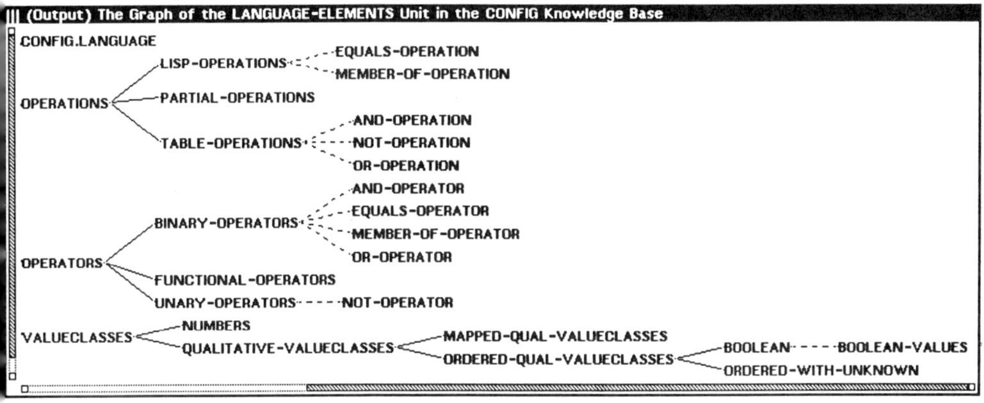

Figure 11. Hierarchy of CONFIG language elements.

The language has lexical representations for the following elements: component variables (and constants), language constants such as numbers and qualitative literals, language operators, and some special symbols ("←", "Try", and "()"). Component variables and constants are defined as attributes of the component class, and the language translator can reference these attributes by their symbolic names. The library builder can create groups of qualitative (or quantitative) literals, called ValueClasses, and Operators (see Figure 12).

There are three classes of CONFIG Statements (see Figure 13). Expression statements are computational forms which, when "evaluated," return some constant value. Expressions use the three basic classes of operators in CONFIG: unary, binary and functional operators (see Figure 11). The expression statements (see Figure 14) use the operators in a syntax based on the traditional arithmetic grammar (i.e., infix notation) plus a Lisp-like notation for functional operators. Binary operators have a numeric precedence that determines the implicit order of the operations in the expressions.

Invocation statements are expressions which return a true/false value whose ValueClass is a member of the BOOLEAN class. Effect statements (see Figure 15) set variables or modes, or activate processes for a component instance, using side effects. They have a special syntax of one of the three forms in Figure 15.

The first form sets a component variable to the value of the *expression*. The second form sets the component's mode to the mode whose name is *mode.name*. The last form activates a process, whose name is *process. name*, for the component.

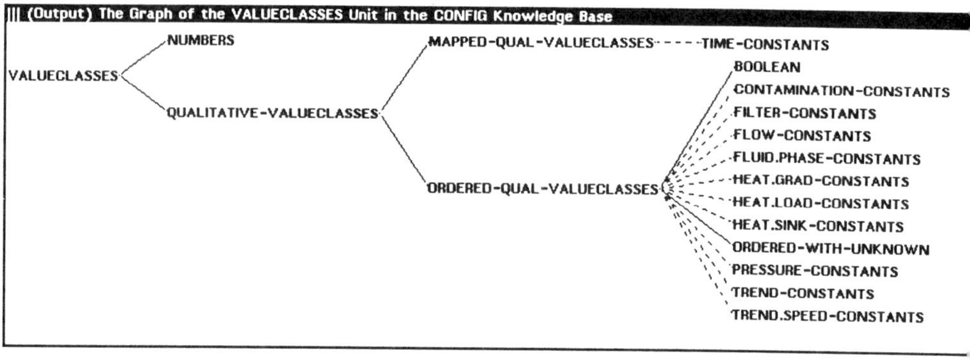

(a)

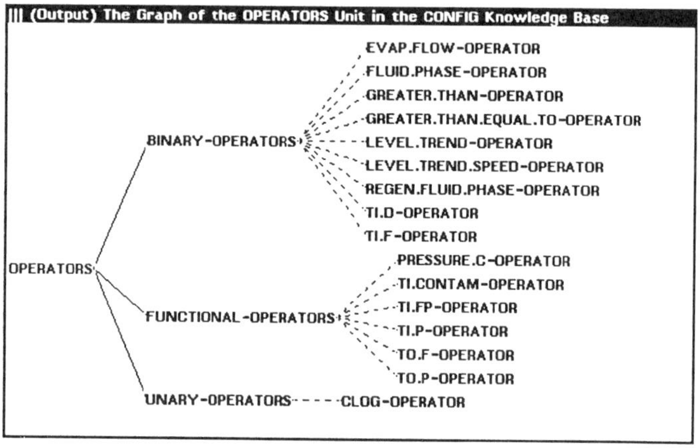

(b)

Figure 12. Thermal language hierarchies. (a) ValueClasses; (b) Operators.

Figure 13. Statements hierarchy.

```
Expression ::= SubExpression | Expression Binary.Op Expression

SubExpression ::= Component.Variable | Component.Constant |
                  Language.Constant |
                  Unary.Op SubExpression |
                  (Functional.Op SubExpression1 SubExpression2 ...)
                  ( Expression Binary.Op Expression )
```

Figure 14. Expression statement syntax.

```
( Component.Variable <- Expression )
( Current.Mode <- Mode.Name )
( Try Process.Name )
```

Figure 15. Forms of effect statements.

Statements, when entered into CONFIG, are first translated into a Lisp-like format and then compiled into Lisp code. During the translation process each statement is checked for semantic, as well as syntactic, consistency and correctness within the context of the component class for which that statement is written. The translator uses the current Library language to parse the expression's operators and constants and determines which operations are to be used, based on the ValueClasses of the operator's arguments. The compiled Lisp code is used to evaluate invocation and effect statements when processes are activated.

Operators describe mappings between one (or more) argument ValueClass(es) into a resultant ValueClass. For example, the real number line can be represented by a SIGN ValueClass of the qualitative literals NEG, ZERO, and POS. The multiplication operator could be described as a mapping of SIGN × SIGN into SIGN, by the following table:

×	NEG	ZERO	POS
NEG	POS	ZERO	NEG
ZERO	ZERO	ZERO	ZERO
POS	NEG	ZERO	POS

```
FLUID.PHASE-OPERATIONS Display/Edit Menu
FLUID.PHASE-OPERATIONS
  Table:
            Arg #2 |   NEG.HIGH   NEG.MED    NO.CHANGE  POS.MED    POS.HIG
            -------+---------------------------------------------------------
     Arg #1: SUB.LIQ  | SUB.LIQ    SUB.LIQ    SUB.LIQ    SAT.LIQ    SAT.VAP
             SAT.LIQ  | SUB.LIQ    SUB.LIQ    SAT.LIQ    SAT.VAP    SUP.VAP
             SAT.VAP  | SUB.LIQ    SAT.LIQ    SAT.VAP    SUP.VAP    SUP.VAP
             SUP.VAP  | SAT.LIQ    SAT.VAP    SUP.VAP    SUP.VAP    SUP.VAP
  Argument.Classes.List:
    FLUID.PHASE-CONSTANTS
      Qualitative.Symbols:
        SUB.LIQ SAT.LIQ SAT.VAP SUP.VAP
    HEAT.GRAD-CONSTANTS
      Qualitative.Symbols:
        NEG.HIGH NEG.MED NO.CHANGE POS.MED POS.HIGH
  Result.Class:
    FLUID.PHASE-CONSTANTS
      Qualitative.Symbols:
        SUB.LIQ SAT.LIQ SAT.VAP SUP.VAP
```

Figure 16. Fluid phase operation table.

It is desirable to be able to define different mappings, called Operations,[4] for the same operator in order to allow the user to describe the *same* operator concept for many qualitative domains. For example, a multiplication operator can be described for the argument sets of SIGN by SIGN, SIGN by PRESSURE, or PRESSURE by PRESSURE. The Process Language allows an operator to have any number of such operations for individual operators (e.g., see Figure 16). Operations can be performed with tables or Lisp functions. They must return symbolic values, all within the same ValueClass, which is called the *range* of the operations.

Operations are used to combine effects of inputs, and results of these combinations of qualitative values are sometimes ambiguous. Yet for any discrete event simulation run, single results of local operations are required. The standard way to deal with uncertainty in discrete event simulation is to identify the distribution of possible result values for an uncertain outcome, and perform a number of simulation runs, with random selections of values of some of the variables. The result of the set of simulation runs is a distribution of output values, to which descriptive statistics is applied. This approach seems well suited for evaluating the overall level of performance of alternative configurations of service resources, where time is the primary type of random variable. Other selection policies, such as selection of the worst-case alternative, are used in failure-oriented design analyses such as failure modes and effects analysis. Whether the approach is statistical or not, methods to deal with ambiguity include identification of local ambiguities and selection methods to eliminate some of the alternatives and explore others.

[4] The Operations subclasses hierarchy can be seen in Figure 11.

In the Process Language, tables can be used to specify how qualitative values combine, and can be used to identify and resolve local ambiguities. Tables are used in a number of qualitative algebras to specify combinations, and ambiguities in particular cells are often resolved with reference to additional information.

Combining quantized magnitudes generally produces more ambiguities than combining signs. For example, one magnitude combination table for three-valued qualitative magnitudes might be as follows:

	LOW	NORM	HIGH
LOW	L	(N, L)?	(N, L, H)?
NORM	(N, L)?	N	(N, H)?
HIGH	(N, L, H)?	(N, H)?	H

Domain-specific quantitative information about landmark values and sizes of qualitative ranges and order-of-magnitude types of assumptions can sometimes be used to select alternatives in the cells, where the (. . .)? denotes the ambiguity. For example, Normal might be the expected result in combinations where the other qualitative value has just passed the limit into High.

Multiple operations can be used when selections are domain specific. For example, multiple operations for the same addition operator could define one table that specifies the expected combination of pressures and another table that specifies the expected combination of fluid phases. More complex operations, using more arguments, can also be used to disambiguate cases. For example, this was done for the resultant pressure in a T-Pipe within the THERMAL library. A two-input, one-output T-Pipe has two pressure values coming from the inputs which need to be "combined" to form an output pressure; this was done using other information in the T-Pipe such as the amount of flow from the input ports.

It is currently a requirement of the Process Language that selections be specified for all ambiguous combination cases. This is a limitation of CONFIG that can be remedied in the future by providing facilities in the Process Language to indicate ambiguities and implement selection policies or note selection assumptions. With this capability in place, many selection approaches, including an array of statistical, qualitative, and quantitative approaches can be implemented.

4.5. Variable Clusters and Relations

The way that variable clusters (VCs) and relations are used to connect components is a unique feature of CONFIG. VCs are used for grouping variables into collections which act as ports, or connections, to other components. For example, a Pipe has two ports, one on each end of the pipe. To designate the flow and pressure of the fluid as distinct variables at each end, the library designer might construct two VCs, called In and Out, both with the variables Flow and Pressure. The Pipe would then contain the four variables: In.Flow, In.Pressure, Out.Flow, and Out.Pressure. Variable clusters provide the mechanism which allows relations to connect a specific set of variables from one component to a specific set of variables in another component. Another example that illustrates the necessity of variable clusters is given by the following:

> Assume a definition of a T-pipe is desired with two inputs and a single output. Assume that this T-pipe has internal variables such as pressure, temperature, flow, and fluid phase. It is not possible to simply connect the outputs of two different components to the same set of input variables in the T-pipe. The T-pipe must be modeled so that it receives two separate sets of inputs (In1 and In2) for these variables and combines them in some fashion to determine what the output values of the T-pipe should be.

Thus VCs allow the library builder to differentiate among multiple inputs (and outputs) for the same set of variables.

Like components, variable clusters can be defined in hierarchies. An example of a hierarchy of VC classes can be seen in Figure 17. Figure 18 shows some variables defined with associated Process Language Value-Classes for variable clusters defined under DOWNSTREAM-VCS. Variables are declared by name *with* a ValueClass, for example, FLOW *with* FLOW-CONSTANTS. This means that the variable FLOW can only take on the values defined in the FLOW-CONSTANTS ValueClass. An example of a variable cluster instance can be seen in Figure 19, which shows the names of the variables concatenated with the name of the variable cluster instance.

Each new variable attribute of a VC instance may have a default value associated with it, which will be inherited by the component(s) that contain the same variable cluster. For example, the D-IN1.FLOW variable has the default value of FULL, and any components that contain this VC instance will inherit that variable and its default value. Many different component classes may share the same variable cluster instances. If a particular component class needs a different default value than that which is defined in the variable cluster, then the default value may be redefined at the component class level. This will affect all children of the component but will not affect any other components that share the same variable cluster. This can even be done for particular component instances without affecting the component class.

A relation is a mechanism that can be used to define types of connections

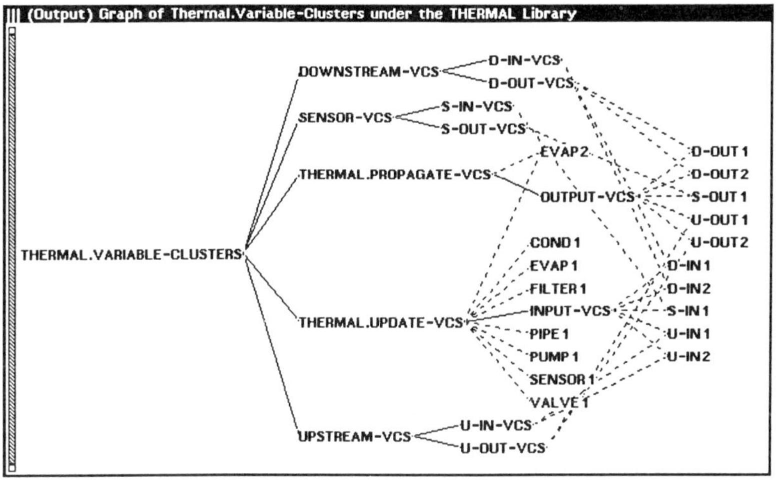

Figure 17. Thermal variable cluster hierarchy.

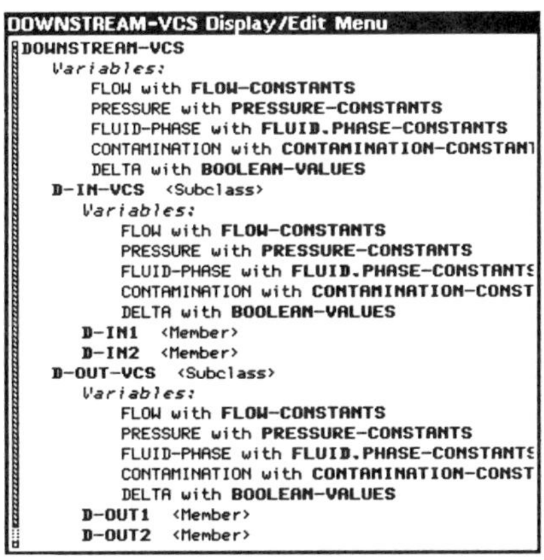

Figure 18. Downstream variable cluster class display.

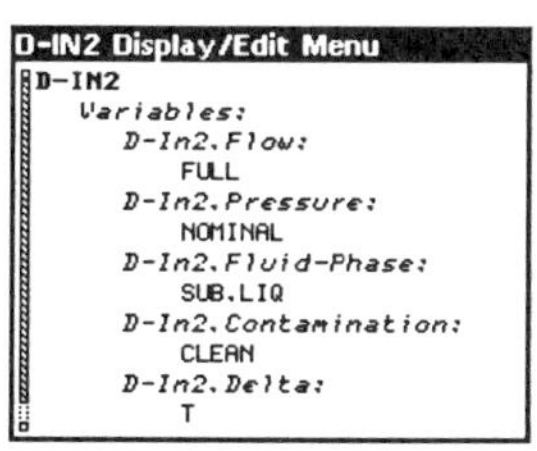

Figure 19. Variable cluster instance display.

between components, in terms of a relation between a variable cluster in a domain component instance and a variable cluster in another component instance called the range component. In discrete event simulation, a relation provides a path for interaction by propagation of information from one component instance to another within a system. In CONFIG, the function of a relation is elaborated, using VCs. When a variable of the domain VC is altered, the change is moved immediately to a corresponding variable in the range VC. The attributes associated with all relations are as follows:

- *Domain* defines the class of components that the relation can have as its domain.
- *Domain Variable Cluster* defines the types of VCs of the domain component that may be related to the range component.
- *Range* defines the class of components that this relation can have as its range.
- *Range Variable Cluster* defines the types of VCs of the range component that may receive the data from the domain component.
- *Include* declares which variables from the domain VC are to be propagated.
- *Exclude* declares which variables from the domain VC are *not* to be propagated.
- *Mappings* declares which variables from the domain VC are to be matched to which variables from the range VC.
- *Transformations* declares any transformations of the values of the variables to be performed during the propagation.

An example of a relation definition is shown in Figure 20. This relation may connect any thermal model object to any other thermal model object; however, it may only connect a variable cluster of the domain component

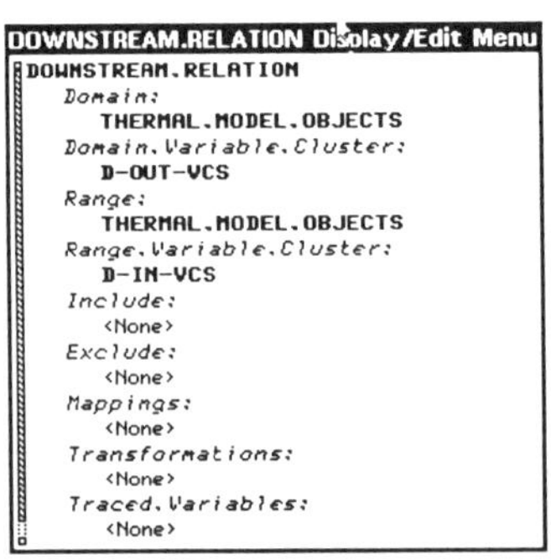

Figure 20. Downstream relation display.

that is an instance of the D-OUT-VCS (which is the class of VCs that define all the characteristics of *downstream output* variable clusters) to a variable cluster of the range component that is an instance of the D-IN-VCS (which is the class of VCs that define all the characteristics of *downstream input* variable clusters).

When a variable cluster is created, the designer is prompted to indicate whether the new cluster is a (1) Propagate VC, (2) Update VC, (3) both, or (4) neither (see Figure 17). When a change occurs in the value of a variable in a Propagate VC during simulation, this value is propagated without delay through relations to connected components. When a change occurs in the value of a variable in an Update VC during simulation, an Update Component event may be scheduled. These two generic VC classes are elements of the CONFIG library and contain all of the necessary functionality to perform the described component updating and data propagation. These mechanisms are described in detail in the next section.

CONFIG provides two types of relations: (1) unidirectional, and (2) bidirectional. A unidirectional relation typically has a Propagate VC defined as its domain VC and an Update VC defined as its range VC. Thus, when a change occurs in the domain VC of the domain component, the change will be propagated to the range VC of the range component. This usually results in the scheduling of an Update Component event for the range component. A bidirectional relation provides the user a simple mechanism for relating two components in both directions. Typically, the VCs used in bidirectional relations should be defined as both Propagate and Update VCs, to correctly propagate data in both directions and schedule Update Component events.

The variable clusters and relations permit the modeler to define types of connections and flows. These VCs can define ports which relations use to connect one component to another, but they can also be used to define hierarchical groups of related variables that help delineate types of connections and flows such as of information (sensor and actuator signals), electrical power, hydraulic, or proximity relations. Future work on creating "generic" component classes may use this facility to group generic hierarchies of variables to create these component classes.

4.6. Control of Simulation and Scheduling

As mentioned in Section 3.2., the CONFIG simulation is based upon a discrete event simulation control structure; that is, CONFIG uses a discrete time clock and a calendar (or queue) of events. Where CONFIG differs most from traditional discrete event simulators is in how events are placed onto the calendar and in the types of events that are represented. Events are placed onto the calendar only when a qualitative change occurs,

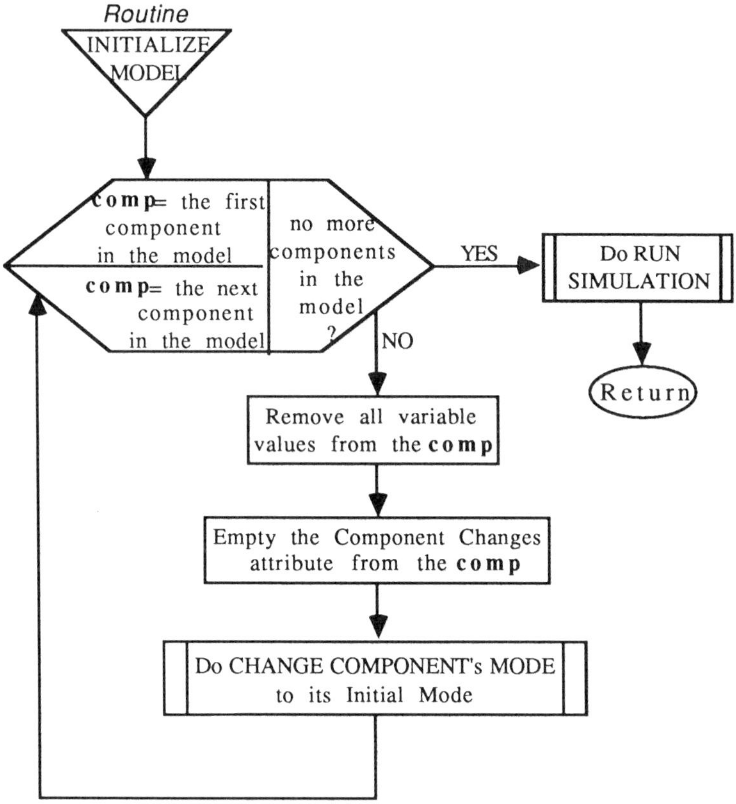

Figure 21. Model initialization flow chart.

and these events describe qualitative changes in continuous systems, rather than events of discrete tasks in systems for queued service scheduling problems. There is also a mechanism for propagating component variable values through the model utilizing relations that connect component instances.

In general, the simulator performs as follows. First, the model is initialized, by initializing each component instance in the model and then running the simulation from this initial set of conditions (see Figure 21). A component is initialized by removing local values of several of its attributes,[5] and then setting the Current Mode to the component's Initial

[5] In particular, all of the component instance's variables and Current.Mode. By removing the local values of component variables, the default values for each variable are automatically put into the variables. These default values are defined by the library builder for component classes, but can be modified for individual component instances within a particular model.

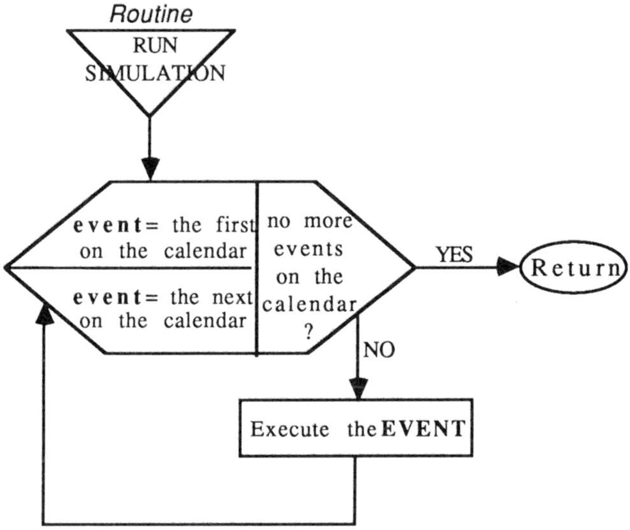

Figure 22. Simulation routine flow chart.

Mode. This initialization of component variables and the Current Mode attribute schedules an Update Component event for every component instance within the model.

Second, a discrete event simulation is performed by executing all of the scheduled events on the calendar (see Figure 22). The simulation ends when there are no more events on the calendar to be processed. At any time during (or after) a simulation, component variables may be changed. These changes may place new events on the calendar, thus allowing the simulation to run with new conditions. This is useful for injecting faults into the simulation and seeing the effects of faults cascade through the system.

In CONFIG, there are two types of events that can be placed on the calendar to drive the simulation: (1) Update Component events, (2) and Effect Execution events (see Figure 23). All of the events on the calendar are associated with a particular component instance; it is this object that the event operates upon. This has the benefit of modularity, which may allow a future implementation on a parallel processing architecture. Within a time slice on the calendar, Effect Execution events have a greater priority than Update Component events and will be executed first. These priorities ensure that the Update Component event will have a complete[6] parameter set that reflects the state of the instance.

[6] All of the effects and associated data propagation for each time slice of the simulation occur for all component instances in the model before Update Component events begin.

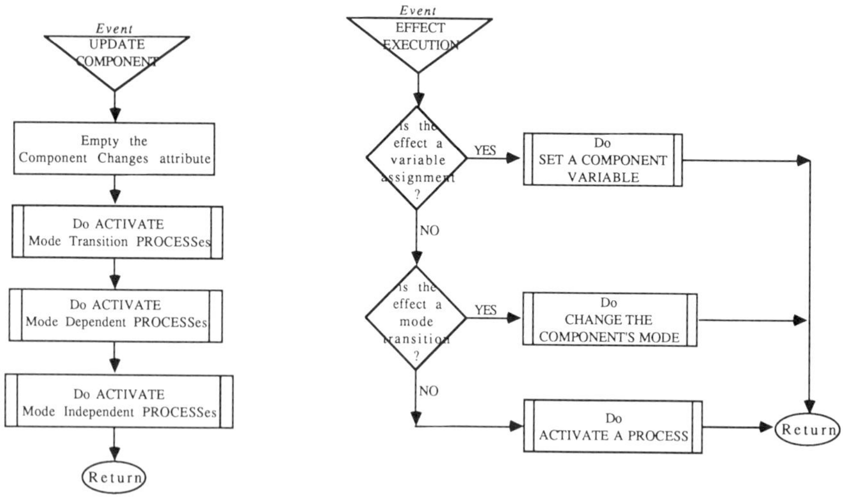

Figure 23. CONFIG event flow charts (a) Update component event; (b) Effect execution event.

The purpose of the Update Component event is to allow a component instance to activate its processes and execute (or schedule) their effects. When an Update Component event is processed, the Component Changes attribute is emptied (thereby allowing updates to be based on new changes), then the component's processes are activated in the following order: mode transition processes of the Current Mode, mode dependent processes of the Current Mode, and finally the mode independent processes. Thus, a change to a new component mode occurs before the mode dependent processes are activated.

Activation of a process results in execution of its effect statements only if all of its invocation statements evaluate to *true* (see Figure 24). If an effect statement has an associated delay, the effect is placed on the calendar (scheduled at *delay* amount after the current time) for later processing as an Effect Execution event. If the delay is zero then the effect is executed immediately. In keeping with the qualitative spirit of the CONFIG system, process delays can be described as qualitative constants. However, since CONFIG has been implemented on top of a discrete event simulation tool, the clock times must be numeric (i.e., integer). Although CONFIG allows the user to specify delays qualitatively, they are translated into integer values. Thus, if cycles or feedback loops in the system model go through at least one component with a process with a non-zero delay, they can be well described.

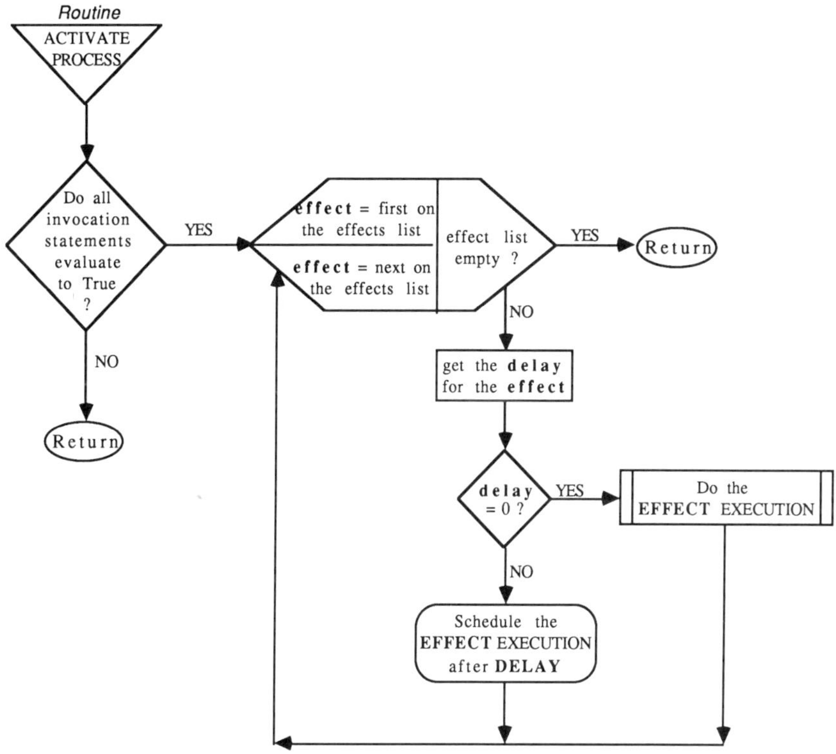

Figure 24. Activate a process routine flow chart.

There are three types of Effect Execution events: (1) setting a component variable, (2) changing the Current Mode of the component, and (3) activating (or "Try"ing) a component process (see Figure 25 for the component attribute setting routines). These Effect Execution events, like Update Component events, are processed on the particular component instance associated with the event. Setting a component variable may have the effect of propagating the data to related components or may prompt an Update Component event. Changing the Current Mode prompts an Update Component event. Process activation within Effect Execution works just like process activation within Update Component (see Figure 24). Since trying a process controls the interruption and rescheduling of trends to new qualitative values and mode transitions, invocation statements that are evaluated include trend direction and duration.

On the first variable or mode change recorded since the last update of a particular component, an Update Component event is scheduled for that

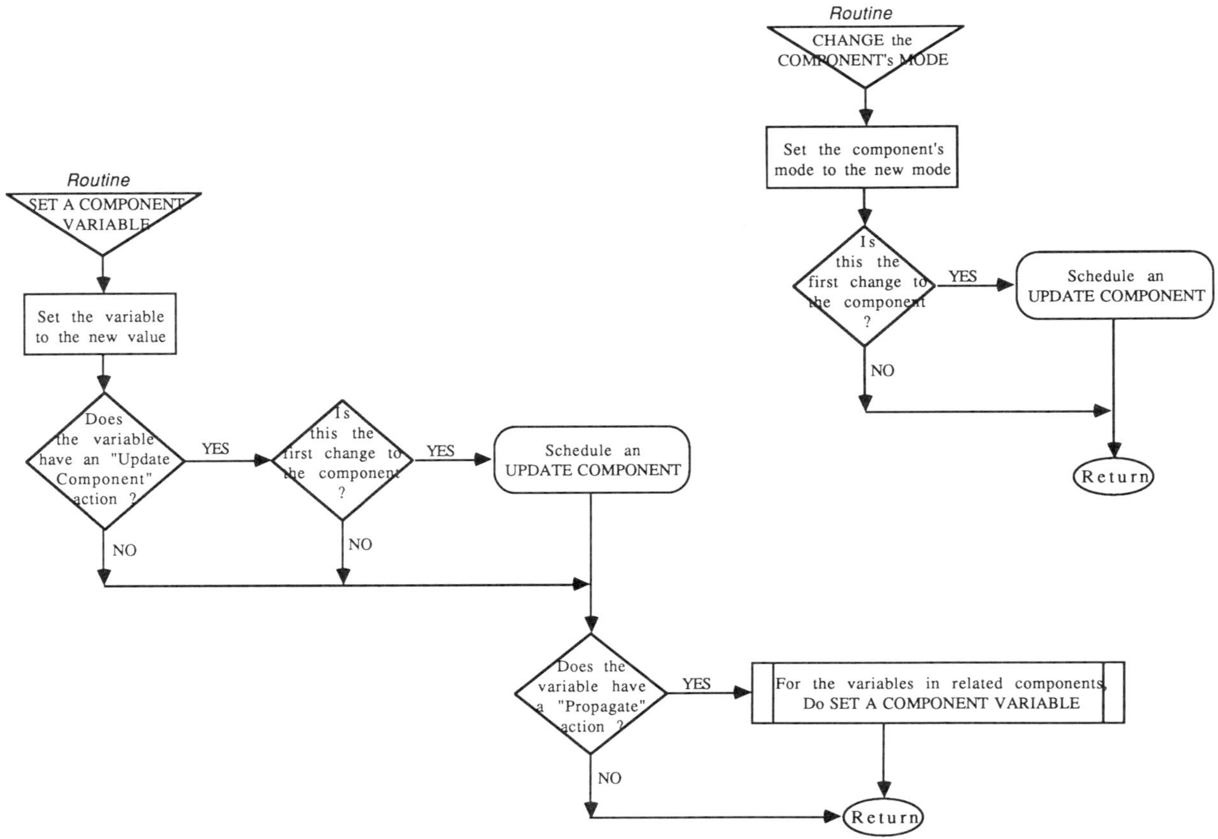

Figure 25. Component attribute routine flow charts. (a) Set a component variable; (b) Change the component's mode.

component in the current time slice. Variable changes can also propagate to related components and cause Update Component events for those components to be scheduled. When change occurs in a Propagate type of variable, the new value is sent immediately to the related variable of all of the components to which the former is connected. This propagation mechanism is basically the standard method used in discrete event simulation to represent and control causal influences between components. When the related component variables are changed, this will usually cause some sort of action to be taken by that component instance, an update for the related component or more propagation (possibly both). In this way, the simulation is driven by propagation of information through relations and by the two events: Effect Execution and Update Component.

In summary, the CONFIG system implements a form of qualitative simulation which uses a modified form of discrete event simulation. In particular, the calendar of the discrete simulator is used to manage Update Component and Effect Execution events. Update Component events are scheduled when a change (either in a variable or in the Current mode) in a component occurs. The Update Component event causes the components processes to activate. the effect statements of these processes may *set* a variable. This may result in another update to the component or propagation of the value to a related component. In the latter case, this may cause an update for a *downstream* component if the related variable is *changed*. In this way, the simulation literally drives itself along.

4.7. System Analysis and Simulations

The purpose of computer-aided simulation is to extend human capabilities to analyze system behavior. Models provide a formal means for describing systems, but often lack the information needed to perform adequate simulation of the physical system. In this object-oriented environment, when a change of design is required, a modification may only need to be made to an element in a library and the user can be assured that all instances in the model of the class definitions will change accordingly.

CONFIG allows the modeler to experiment with both the library design definitions of the components and the specific component configurations set forth in one or more system models. Such experiments may be analyzed or compared by various forms of interactive and batch information. In addition to the analysis information provided by CONFIG, the underlying discrete event simulation tool also provides common statistical information by means of *data collectors*. Several types of data collectors are provided which support both interactive graphical and report-oriented methods of display.

It is CONFIG's analysis information that provides a robust environment for experimentation. The user can easily compare the difference between particular model perturbations by using one or more of the analysis tools. Investigation of component redefinitions require no different approach to analysis. The only difference is the motivation behind each type of analysis. Redefinition analysis is motivated by the user's desire to understand and/or modify the design of the system components, whereas model perturbation analysis is motivated by the need to gain understanding of the system dynamics. The manner in which CONFIG accommodates both motivations and allows the user to move freely between them is a significant feature of CONFIG.

Analysis information may be produced in various manners. Interactive visual analysis can be performed by running a simulation with a portrayal of dynamic activity via bitmap or active trace images. The simpler bitmap images move a variable name along the relations image on the screen whenever a value of that particular variable gets propagated through the relation. The more informative active trace images move with propagation, just as bitmap images do, but change to show the new value that is being propagated rather than just a static variable name. Both types of trace images are associated with particular relation types and may be turned on or off for any number of relations depending on the analysis desired by the end user. An important feature of these trace images is the ability to associate trace speeds with each image. This is useful when the user wishes to slow down the simulation. Interesting simulations may be obtained by assigning different trace speeds to different variables propagated across different relations.

The debug facility allows the user to specify one of several levels of debugging for any class of components. To see a textual description of all the events that occur during a simulation, the user can set the debug level appropriately on the root component class in a library. To see only the information associated with a particular instance of a component, the user can set debug for the desired component instance only. The user may choose to view the information generated by this facility in an interactive mode, or can create a batch log file of the entire simulation, providing a permanent record that may be perused later. These files of simulation history provide an automatic means for comparing the results of simulations. For example, the simplest form of comparison would be to point out where and how two different changes may or may not produce different simulation results. There are many tools already available that perform such tasks on text files.

In most physical systems, the operator of the system is usually only provided with sensor information on which he or she must make all diagnostic decisions. Thus, CONFIG supports the recording and display of information provided by the sensor class of components. A log file of sensor information plus other component information obtained (such as mode changes) from a simulation may be used to support the development and test of diagnostic rules for intelligent decision making or procedures for system operation. In addition, this type of analysis will also help determine instrumentation issues. That is, it may become evident that certain sensors are not needed in the diagnostic process and thus might be eliminated, or that the addition of a particular sensor may provide crucial information needed for a more accurate diagnosis.

To help the user look for particular scenarios and analyze the results of such scenarios, syndromes may be defined in terms of symptom patterns. CONFIG can automatically monitor the simulation and determine which symptoms and syndromes are satisfied. Upon satisfaction of a syndrome, CONFIG will execute the desired action specified by the analyst.

5. Conclusions and Future Work

In this chapter, an approach to integrating qualitative and discrete event simulation and modeling has been presented. The current implementation of CONFIG is a usable working prototype.[7] CONFIG is currently in limited field test, being used to investigate the effects of failures in reconfigurable computer networks with various communication protocols and to model a portion of the Space Shuttle Remote Manipulator System. The objective of the Remote Manipulator System modeling activity is to support development of a diagnostic expert system. Now that the first task of developing a usable prototype that combines the technologies has been accomplished, a number of areas of development need to be pursued.

Additional approaches and capabilities need to be developed to provide libraries of generic models of types of components, faults, processes, and relations. More work is needed to provide capabilities for modeling procedures and functions. Additional approaches and capabilities are needed to support model composition, translation and abstraction, to exploit component groupings and hierarchies.

Approaches for representing and controlling qualitative ambiguity that are already available in the qualitative modeling literature and discrete event literature need to be more fully explored. The process language capabilities for defining operations will need to be augmented to permit the

[7] CONFIG is covered by a pending patent application by the National Aeronautics and Space Administration.

designer to mark ambiguities and implement selection and pruning policies. Test implementations in the Process Language of various approaches to qualitative values and algebras such as orders of magnitude (Raiman, 1986; Mavrovouniotis and Stephanopoulos, 1987) and numerical intervals (Simmons, 1986) are needed. Use of global criteria for pruning possibilities (Struss, 1988; Lee and Kuipers, 1988) needs to be explored. Additional capabilities for trend interruption and rescheduling need to be explored, including methods for identification and analysis of potentially interacting try-process events on the event list to identify possible reversals of event sequences.

Additional capabilities need to be developed to automate and assist a range of engineering and operations tasks: model-based development and validation of fault management expert systems, failure modes and effects analysis, and on-line model-based fault management.

Acknowledgment

The authors wish to thank Jodi Seaborn for developing the thermal library and model that were used as a source of examples for this chapter. Kenneth Forbus, Eduardo Morales, and Héctor García provided helpful comments on an earlier draft of this chapter.

References

Abbott, K. H. Robust operative diagnosis as problem solving in a hypothesis space. In *Proc. AAAI-88*, St. Paul, MN, San Mateo, CA: Morgan Kaufmann, 1988, pp. 369–374 (1988).

Davis, R., and Hamscher, W. Model-based Reasoning: Troubleshooting. In *Exploring Artificial Intelligence* (H. Shrobe and AAAI, eds.), San Mateo, CA: Morgan Kaufmann (1988).

de Kleer, J. The origin and resolution of ambiguities in causal arguments. *Proc. IJCAI-79*, Tokyo, Japan, San Mateo, CA: Morgan Kaufmann, pp. 197–203 (1979).

de Kleer, J., and Brown, J. Assumptions and Ambiguities in Mechanistic Mental Models. In *Mental Models* (D. Gentner and A. Stevens, eds.), Hillsdale, NJ: Lawrence Erlbaum (1983).

de Kleer, J., and Brown, J. A qualitative physics based on confluences. *Artificial Int., 24,* pp. 7–83 (1984).

Fishman, G. S. *Principles of Discrete Event Simulation*. New York, NY: Wiley (1978).

Forbus, K. Qualitative Process Theory. *Artificial Int.*, 24, pp. 85–168 (1984).

Forbus, K. Qualitative Physics: Past, Present, and Future. In *Exploring Artificial Intelligence* (H. Shrobe and AAAI, eds.), San Mateo, CA: Morgan Kaufmann (1988).

Hayes, P. The Naive Physics Manifesto, In *Expert Systems in the Microelectronic Age* (D. Michie, ed.), Edinburgh, Scotland: Edinburgh Univ. Press (1979).

Kuipers, B. J. Qualitative simulation. *Artificial Int., 29,* pp. 289–338 (1986).

Lee, W. W., and Kuipers, B. J. Non-intersection of trajectories in qualitative phase space: A global constraint for qualitative simulation. *Proc. AAAI-88*, Saint Paul, MN, San Mateo, CA: Morgan Kaufmann, pp. 286–290 (1988).

Malin, J. T., and Lance, N. Processes in construction of failure management expert systems from device design information. *IEEE Trans. on Systems, Man, and Cybernetics, SMC-17*, pp. 956–967 (1987).

Mavrovouniotis, M., and Stephanopoulos, B. Reasoning with orders of magnitude and approximate relations. *Proc. AAAI-87*, Seattle, WA, San Mateo, CA: Morgan Kaufmann, pp. 626–630 (1987).

Pan, J. Y. Qualitative reasoning with deep-level mechanism models for diagnoses of mechanism failures. *Proc. 1st Conf. Art. Int. Applications*, Denver, CO, Los Angeles, CA: IEEE Computer Society, pp. 286–290 (1984).

Pearce, D. A. The induction of fault diagnosis systems from qualitative models. *Proc. AAAI-88*, St. Paul, MN, San Mateo, CA: Morgan Kaufmann, pp. 353–357 (1988).

Pople, H. Heuristic methods for imposing structure on ill-structured problems: The structuring of medical diagnostics. In *Artificial Intelligence in Medicine* (P. Szolovits, ed.), Boulder, CO: Westview Press (1982).

Raiman, O. Order of magnitude reasoning. *Proc. AAAI-86*, Philadelphia, PA, San Mateo, CA: Morgan Kaufmann, pp.100–104 (1986).

Simmons, R. "Commonsense" arithmetic reasoning. *Proc. AAAI-86*, Philadelphia, PA, San Mateo, CA: Morgan Kaufmann, pp. 118–124 (1986).

Struss, P. Global filters for qualitative behaviors. *Proc. AAAI-88*, St. Paul. MN, San Mateo, CA: Morgan Kaufmann, pp 274–279 (1988).

Williams, B. Qualitative analysis of MOS circuits. *Artificial Int., 24*, pp. 281–346 (1984).

Williams, B. Doing Time: Putting qualitative reasoning on firmer ground. *Proc. AAAI-86*, Philadelphia, PA, San Mateo, CA: Morgan Kaufmann, pp. 105–112 (1986).

Zeigler, B. P. *Multifacetted Modelling and Discrete Event Simulation.* Orlando, FL: Academic Press (1984).

3 An Expert System for Diagnosis of a Sequential, PLC-Controlled Operation

DOUGLAS R. MYERS AND
JAMES F. DAVIS

Ohio State University
Department of Chemical Engineering
Columbus, Ohio

CHARLES E. HURLEY II

Westvaco
Bleached Board Divison
Covington, Virginia

I. Introduction

1.1. Diagnosis of PLC-Controlled Operations

Programmable Logic Controllers (PLCs) are logic-based computers heavily used in the manufacturing industries for the control of machine and process operations (Jones and Bryan, 1983; Otter, 1988; Pessen and Huebl, 1979). Modern-day PLCs are flexible control tools offering the ability to integrate on/off, analog, and positioning controls. Furthermore, the "programmable" feature makes it possible to program and/or alter complex logical algorithms readily. Covering a wide spectrum, PLC-controlled operations include such applications as batch processing finished product handling and wastewater treatment in the chemical/petrochemical industry; injection/blow molding, metal casting and energy management in the discrete-part manufacturing industry; and digesting, coating, and wrapping in the pulp and paper industry (Jones and Bryan, 1983). Providing primarily a decision-making device, the PLC receives information from field sensors, performs logical tests on the input data, carries through

ISBN 0-12-480575-2

logical computations, and specifies output commands that can be used to drive various field devices.

For all PLC applications, diagnosis plays a key role in day-to-day operations. Field sensing devices can be in error, process conditions can change, mechanical components can wear, and malfunctions can occur in the PLC itself. Because an intimate relationship exists between the functionality of the physical operation and the functionality provided by the PLC program code, diagnosis necessarily involves simultaneous considerations of the PLC code, the operability of the PLC itself, and the operability of the field devices. However, these considerations can often be numerous and complex, in which case diagnostic tools are needed to resolve the situation expeditiously.

In this discussion, we focus on those PLC applications that involve sequential operations. Two diagnostic states are recognized. The first is the *dead operating state* scenario in which a malfunction(s) causes an operation to come to a complete stop. Generally, the cause is a faulty field device which results in an unsatisfied permissive in the PLC logic and prevents continued operation. The second is the *degrading performance state* in which the operation has noticeably deteriorated, but does not stop. In this scenario, the PLC must be receiving appropriate field measurement data necessary to satisfy its logic permissives. Therefore, the cause is either a malfunctioning device which is not associated with a PLC logical test or a degradation in the operation which does not affect an existing logical test. Diagnostically, these two scenarios are distinctly different. In the dead state, data is static and the diagnosis emphasizes the states of the PLC logical tests at the time the machine halts. On the other hand, with a degrading performance situation, the PLC logical tests continue to change since the machine continues the run. The diagnosis, therefore, emphasizes the use of observed information.

While recognizing the existence of distinct differences in the diagnosis of the two scenarios, this presentation discusses only the dead machine diagnostics. Broadly speaking, the diagnosis problem under consideration is then defined as the identification of a malfunctioning field device which, because of an unsatisfied logical test in the PLC code, prevents the operation from continuing.

1.2. Dead State Diagnostics

In examining dead state diagnostics and methods for resolution, it becomes apparent there is a range of complexity in diagnostic situations which affects the ability of the currently used methods and tools to identify malfunctions. The applications that are the simplest to diagnose involve discrete functional steps that require single (very few) input data pieces and output a single field device execution command, are dependent only on the immediately preceding functional step (linear sequence), and contain straightforward logic in processing the input information. The diagnostic complexity increases rapidly when the functionality involves the logical processing of many input data pieces and makes simultaneous determinations on multiple field operations. Furthermore, in complex operations a given functional step often depends upon the completion of a number of preceding steps (not necessarily the step immediately preceding).

Troubleshooting manuals and PLC diagnostic software are the two major types of conventional tools currently used. Manuals are based on an enumeration of data/malfunction relationships, while PLC software is designed for accessing the PLC logical test values and is useful for identifying the logical tests which may have stopped the operation. Both types of tools readily lead to a diagnosis for the simpler applications where often a one-to-one correspondence exists between the logical tests stopping the operation and the malfunctioning device. These conventional approaches, however, do not interpret data in terms of identifying malfunctions. As a result, they are ineffective in dealing with the large number of symptomatic combinations that can occur in complex operations with many interacting steps. In these situations, the conventional methods offer useful data, but the diagnostic emphasis becomes one of selecting and interpreting data to arrive at a diagnostic conclusion, the typical role of an expert diagnostician. Because of the strong need for expert intervention, an expert system offers an approach for automating the diagnosis of these complex operations.

1.3. An Expert System Approach to Automated Diagnosis

In this chapter we describe the critical aspects of a diagnostic expert system called *WRAPITUP*. WRAPITUP has been developed for a PLC-controlled Roll-Wrapping Machine (RWM) which receives, wraps, and labels large paper rolls produced in a paper mill. The diagnostic expert system is a fielded system currently in direct computer communication with an Allen-Bradley PLC-III (Myers *et al.*, 1988, 1989). The system has been thoroughly performance tested in terms of accuracy of diagnostic conclusions and computational speed and has successfully diagnosed a number of actual RWM malfunctions.

The RWM consists of several major stations: the positioning station, the waiting station, the wrapping station, and so forth. To provide a source of specific examples for describing the development of WRAPITUP, we will focus on the positioning station shown in Figure 1. Briefly, the positioning

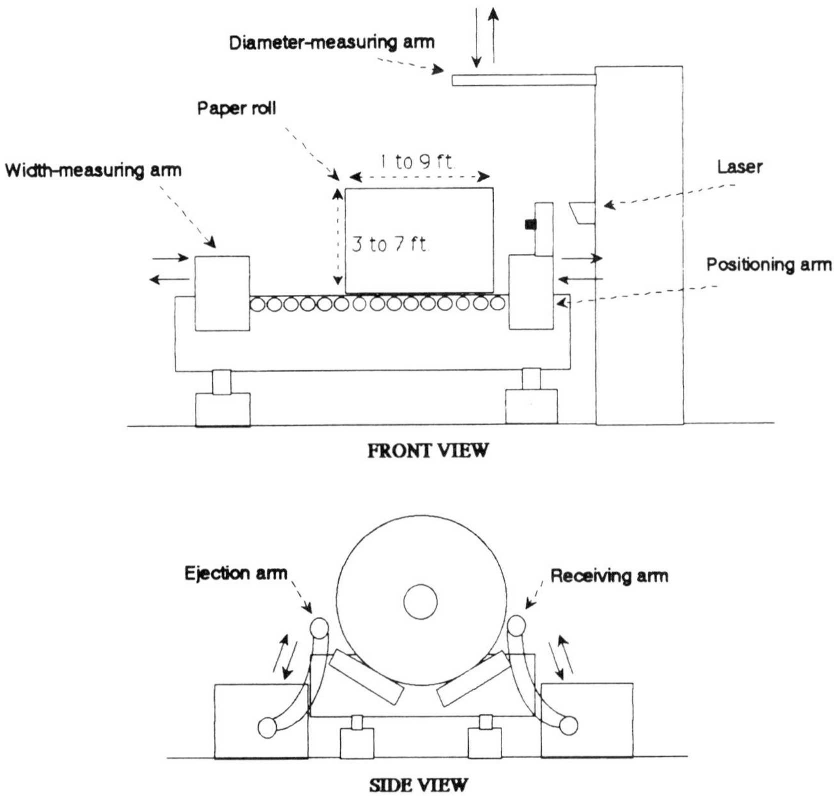

Figure 1. Schematic of the RWM positioning station.

station is the first station that sees a paper roll. This station is responsible for receiving, weighing, measuring the diameter, measuring the width, determining the desired wrap for the roll, positioning the roll for downstream processing, and finally ejecting it from the station.

Functionally, the RWM can be categorized as a complex operation because each functional step (e.g., measuring the width) involves a large number of field device actions and relies upon a number of accompanying sensors. Furthermore, the logic used to coordinate the multiple machine movements not only is extensive but involves a substantial number of nested subroutines. Additionally, there are extensive functional interactions caused by various machine movements depending on the completion of previous steps as executed by the PLC code.

In describing the development of the diagnostic expert system, this chapter emphasizes the characterization, organization and implementation of the knowledge, and the problem-solving strategy. Specifically, it is shown how a combined functional understanding of both the physical process and the PLC program can be used as an effective leverage for reducing diagnostic search. As a result of compiling the PLC program and physical process functionalities into a diagnostically useful form, the resultant expert system is robust and computationally efficient.

2. Programmable Logic Controllers

2.1. Perspective

A simplified schematic of a PLC-controlled operation is shown in Figure 2. Via its input/output (I/O) rack, the PLC receives input information from a variety of field devices such as limit switches, pushbuttons, selector switches, photoeyes, and other types of signal-producing devices. Based on this input information, the PLC processor then makes logical control decisions that lead to output actions such as turning on motors, energizing solenoid valves, and so forth.

The control decisions made by the PLC processor reflect the logical sequencing of control actions. Before the development of the PLC, these control decisions were hardwired using relay control logic. In these hard-

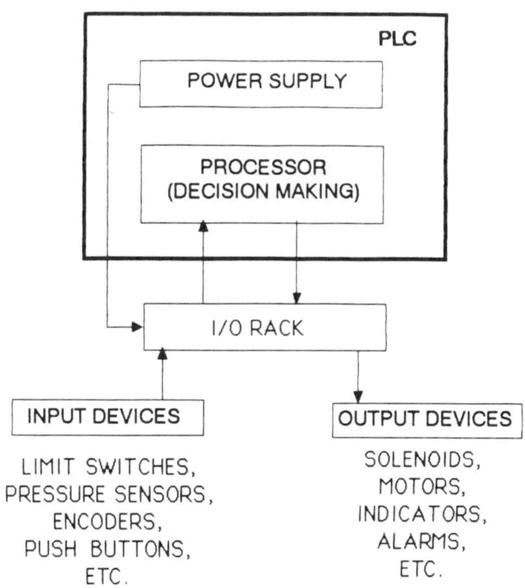

Figure 2. Schematic of a PLC process.

wired systems, modifications to the relay logic often involved a significant amount of rewiring. However, with the development of the PLC, the control decisions are specified in a user-defined program interpreted by the PLC processor. When a change in the control decisions is desired, only the program itself has to be modified. Rewiring the entire process is no longer necessary.

The language commonly used to program and operate a PLC is known as *ladder logic*. Ladder logic is so named because it consists of *rungs*, each of which represents a set of logical tests and associated output commands. While in operation, the PLC uses a scanning process which involves scanning all the input and output devices attached to the I/O rack. Since the PLC operates on digital data, I/O modules within the rack convert incoming electrical signals into digital data and likewise transform outgoing digital PLC commands into appropriate electrical signals. The digital information obtained during the I/O scan is stored in memory for interpretation by the PLC program. As part of the scanning cycle, the main program ladder logic is also scanned in total, beginning with the top rung. The output data is updated based on the commands executed in the program. Depending on the type of PLC, the input/output data are either continuously scanned and updated as the PLC scans the program rungs or are updated after part or all of the program is scanned (Jones and Bryan, 1983). The PLC scanning cycle repeats on the order of once every 1 to 100 milliseconds, depending on the configuration of the PLC system (Jones and Bryan, 1983).

2.2. Program Example

The decision making occurs when the PLC scans the ladder logic. Figure 3 shows a simple PLC ladder logic program consisting of two rungs, labeled 1Ø1 and 1Ø2. Output commands are located on the far right-hand side of the rung and logical tests are placed along the rungs to the left. During the program scan, the PLC examines the logical tests on each ladder rung. If these logical tests are all satisfied, a path from the left side of the rung to the right side is made and the output commands on the right side are invoked. If no path can be made, its output commands cannot be invoked and the PLC scan proceeds to the next rung.

The logic tests are usually performed on input or output data contained within the PLC. These data take the form of bits, having binary values of 1 or 0 (on or off), or *words*, which can be used for both integers and floating-point numbers. The PLC stores these as bit addresses and word addresses, respectively. The "I:ØØØØ/ØØ" listed above rung 1Ø1 in Figure 3 is a bit address, with the prefix "I" indicating that it is an input value. Similarly, the "O:ØØ1Ø/ØØ" at the end of this rung is an output bit address, indicated by the "O" prefix.

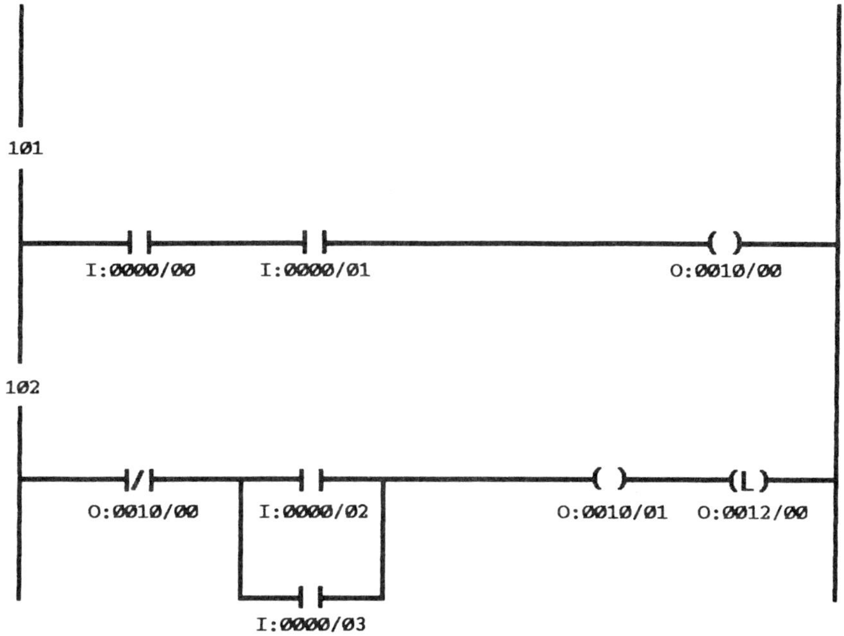

Figure 3. A PLC ladder logic program example.

Many of the simple logic tests are performed on binary bit address data. All of these tests are represented by specific symbols in the ladder logic. As examples, two of the more common logic tests are represented in the PLC language by the following symbols:

-| |- *Examine On*. The test is true if it has a value of "1."
-|/|- *Examine Off*. The test is true if it has a value of "0."

Output commands given on the right side of the logic rungs are used to invoke output actions. Some of the PLC program symbols for the more common output commands are:[1]

-( )- *Output Energize*. Should the logic rung indicate a true path to this command, the associated output bit address will be given a value of 1 as long as the true path exists. Whenever the conditions on the rung become false, the output bit address goes to 0.

[1] These output command symbols are often referred to as *coils*. This term reflects the development of PLCs as a programmable tool to replace hardwired relay logic.

-(L)- *Output Latch.* Once a true path reaches this command, the associated output bit address is given a value of 1. The output bit address will remain latched (remain a value of 1), even if the conditions on the rung become false. The output command will remain latched until it is explicitly unlatched.

-(U)- *Output Unlatch.* Once a true path reaches this command, the associated output bit address is given a value of 0. Latched bit addresses can be unlatched by this command only.

For an example using the logic tests and output commands, refer again to Figure 3. Rung 1Ø1 shows that two *Examine On* tests in series are equivalent to an AND statement, while rung 1Ø2 shows that two of the tests in parallel are equivalent to an OR statement. The logic and output commands represented by these rungs can be stated as follows:

RUNG 1Ø1:
If input bit address I:ØØØØ/ØØ is on (value of 1) AND input bit address I:ØØØØ/Ø1 is on, then energize output bit address O:ØØ1Ø/ØØ (assign it a value of 1).

RUNG 1Ø2:
If output bit address O:ØØ1Ø/ØØ is not energized (value of zero) AND either input bit address I:ØØØØ/Ø2 OR I:ØØØØ/Ø3 are on, then energize output bit address O:ØØ1Ø/Ø1 and latch output bit address O:ØØ12/ØØ.

Although the PLC involves simple relay logic operations based on binary bit address data, most PLCs have added functionality for more complex decision making. Timers, counters, and arithmetic operations can be included on the logic rungs along with instructions for manipulating word and file address data (*Allen Bradley Bulletin*, 1983; Jones and Bryan, 1983; Otter, 1988; Pessen and Huebl, 1979). PLC languages typically also include various means for modifying the flow of program control such as jumping to other parts of the ladder logic code (subroutines are called in this manner). For example, a ladder logic rung may test whether one word address value is greater than another. If true, then an output command may write a data file to a portion of the PLC memory, which in turn may be read by another PLC subroutine. Each ladder logic rung can use various combinations of these more complex types of decision making.

2.3. Diagnostic Complexities in Sequential Process Operations Using PLCs

The complexity of PLC applications can range from relatively simple processes involving a few hardware items and a PLC program of several rungs to highly complex processes involving hundreds of hardware devices and a program consisting of several thousand rungs. Likewise, depending on the complexity of the particular application, the functional relationships between sequential steps and a physical process may range from very simple, independent, sequential functions to highly complex, interdependent, nested functionalities. As a result of the interplay between both the scope and the functional interdependence of the application, diagnosis may range from a relatively simple task to a very difficult one.

As an example of a simple sequential operation, consider a small portion of an interlock system in which a pump motor M1 cannot be turned on until an upstream valve V1 is opened. An analog device measures the valve position and, via an A/D converter, sends a discretized signal to the PLC indicating the percent open on a scale of 0 to 100. The PLC program tests whether the valve position is greater than 40. If true, the output command at the end of a ladder logic rung energizes the output bit address which turns on the pump motor.

In the event the operation stops before the pump is turned on, the first piece of symptomatic information is the observation that motor M1 has not turned on. With the relationship that M1 is controlled by a PLC requiring the V1 valve position, the observation that M1 is not on localizes the diagnostic space to the valve system (the valve, valve position measuring device, connections to I/O rack, A/D converter module, etc.) and the motor drive system (I/O module, connections to the I/O rack, motor, etc.). Given that other process interlocks operate independently of V1 and M1, other portions of the process do not have to be considered in the diagnosis. The specific relationship between V1 and M1 as dictated by the PLC must be understood to carry the diagnosis through to this point.

With the diagnosis focused specifically on the motor drive and valve systems, the input word address value in the PLC corresponding to the valve position becomes a useful piece of information for continuing the diagnosis. If this word address value is less than 40, then the malfunction must exist somewhere in the functionality of the valve system. If, on the other hand, the valve position value is greater than 40, then the diagnosis must focus on the motor system. In this example, knowing the simple relationship between the PLC program and the process allowed identification of the valve position value for quickly isolating the malfunction to a few potential hardware items.

As a diagnostic scenario, this interlock example is straightforward once the V1 position is identified as a crucial piece of data. Functionally, turning on the motor requires no complicated logic and depends only on opening the valve, the step immediately preceding. Additionally, the M1 step is functionally simple, involving a single input (V1), and results in a single output command (turning on M1).

The automated roll-wrapping machine (RWM), for which the expert system was developed, is an example of a complex PLC operation as pointed out in the introduction. The RWM process is complex for the opposite reasons that the interlock example above is considered simple. First, the RWM involves sequences of complex events which entail multiple inputs, outputs, and field devices. Each of the many functional operations executed at the various stations involves precise execution of combinations of physical hardware devices. Furthermore, the RWM is designed to both move and process the paper rolls as fast as possible. As a result, where possible, portions of the process are carried out in parallel. Additionally, many functional sequences depend upon the completion of multiple steps. Careful timing of these multiple process steps is critical since many of the steps interact. For example, a robot arm measuring the roll must be out of the way before another robot arm moves the paper roll. Lastly, the RWM must accommodate a variety of roll sizes and wrap types that are different from roll to roll. The PLC logic for automatically detecting roll dimensions, selecting wrapper types, and physically maintaining a precise wrap quality is, therefore, quite complex.

3. The Dead Operating State Diagnostic Scenario

3.1. Dead Operating State

During the operation of a process controlled by a PLC, a condition may occur where one or more functions of the process come to a complete halt. We refer to this total stoppage of an operation as a *dead operating state*. This dead state occurs when a single permissive or combination of permissive within a portion of the ladder logic are not satisfied, therefore preventing the PLC from carrying out a particular function. The PLC is still functioning[2] and attempting to scan its program, but the proper conditions for continuing the process are not seen by the PLC. The ultimate cause of this dead state is, therefore, some actual hardware malfunction external to the PLC, such as a faulty limit switch, a bad input/output card, an inoperative solenoid valve, and so forth.

If the operation involves major process functions that are essentially independent of one another, then a dead operating state may affect only one of the functions. In this case, that particular function would stop while the other process functions would progress normally. Generally speaking, though, many operations controlled by PLCs involve sequences of func-

[2] We are making a reasonable assumption that a malfunction does not exist with respect to the logic in the PLC itself. We do, however, consider malfunctions in the primary components of the PLC.

tions that depend on the completion of preceding functions. The inability to complete one step in the sequence will cause remaining steps in the process to stop and wait for the dead part of the process to complete. Since the entire process ultimately comes to a halt, the PLC data associated with the entire process are static and serve as a diagnostic "snapshot" of the process. This is often the case with the RWM. Since the paper roll has to travel sequentially to various stations in the wrapping process, a malfunction in one part of the process will hold up the rest of the process.

The dead machine scenario is contrasted with an operation that does not come to a complete stop, but instead degrades. We refer to this scenario as a *degrading performance state*. It occurs when a malfunctioning hardware item causes the process to exhibit a deteriorated behavior. Often the cause is mechanical wear. For example, the robot arms of the RWM may not be moving smoothly. A dead state does not occur because the PLC ladder logic permissives are still being satisfied. We recognize the degrading performance situation, but we limit our discussion in this chapter to dead state diagnostics.

3.2. Causes of a Dead Operating State

In most sequential operations, the programmer designs the PLC program to incorporate permissives based on the process data (i.e., feedback from the process) for at least two reasons. The primary reason is to ensure that a particular step in the process sequence has been completed before beginning some other step. For example, the PLC on the RWM checks if the paper roll has been ejected from a station before accepting a new roll. Another important reason, however, is safety. This can be either safety for the equipment itself or safety for the operators involved with running the process. For example, the RWM uses photoeyes at various points in the process to ensure that a path is not blocked by either a human or another object before carrying out a process step.

When a ladder logic permissive is unsatisfied because of unexpected data from a particular measuring device, there are generally two types of causes. One cause may be that the data are incorrect because of a malfunctioning measurement device. For example, the PLC program for the RWM checks to make sure a mechanical ejection arm is physically all the way down by testing the status of a down limit switch. If the limit switch is faulty, it may give a false reading even if the ejection arm is all the way down. The second type of cause occurs when a measurement device is reading correctly, but a malfunction in the operation is causing the unexpected data. For example, the down limit switch may be indicating correctly that the ejection arm is not down. The problem is then some malfunction in a prior event which has prevented the ejection arm from going down. This example illustrates one of the most difficult dead state scenarios. Although one input data permissive (ejection arm down limit switch) is apparently preventing the PLC from continuing the process, the actual cause is elsewhere in the process.

In addition to testing input data values from measuring devices, the PLC ladder logic may also test permissives that are internally set by the PLC, such as output bit addresses. A failure in an internal test may also result in a dead state. The cause will likely be associated with a malfunctioning hardware item, which, in another portion of the ladder logic program, affected how that output bit was determined. Again, because the cause of an unsatisfied permissive is in another portion of the program, the diagnosis of the malfunction can be quite difficult.

3.3. Types of Hardware Malfunctions

In a dead state scenario, we generally assume that the cause of the malfunction is associated with devices external to the PLC. The types of hardware malfunctions can be faulty field devices such as a limit switch giving a false reading or a solenoid that is not activating. Hardware malfunctions can also be associated with devices that participate in exchanging information between the I/O rack and the process. These malfunctions can include faulty I/O modules, bad proportional amplifier cards, or even disconnected electrical wiring. Additionally, malfunctions may be caused by circumstances external to the physical hardware devices themselves. For instance, loss of electrical power or hydraulic pressure can also prevent process functions from being completed.

While it is quite reasonable to assume that hardware malfunctions occur external to the PLC, there are rare occasions when either a software or hardware malfunction associated with the PLC itself may be responsible for a dead operating state. A software problem occurs when a programming error in the ladder logic unintentionally causes a dead state. We assume that such software programming errors are not present since they belong to the realm of ladder logic program design and testing. However, hardware problems associated with the PLC itself are considered part of dead state diagnostics. Key status bits set within the PLC can be used to determine whether a PLC hardware fault has occurred.

4. Diagnosis

4.1. Characteristics of the Diagnostic Task

Generally speaking, once a PLC-controlled operation is in a dead state, the diagnostic goal is to identify the malfunctioning hardware device(s) ultimately responsible. The identification of the malfunction must be at an appropriate level of detail so that a hardware device can be repaired or replaced. As a result, the collection of all hardware devices in the entire process constitutes the diagnostic *search space*. A search space for a typical PLC operation may include various I/O modules, sensors, solenoid valves, motors, hydraulic lines, etc. With the potential for very large search spaces, the diagnostic task becomes one of effectively and efficiently searching this space for the responsible malfunctioning device. Even for relatively simple processes, the search space of all possible malfunctions can often be quite large. As a result, an efficient diagnostic strategy is needed to focus the search so that only a few potentially malfunctioning hardware items have to be considered.

From our investigation, it has become clear that a predominant aspect of diagnosis is selecting pertinent symptomatic information to effectively focus the diagnostic task based on the understanding of process functionalities (Davis *et al.*, 1987; Finch and Kramer, 1987; Ramesh *et al.*, 1988; Shum *et al.*, 1988). For PLC operations, the process is comprised of a sequence of many functions each of which accomplishes a specific purpose. Each functional step consists of the following elements:

- One or more input devices which supply information to the PLC.
- A control scheme implemented by a portion of the PLC ladder logic program.
- One or more output field devices which act together, as directed by the PLC, to accomplish the overall function.

The key benefit of a functional understanding is the ability to collect individual hardware devices into groups or classes. In turn, these groups can be further collected into broader classes representing more generally defined process functions. For example, the RWM has a generally defined process function called *width measurement*. This function comprises various subfunctions including beginning the width arm motion, controlling the speed of the width arm as it impacts the roll, taking the actual width measurement, and retracting the width arm. Associated with each of these subfunctions are collections of hardware components that work together to accomplish the process function. From a diagnostic viewpoint, consideration of specific process functions translates into an effective strategy since many of the individual devices associated with the function can be simultaneously considered in one context.

As stated earlier, localizing the focus of the dead state diagnosis to a single process function, or even just a few related functions, considerably reduces the search space. Only relevant process functions and their associated hardware then have to be considered. As an example of this functional approach to diagnosis, again consider the roll-wrapping machine. The positioning station, the first major station, has the primary function of determining wrap requirements for a particular paper roll and preparing the roll for the subsequent wrapping processes. In carrying out this broadly defined function, several distinct subfunctions must be accomplished in sequence: receiving a paper roll from a lifting scoop, measuring the width, diameter and weight of the roll, reading the roll's bar code identification, checking the wrapping information with the plant mainframe computer, sending the wrapping information to the next station on the roll-wrapping machine, positioning the roll (so that the wrapper will fit evenly when applied at a later station), and finally ejecting the roll to the next station. Each function is accomplished via its own portion of the PLC program along with a variety of input/output hardware.

Now, visualize a dead state diagnosis, in which the operation halts with a paper roll physically located on the positioning station, but with no roll measurements having been taken. By knowing the sequence of functional steps, the diagnosis can immediately be directed to either the receiving or the width-measuring functions.

With this example, we see the advantage of approaching the diagnosis from a functional viewpoint. While investigating the receiving function, the diagnosis is collectively considering all subfunctions and associated hardware items which support it. If a function can be eliminated from consideration during diagnosis, then all subfunctions and associated hardware devices are also removed from consideration. On the other hand, if a process function was not completed, and therefore is believed to be associated with why the operation stopped, then only subfunctions and supporting hardware items associated with that function need to be further considered. Broadly speaking, the diagnostic process is one of selecting appropriate symptomatic information that supports or rejects malfunction hypotheses about various levels of process functions.The most effective way is to proceed in a direction of functional generality to functional detail.

4.2. Physical Process Functionality and PLC Program Functionality

Visual symptoms, along with a basic understanding of the sequence of process functions, are often sufficient for localizing the general functional levels that are likely causes of the dead operating state. However, further refinement with visual information alone is generally impossible because of the integration of the process and the PLC functionalities. For the diagnosis to progress, there must be an understanding of both:

- The PLC program functionality, i.e., how the PLC is programmed to accomplish the desired process functionality.

- The physical process functionality, i.e., how the input and output field devices associated with a process function operate in conjunction with the PLC.

With respect to the PLC program, it is extremely important to understand the relationship between the PLC ladder logic and how the process functions are executed. With respect to the previous receiving/width-measurement example, it is diagnostically important to know that the PLC program encompasses the following decisions and actions that are not immediately evident from consideration of the process alone:

(1) Before a roll is received, the PLC must perform a "handshake" with the lifting scoop[3] thereby giving the lifting scoop permission to deliver a roll to the positioning station.
(2) When a roll is "seen" approaching the positioning station by several photoeyes, the PLC energizes a system to retract a receiving arm that gently receives the roll.
(3) The successful completion of the receiving arm retraction is checked by two limit switches before any measuring process begins.
(4) Before the robot arm for measuring the roll width is allowed to move, a brake normally locking the robot arm in place must be released.
(5) The PLC sets a destination for the width-measuring arm and gives it permission to move. A PLC subroutine employs an algorithm to control the speed of the arm. The subroutine algorithm receives the width arm position via an encoder and dictates an appropriate signal to a proportional amplifier which controls the speed of a hydraulic motor. The PLC stops the width arm when its current position equals its destination.

Since the PLC plays such an integral role in controlling the process, understanding how the PLC functionally participates is essential for effective diagnosis. Without the appropriate knowledge of the PLC program, the diagnosis can easily be misdirected. In the example described, a diagnostician unfamiliar with the PLC operation may suspect a problem with the width-measuring arm and have a mechanic disassemble portions of the hydraulic system that controls the arm. However, the expert, familiar with the PLC logic code, would check to ensure that the receiving function had been successfully completed (since this is a prerequisite for beginning the width-measuring function).

[3] The lifting scoop system is controlled by an entirely different PLC than the one that controls the roll-wrapping machine, thus making an electronic "handshake" necessary.

While an understanding of the PLC functionality is necessary, it is also not sufficient in itself for carrying through the diagnosis. This PLC understanding must be integrated with an understanding of the physical process functionality. It is important to know the purpose of the physical process functions, where they are located in the operation, and how they are accomplished using various hardware components. Furthermore, it is essential to understand what types of information are transferred between the physical process and the PLC and how this transfer occurs. Continuing with the same example, suppose that the PLC is giving a command to move the width-measuring arm, yet the width-measuring arm does not move. An understanding of the physical components in the process is needed to know that:

(1) The PLC first initiates the width-measuring step by releasing a width arm brake and setting a permissive enabling a subroutine for width arm positioning.
(2) A digitally encoded output is sent from this subroutine to a proportional amplifier which controls a proportional hydraulic solenoid value that actuates a hydraulic motor on the width arm. While the PLC dictates the signal to the proportional amplifier for controlling the acceleration/deceleration speeds, the proportional amplifier itself acts as a local controller for the solenoid valve.
(3) The motion of the width arm is complete when a limit switch sees that the width arm has forced the paper roll over to a mechanical stop.

4.3. Using Symptomatic Data in Diagnosis

Since a diagnosis proceeds by determining the functional status of the operation, symptomatic information, therefore, drives the diagnosis. This symptomatic information can be grouped into three general types:

- *Process Observations:* This information includes the halted status of various physical components in the process. Process observations can range from general information, such as which major functions of the entire process are stopped, to more specific information, such as whether a particular robot arm is all the way up. Although such process information is obtained visually, there are exceptions. For example, squeezing a flexible hydraulic line can establish the presence of hydraulic pressure.
- *Internal PLC Data:* This information includes all input/output bit and word addresses as they existed in the PLC at the time the operation stopped.
- *Data External to the PLC:* This information includes field measurements taken on the physical process itself.

To be useful diagnostically, these various types of symptomatic process data need to be interpreted within the functional context of the process. This means that process information cannot be treated merely as a collection of numbers, values, or states. Instead, this information is interpreted to represent a relationship between the observed data and the process functionality. The various relationships are used to focus the diagnostic search by ruling out unlikely possibilities and establishing likely ones. For example, output bit address O:Ø15Ø/Ø2 may have a dead state value of 1, but for diagnosis this piece of information is interpreted to mean that the paper roll has been properly received. We see from this example that an understanding of the PLC program and the process establishes a relationship between the raw process information and the functional status of the dead operation.

The three types of symptomatic information are employed at different phases throughout a dead state diagnosis. First, general observations are initially used to narrow the diagnostic focus to a particular broadly-defined portion of the overall process. The data interpretation at this level involves relating observable events of the process to the normal overall sequence of process functions. (An operator with only a moderate familiarity with the process usually can carry the diagnosis through to this level.) For example, the observation that the width-measuring arm has not moved from home position means that the width-measuring function was not executed. The first phase of the diagnosis, therefore, is one of focusing on a broad functional portion of the process.

In the next phase, the PLC/process functionality associated with this specific portion is examined. Generally speaking, this phase is characterized by a level of consideration that involves understanding what has been observed in relation to the PLC program. For instance, observing that the width-measuring arm has not moved may focus the diagnosis to only those portions of the PLC program involved with the width arm function. However, at this level of detail, further interpretation involves examining the status of "crucial" internal PLC data in relation to process functionality. These crucial data comprise bit or word address values that are used to determine which specific process functions the PLC completed and which were not completed or executed at all. For example, an output bit address may be identified and examined to determine whether the PLC has attempted to release the width arm brake.

In the final diagnostic phase, field tests or more specific observations of the process further focus the diagnosis to a few potentially malfunctioning hardware systems. For instance, a voltmeter might be used to determine whether the width arm is receiving the PLC signal and is physically being released.

5. Diagnostic Methods

5.1. Conventional Approaches

Given the wide range of complexity in PLC-controlled processes, in virtually all practical applications there is some level of need for diagnostic tools for efficient diagnosis and reduction of downtime. The complexity of the diagnosis, however, determines the level of sophistication needed in the tool or, in the case of an existing tool, its effectiveness. The evaluation of a tool reduces to an examination of its ability to identify crucial pieces of symptomatic data for focusing the diagnostic search.

Conventional diagnostic tools can be characteristically categorized as tools that attempt to enumerate relationships between dead state symptoms and malfunctioning devices or tools that offer sophisticated representations of data that aid the diagnostician. These tools are provided to the diagnostician through a variety of media, such as troubleshooting manuals in book or electronic form, control panel indicator lights, text and graphics interfaces to the PLC program, and, more recently, bit mapping fault detection techniques. For PLC applications with lower levels of complexity in terms of the number of field devices and especially functional interdependency, these tools are very effective. The primary reason is that the crucial data are few and can be directly associated with a malfunction. The relationships, therefore, can be either enumerated *a priori* or readily determined by the diagnostician. However, for those applications that involve a large number of functional interactions, these tools do not provide the information for resolving them. The combinations of effects are too many to enumerate exhaustively or to remember.

5.1.1. Troubleshooting Manuals

Whether available in a book form or on-line in electronic form, the troubleshooting manual is a common diagnostic tool developed for an operator or someone not familiar with interpreting internal PLC data. As such, the troubleshooting manual is useful in PLC applications where visual data and easily obtained process data are sufficient for localizing the source of the dead state. Since rather direct relations between data and malfunctions must exist, manuals are most effective when the PLC program functionality involves simple on-off relationships between hardware components. In general, a troubleshooting manual represents a compilation of these functional relationships, so that troubleshooting does not require accessing and evaluating PLC address data. Normally organized into a type of flowchart, the troubleshooting manual directs the user to check various pieces of hardware in relation to visual process symptoms.

For example, the manual may suggest, "if the wash-cycle motor does not engage, check the operation of the level indicator." In this manner, the troubleshooting manual is effective for representing direct relationships between process symptoms and hardware devices.

In complex PLC applications with highly interdependent process functions, simple relationships between process symptoms and hardware malfunctions often do not exist. Functional interpretation of PLC data is needed in addition to the observations and process field data to establish the data/malfunction relationships. Without the additional diagnostic knowledge provided by the interpretation of internal PLC data, a troubleshooting manual that enumerates direct relationships results in the time-consuming examination of a large number of hardware devices.

5.1.2. Software Tools for Diagnosis

Data Display Method. In more complex diagnostic situations where access to internal PLC data is necessary, methods of examining crucial PLC data are required. The most direct method is simply to access the PLC internal database and display the address values, but this can often be inconvenient when only a few key data pieces are frequently needed. To facilitate the investigation of often-used data, several different software approaches have been developed. One approach involves using status indicator lights corresponding to key data bits in the PLC program. Programming additional ladder logic can be used to turn lights on or off, where these indicator lights correspond to a set of conditions in the ladder logic being true or false. Alternatively, external software is also available for providing indicators of various pieces of PLC data. Without increasing the size of the PLC program, this software can access the data in the PLC memory and present it via a display terminal.

While methods for displaying PLC data can be quite sophisticated, it is important to recognize they are useful only as diagnostic tools. These methods do not determine what PLC data is useful, nor do they interpret how this information is to be used in the diagnosis. These decisions are left to a person familiar with the PLC process functionality. To illustrate, consider the valve/motor interlock example previously discussed. A needed piece of diagnostic information to be displayed was the value of the valve position as seen by the PLC. Even with this PLC address being displayed, the diagnostician must know the functional relevance of this information, i.e., the motor cannot be turned on until the value is greater than 40.

As with troubleshooting manuals, for PLC applications where the status of key PLC data is functionally related to a single specific process function, the data display methods can be used effectively by the diagnostician to focus the diagnostic search to a few hardware devices. This is often the case with PLC interlock applications. However, in more complex applications, especially those that use data addresses for more than one purpose or context, the status of PLC data alone cannot effectively focus the scope of the diagnosis. In these situations, observations, field data, and a variety of PLC data are required.

Data Mapping Methods. Data mapping is a more sophisticated software method for examining internal PLC data that provides a form of *automated fault detection.* This software compares a "map" of normal PLC data obtained for a particular scan cycle to a data map for a dead state, and alerts the user to any discrepancies. In some systems, the user can determine which set of individual data addresses should be examined for data map comparisons. Like the data status indicator approaches, these fault detection methods can display data values, but provide the added function of alerting the user to multiple data discrepancies. This tool is particularly useful where a distinct data "map" exists for each hardware malfunction. However, when PLC addresses are used in multiple contexts or when many discrepancies are obtained, the malfunction is still not easily identified. Furthermore, a unique data map does not exist for every malfunctioning hardware item, since malfunctions can be caused by physical events occurring in the process, which cannot be detected by PLC data alone. Automated fault detection systems can offer a useful PLC data analysis technique, but again a functional interpretation of all three types of symptomatic information is still required for an effective diagnosis of complex process.

6. Expert System Development

6.1. Perspective on an Expert System Approach

As the complexity of the diagnostic application increases, conventional methods require more and more human interaction for identifying, interpreting and making decisions about the observed data in an effort to troubleshoot a malfunction. The diagnostic emphasis is seen to shift from one of simply obtaining data which readily indicate a malfunction to one of interpreting the symptomatic process information for focusing the diagnostic search. In these situations, the expert diagnostician plays the key role by bringing an in-depth understanding of the PLC program and the physical process together into an effective diagnostic strategy. The essence of this strategy is a problem decomposition that leverages a functional understanding for efficient search.

While the diagnosis (with only conventional diagnostic tools available) requires an expert who understands the PLC/process functionality and can interpret in some systematic fashion the available data (visual, PLC, and field test), the expert is literally incapable of remembering all detailed interactions that can occur and the data that delineate them. As a result, large complex PLC operations involving thousands of ladder logic rungs, thousands of PLC data addresses and hundreds of hardware devices, require a great deal of time to diagnose even for the expert. The primary reason is that the crucial PLC data, which delineate the various functions, are identified by reconstructing (thinking through) the PLC logic for each function under consideration. This time-consuming process is used even for repeat malfunctions. Thus, the expert diagnostician is very effective in decomposing the problem and deciding what portions of the operation are most likely malfunctioning, but is ineffective in cataloging the large number of PLC addresses and expected values that are used to examine the state of the operation.

Recognizing that the diagnosis proceeds effectively with functional decomposition knowledge and interpretive knowledge associated with crucial symptomatic data, an expert system provides an approach for capturing this knowledge and automating the diagnosis. This approach offers several distinct benefits over the current methodologies:

(1) An expert system documents the functional decomposition knowledge and diagnostic strategy that is leveraged for efficiency. In an electronic form this problem-solving methodology can be made available to the operation.
(2) The expert system can assimilate all three types of symptomatic data into its problem solving.
(3) With respect to the PLC data, the expert system provides a means of cataloging those crucial data and, therefore, preventing reidentification during each diagnosis.
(4) The expert system has the ability of carrying through a complete diagnosis independent of the expert.
(5) As a computer implementation that takes advantage of a cataloging capability, the expert system can perfom the diagnosis much faster (not better) than the diagnostician.

6.2. *The Task Analysis*

An analysis of the diagnostic process reveals several aspects of the expert's approach which constitute the expertise used in solving the problem. First, a strategy is used in which the problem is considered in terms of broad process functionalities. The solution then progresses systematically through more and more specific considerations. Each consideration is associated with a more specific function than the previous one. It is clear that the expert makes use of a hierarchical strategy based upon a functional decomposition of the process to narrow down the number of possible causes of the malfunction, i.e., focusing from more general diagnostic possibilities to those that are more specific (Ramesh *et al.*, 1988). This decomposition is in the form of hypotheses about malfunctions and is organized in such a way that, if a detailed hypothesis is true, then higher level hypotheses must also be true, i.e., if there is a malfunction in the weight measurement system, then the measurements systems and the positioning station must be malfunctioning.

Another aspect of this problem approach is that the hierarchy of malfunction hypotheses can be considered separately from the way in which each hypothesis is examined. In other words, hypothesis testing is an individualized activity associated with each hypothesis and a test used to examine a hypothesis is distinctly separate from those applied to other hypotheses. The hierarchy of hypotheses therefore provides an organizational structure for the tests of the hypotheses. The effect is a modularized system with respect to determining the validity of each hypothesis.

An individual test for a hypothesis consists of a pattern of crucial data pieces which serve to establish or reject the likelihood of a malfunction with respect to the associated function. As discussed earlier, these data pieces included in this hypothesis testing pattern are some combination of visual data, PLC data, and field data. The expert knows what data are needed, where the data are found, and how to construct the data to test for each hypothesis.

The entire process of focusing the diagnosis by hierarchically considering malfunction hypotheses in terms of specific process symptoms is a specific type of problem-solving referred to as *classification* (Chandrasekaran and Punch, 1987; Davis *et al*, 1985, 1987). Recognizing that PLC process diagnostics are predominantly classificatory in nature, in the next sections we discuss the details of capturing the knowledge organization and problem-solving strategy used to develop the expert system for the roll-handling process.

6.3. Selection of the Expert System Implementation Language

The objective of the expert system is to capture not only the knowledge of the PLC and the physical operation, but also a means of using the knowledge effectively to reach an accurate diagnostic conclusion in a very efficient manner. The relevant knowledge that the system leverages for an effective solution is, therefore, specific knowledge about the domain, to be sure, but also the underlying ways of organizing the knowledge and the solution strategies for manipulating the knowledge. We refer to the process of identifying the underlying organizational and problem-solving strategies as the *task analysis*.

As described in the previous section, the task analysis identifies Classification as the predominant problem-solving task in PLC dead state diagnosis. The Classification is achieved through a hierarchical decomposition of the process in terms of functions. The diagnosis progresses from general considerations to detailed considerations by first establishing or rejecting hypotheses about general malfunctioning systems. If established, these general malfunctions are considered in greater detail. If not established, then large portions of the process are excluded from consideration, thereby greatly reducing the number of detailed considerations. Furthermore, each malfunction hypothesis is associated with its own test for determining whether it is a likely possibility or not (establish or reject, respectively). The test consists of combinations of symptoms obtained visually, from the PLC address database and from field measurements.

The recognition that the underlying problem-solving in Classification has led to the selection of CSRL (Conceptual Structures Representation Language), a paradigm-specific programming language which supports Classification. Developed at the Ohio State University Laboratory for Artificial Intelligence and now commercially available through Battelle Memorial Institute, CSRL facilitates the building of classificatory systems by providing directly the appropriate knowledge organizational structures and inferencing mechanisms (Chandrasekaran, 1984, 1986; Davis, 1987; Davis *et al.*, 1987).

Specifically, CSRL provides the programming structures for capturing the decomposition knowledge and the hypothesis-testing knowledge, the two primary knowledge types used in Classification. With respect to the decomposition knowledge, CSRL offers a programming framework for graphically generating a hierarchy of nodes, each of which can be associated with some decomposition concept. In the case of the PLC diagnosis, the concept is function or system. Arcs connecting CSRL nodes represent related layers of functional detail. The organization strategy of focusing the diagnosis through successive layers of process detail is achieved by organizing the nodes to reflect more general functions at the top and successively more detailed functions as the hierarchy is traversed from the top to the bottom. For PLC diagnostic applications the hierarchical decomposition is based upon what we refer to as a *Sequential-Functional* decomposition. The resulting hierarchy corresponds to a diagnostic search space explicitly organized to leverage knowledge for rapidly focusing the diagnosis.

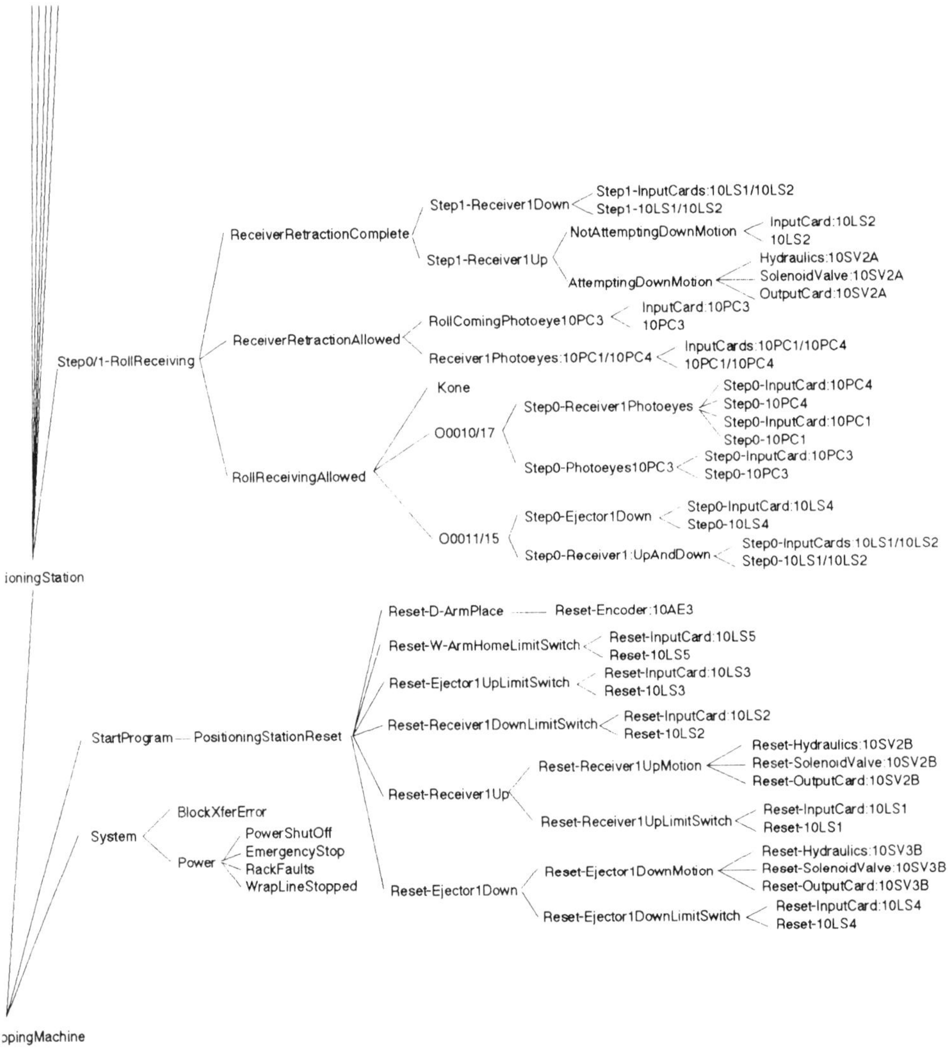

Figure 4. CSRL diagnostic hierarchy for WRAPITUP.

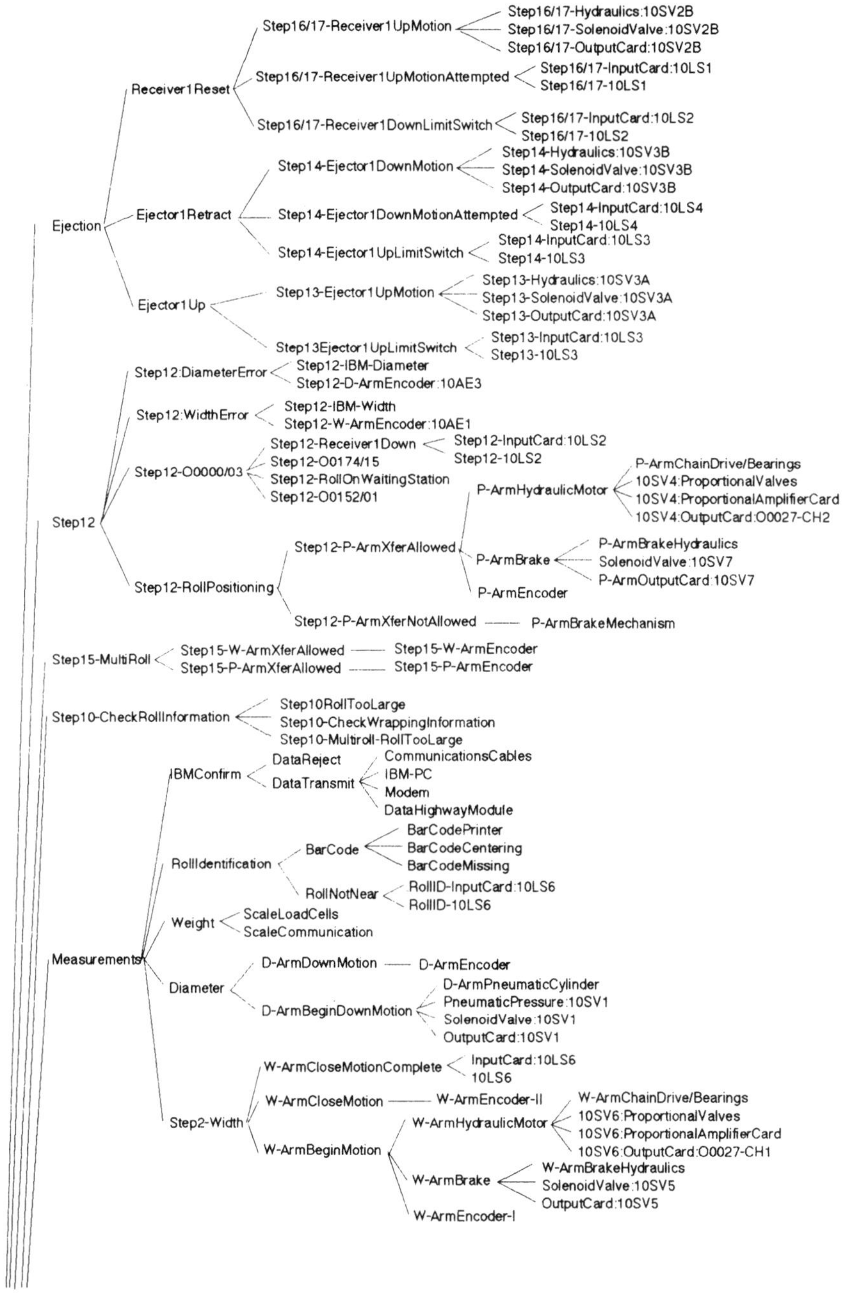
Ejection
Receiver1Reset
Step16/17-Receiver1UpMotion
Step16/17-Hydraulics:10SV2B
Step16/17-SolenoidValve:10SV2B
Step16/17-OutputCard:10SV2B
Step16/17-Receiver1UpMotionAttempted
Step16/17-InputCard:10LS1
Step16/17-10LS1
Step16/17-Receiver1DownLimitSwitch
Step16/17-InputCard:10LS2
Step16/17-10LS2
Ejector1Retract
Step14-Ejector1DownMotion
Step14-Hydraulics:10SV3B
Step14-SolenoidValve:10SV3B
Step14-OutputCard:10SV3B
Step14-Ejector1DownMotionAttempted
Step14-InputCard:10LS4
Step14-10LS4
Step14-Ejector1UpLimitSwitch
Step14-InputCard:10LS3
Step14-10LS3
Ejector1Up
Step13-Ejector1UpMotion
Step13-Hydraulics:10SV3A
Step13-SolenoidValve:10SV3A
Step13-OutputCard:10SV3A
Step13Ejector1UpLimitSwitch
Step13-InputCard:10LS3
Step13-10LS3
Step12
Step12:DiameterError
Step12-IBM-Diameter
Step12-D-ArmEncoder:10AE3
Step12:WidthError
Step12-IBM-Width
Step12-W-ArmEncoder:10AE1
Step12-O0000/03
Step12-Receiver1Down
Step12-InputCard:10LS2
Step12-10LS2
Step12-O0174/15
Step12-RollOnWaitingStation
Step12-O0152/01
Step12-RollPositioning
Step12-P-ArmXferAllowed
P-ArmHydraulicMotor
P-ArmChainDrive/Bearings
10SV4:ProportionalValves
10SV4:ProportionalAmplifierCard
10SV4:OutputCard:O0027-CH2
P-ArmBrake
P-ArmBrakeHydraulics
SolenoidValve:10SV7
P-ArmOutputCard:10SV7
P-ArmEncoder
Step12-P-ArmXferNotAllowed
P-ArmBrakeMechanism
Step15-MultiRoll
Step15-W-ArmXferAllowed
Step15-W-ArmEncoder
Step15-P-ArmXferAllowed
Step15-P-ArmEncoder
Step10-CheckRollInformation
Step10RollTooLarge
Step10-CheckWrappingInformation
Step10-Multiroll-RollTooLarge
Measurements
IBMConfirm
DataReject
DataTransmit
CommunicationsCables
IBM-PC
Modem
DataHighwayModule
RollIdentification
BarCode
BarCodePrinter
BarCodeCentering
BarCodeMissing
RollNotNear
RollID-InputCard:10LS6
RollID-10LS6
Weight
ScaleLoadCells
ScaleCommunication
Diameter
D-ArmDownMotion
D-ArmEncoder
D-ArmBeginDownMotion
D-ArmPneumaticCylinder
PneumaticPressure:10SV1
SolenoidValve:10SV1
OutputCard:10SV1
Step2-Width
W-ArmCloseMotionComplete
InputCard:10LS6
10LS6
W-ArmCloseMotion
W-ArmEncoder-II
W-ArmBeginMotion
W-ArmHydraulicMotor
W-ArmChainDrive/Bearings
10SV6:ProportionalValves
10SV6:ProportionalAmplifierCard
10SV6:OutputCard:O0027-CH1
W-ArmBrake
W-ArmBrakeHydraulics
SolenoidValve:10SV5
OutputCard:10SV5
W-ArmEncoder-I

Fig. 4 (*Cont.*)

To accommodate the hypothesis-testing knowledge, CSRL provides the programming facilities for attaching to each node the patterns of data and their values which serve to establish or reject the particular hypothesis. These data pattens are referred to as *knowledge groups*. Since the expressed purpose of each node is to determine the likelihood of its malfunction hypothesis, CSRL requires an independent test or knowledge group for each node. The broad effect is a hierarchy of independent expert systems which are coordinated to arrive at an overall diagnostic conclusion. The process of establishing nodes that represent successive layers of diagnostic detail is referred to as *establish-refine*.

6.4. Sequential-Functional Hierarchy

Figure 4 shows the CSRL diagnostic hierarchy for WRAPITUP, the RWM application. The nodes in the hierarchy are organized into a *Sequential-Functional* hierarchy that emphasizes not only the functionality of the RWM but also the sequence in which the functions are performed. Each node in the hierarchy is associated with a particular sequential function of the process. Nodes higher up in the hierarchy represent more general process functions, while nodes further down represent more specific process sub-functions.[4] Conceptually, each node represents a hypothesis concerned with determining whether or not the associated sequential function has been properly executed. Knowledge within each hypothesis node is specifically directed toward this determination. If this function has been properly executed, then it is not relevant to the diagnosis and the hypothesis node is ruled out from consideration. By the conceptual definition of a hierarchy, when a hypothesis node is rejected, all of its subnodes are also rejected. On the hand hand, if the knowledge within a hypothesis node determines that its sequential functions has not been properly executed, then this process function must be relevant to the diagnosis. The diagnosis is then refined by considering its subnodes.

For example, consider a malfunction scenario in which the "Measurements" node in Figure 4 is established, indicating the likelihood of a malfunction with the measurement function of the RWM. The diagnosis is refined by pursuing the subnodes: "Step2-Width," "Diameter," "Weight," "RollIdentification," and "IBMConfirm." These subnodes represent more specific measurement subfunctions performed by the RWM. The ordering of the nodes is determined by the order in which these functions are actually carried out. These nodes are then explored one at a

[4] Although we refer to CSRL nodes being organized in a top-down fashion, the nodes are actually displayed in a left to right fashion for more convenient viewing.

time, until one or more are established. This establish-refine process is repeated.

For a given diagnostic session, this process of establishing nodes followed by refining subnodes continues until a tip node, a node with no more subnodes, is established. In the WRAPITUP hierarchy shown in Figure 4, tip nodes are located farthest to the right side of the hierarchy. Instead of corresponding to process functions, these tip nodes correspond directly to potentially malfunctioning hardware devices. Establishing a tip node identifies a particular hardware device as the cause of the dead state. The level of device detail is that which is appropriate for determining the corrective action. Typically, the corrective action is either replacing the device or further testing and repair. The current expert system hierarchy as shown in Figure 4 consists of 188 total hypothesis nodes, 113 of which are tip nodes corresponding to specific malfunctioning hardware items (Myers *et al.*, 1988).

6.5. Hypothesis-Testing Knowledge

6.5.1. Knowledge Groups

Whereas the diagnostic hierarchy as shown in Figure 4 graphically depicts the relationships between the various hypothesis nodes, decision making is performed within each knowledge group to either establish or reject that node. In this sense, each CSRL node is referred to as a *specialist*, since each node essentially contains specialized local knowledge for establishing or rejecting itself. Each specialist has its own *knowledge groups* which contain this local knowledge.

As an example of how knowledge groups are used within a CSRL specialist, refer to Figure 5. This figure shows the information contained in the "D-ArmBeginDownMotion" specialist (node). This node considers the hypothesis of whether the process function for initiating the downward motion of the diameter-measuring arm has been successfully completed or not. Within this specialist there are three sections: *Declare*, *Kgs*, and *Messages*. The Declare section lists the parent node (called a Superspecialist) and all of the children nodes (Subspecialists) associated with the "D-ArmBeginDownMotion" node. The purpose of the Declare section is to establish the nodal relationships graphically depicted in Figure 4.[5]

The Kgs section lists the *knowledge groups* associated with the node. In this particular example, there are three data interpretations needed to

[5] CSRL generates the information in the *Declare* section automatically when a hierarchy is constructed.

```
(Specialist
  D-ArmBeginDownMotion       (* specialist for D-ArmBeginDownMotion)


(Declare
    (Subspecialists
       OutputCard:10SV1 SolenoidValve:10SV1
       PneumaticPressure:10SV1 D-ArmPneumaticCylinder)
    (Superspecialist
       Diameter))

(Kgs
    (D-ArmBeginDownMotion   Table

      (match

         (EQ 1 (DB-FETCH (QUOTE BitAddressValues)
                         (QUOTE 0000:0024/15))
               (* diameter-measuring arm down solenoid 10SV1)

         (GT (DB-FETCH (QUOTE WordAddressValues)
                       (QUOTE N010:0021))
               (* diameter arm home position constant)

             (DB-FETCH (QUOTE WordAddressValues)
                       (QUOTE N010:0020))
               (* diameter arm position))

         (AskYNU? "Is the diameter-measuring arm
                   all the way up at home position")

      with

      (if      T  F  T    then  3
      elseif   T  T  F    then -3
      else     F  ?  ?    then -3))

(Messages   (Establish (SetConfidence self D-ArmBeginDownMotion))))
```

Figure 5. CSRL specialist for D-ArmBeginDownMotion.

establish the "D-ArmBeginDownMotion" node as a likely hypothesis. Stated descriptively, these are:

(1) For this hypothesis to be established, the PLC must be attempting to move the diameter arm downward by energizing pneumatic solenoid valve 1ØSV1 whose PLC output address is OØØØ:ØØ24/15 by giving it a value of 1. If this bit address is a 0, then the PLC has not yet attempted to move the diameter arm downward. A "0" value would rule out this hypothesis.

(2) If the PLC is attempting to energize the diameter arm, then the PLC program tests whether the home position constant of the diameter arm is greater than the position as measured by an encoder. The home position constant is stored in PLC word address NØ1Ø:ØØ21 and its current position is stored in word address NØ1Ø:ØØ2Ø. If the home position constant is greater than the current position of the arm, then the diameter arm has moved from home position. If the value of this test is True, then the current hypothesis can be rejected.

(3) To corroborate, it can be visually observed whether or not the diameter arm has indeed moved.

We see from this example that the diagnosis has specifically focused (via a higher-level diagnostic node) onto the fact that the problem must exist in the diameter-measuring function. In this context, only three data interpretations are required to determine the likelihood of a diameter arm movement malfunction. The selection and interpretation of these particular data with respect to the hypothesis being pursued represent a local modularization of the knowledge.

Referring back to the Kgs section in the "D-ArmBeginDownMotion" specialist shown in Figure 5, there are three expressions corresponding to the three data interpretations needed. The first expression determines whether bit address value OØØØ:ØØ24/15 is equal to 1. The second expression determines whether the word address value NØ1Ø:ØØ21 is greater than NØ1Ø:ØØ2Ø. The third expression asks the user, "Is the diameter-measuring arm still all the way up at home position?"

Below these three expressions is a data-matching table that summarizes the likelihood of the hypothesis in terms of combinations of the values of these three tests. If the values are True, False, True, respectively, then the knowledge group is assigned a confidence value of +3, an arbitrary measure that indicates the hypothesis is *very likely*. On the other hand, if the values are True, True, False, or F ? ?, then the knowledge group is assigned a confidence value of −3 which means *highly unlikely*.[6] These symptom combinations are pre-enumerated[7] (discussed in more detail later). There is no restriction to the number of expressions in the knowledge group or the number of matching patterns.

As illustrated, CSRL provides the capability of assigning various confidence values to a knowledge group via a matching table. These confidence values can vary on a discrete scale of −3,−2,−1,0,+1,+2, and +3,

[6] A question mark "?" within a matching expression simply means that the value of that expression is ignored.

[7] Only relevant combination of symptom-matching patterns are pre-enumerated in Figure 5. Such patterns as T T T or T F F are not possible in the context of "D-ArmBeginDownMotion."

corresponding to a qualitative range from *very unlikely* to *very likely*. We have found that in the case of the roll-wrapping machine, the availability and certainty of the various types of process information used in the diagnostic allow hypothesis nodes to be either absolutely established (+3) or absolutely rejected (−3).

The final part of the specialist's structure is the "Messages" section (as shown in Figure 5). This section sends a message to the specialist to assign itself the same confidence value as the knowledge group. CSRL establishes any hypothesis whose confidence value is +2 or +3, and passes problem-solving control onto a subspecialist. If a hypothesis is rejected, control is passed to another branch of the hierarchy. For example, if the specialist node "Step 0/1-RollReceiving" in Figure 4 is rejected, control is passed on to the "Measurements" node.

6.6. Developing Hierarchies and Knowledge Groups

With the presentation of a pattern-matching representation and strategy for testing malfunction hypotheses, it is important to appreciate the source of this knowledge and the computational advantages of using the knowledge in this form. An examination of the diagnostician's strategy for establishing or rejecting a given hypothesis reveals that the hypothesis is used to identify a chunk (sometimes large) of the PLC program associated with the particular function. Since the objective is to determine if that particular functional step has been completed, the expert traces through the ladder logic and tracks the values of logical tests in an attempt to locate those tests that establish whether or not the step has indeed been completed. This process of tracking through the logic code in the context of a specific malfunction hypothesis results in the identification of a few "crucial" tests for establishing the completion of a step. The difficulty with this approach for the expert is that without these crucial tests properly cataloged, they cannot be remembered and this time-consuming process must be repeated even for reoccurring malfunctions.

Conceptually, the PLC program represents a complete and accurate behavioral description of the operation (a deep behavioral model) (Chandrasekaran and Mittal, 1983a; Kramer and Palowitch, 1985; Kramer, 1987a; Rich and Venkatasubramaniam, 1987). As such, it could be used directly in the expert system to generate the crucial tests by a process of simulation. However, to generate the tests for each hypothesis at run time is computationally inefficient when the tests, once generated, do not change. As a result, the expert system implementation recognizes that the only tests required to establish or reject a hypothesis are a very few crucial tests that can be identified by a single simulation. Furthermore, with the crucial tests identified, the hierarchy of malfunction hypotheses provides an appropriate index for cataloging the tests.

The pattern matching data associated with each node are thus data compiled from a simulation of the specific functional step. As a compiled form of the behavioral model (PLC logic), the conclusions reached are as robust as the behavioral description (Ramesh *et al.* 1988). Using the deep model, malfunctions were simulated for all of the various hardware items on the RWM positioning station, without having to invoke actual malfunctions on the real process. The results were then compiled into the Sequential-Functional hierarchy. Compiling diagnostic knowledge in this form is computationally efficient since the need for run-time simulation is eliminated. The leverage provided by the hierarchical decomposition in rapidly reducing the search space, together with the computational efficiency of the compiled data tests, leads to a very robust, high-performance system.

As an illustration of how the hierarchy and the compiled knowledge are engineered via a functional simulation of malfunctions, refer to Figure 6. The simulation shown involves selecting a failure mode of a hardware item and then determining how this hardware failure (malfunction) results in a dead state. In this example, suppose that solenoid valve 1ØSV3A fails, i.e., the valve does not open when given an electrical signal, and therefore cannot supply hydraulic pressure. This valve normally supplies hydraulic pressure to ejector #1 hydraulic cylinder which forces the ejector arm up. Therefore, a malfunction associated with solenoid valve 1ØSV3A results in the failure of the up motion of ejector #1 when the PLC gives the output command to energize 1ØSV3A. Knowledge of physical process functionality indicates that the downward position of the ejector arm will activate a down limit switch 1ØLS4 (PLC address I:ØØ23/Ø4 = 1) and that an up limit switch 1ØLS3 will not be activated (PLC address I:ØØ23/Ø3 = 0). At this point in the simulation, knowledge of the PLC program is needed to determine if and how a dead state can occur in this scenario.

In analyzing this scenario, it must be determined how, in the context of the solenoid valve failure, data from the ejector #1 up and down limit switches participate in the RWM program. A cross reference (part of the PLC program documentation) lists all of the logical tests and the rung locations for every PLC data address used in the program. Figure 6 shows that 13 different cross references are involved: seven for address I:ØØ23/Ø4 (ejector #1 down limit switch 1ØLS4) and six for address I:ØØ23/Ø3 (ejector #1 up limit switch 1ØLS3). For example, bit address I:ØØ23/Ø3 occurs in examine-on tests on rungs RS-143, RS-184, and RS-223 and in examine-off tests on rungs RM-264, RS-148, and RS-223.[8]

[8] "RM" stands for a rung in the main program and "RS" stands for a rung in a subroutine.

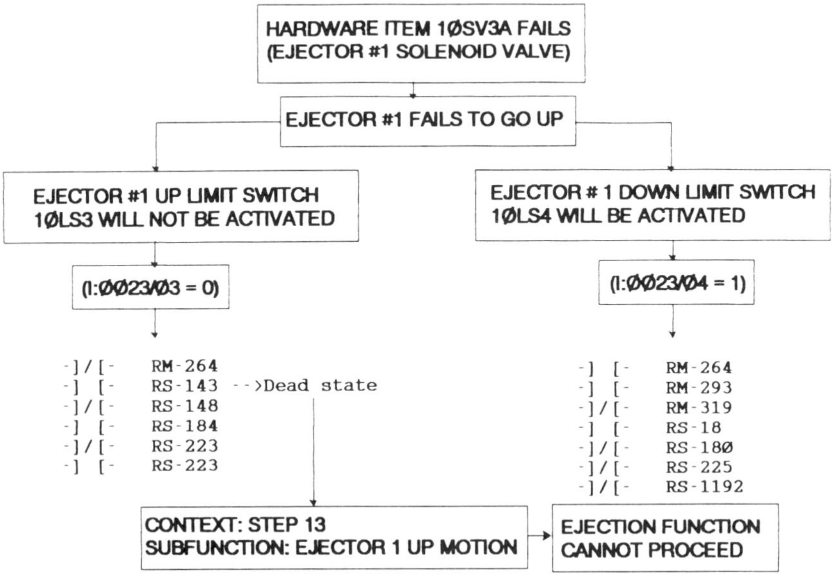

Figure 6. Simulation example for knowledge compilation.

The program segment associated with each of the test locations is analyzed to determine whether a dead state will be induced in the context of the current scenario. By understanding how the PLC scans the program and the specific program functionality of the RWM, it is determined that of all 13 program references to I:ØØ23/Ø3 and I:ØØ23/Ø4, only the -| |- RS-143 (examine on) test will result in a dead state. Figure 7 shows the PLC subroutine called "Step 13" which begins at rung RS-143 and ends at RS-147.

The test on rung RS-143 is an examine-on for I:ØØ23/Ø3, which, if true, will unlatch address O:Ø15Ø/15 (an internal output bit for Step 13) and latch address O:Ø15Ø/16 (a bit for Step 14). The next rung, RS-144, returns (RET) the subroutine to the main program if O:Ø15Ø/16 is 0 (i.e., examine-off is true). The main program is designed to keep calling the Step 13 subroutine until the Step 13 bit address O:Ø15Ø/15 has been unlatched by RS-143. Therefore, a dead operation state occurs if the conditions on RS-143 are not met, i.e., the examine-on test for I:ØØ23/Ø3 is not true. In addition, immediately prior to calling the Step 13 subroutine, the PLC latches the output address for ejector #1 up motion hydraulic solenoid valve 1ØSV3A. Therefore, the overall conclusion of the malfunction simulation of a faulty solenoid valve is:

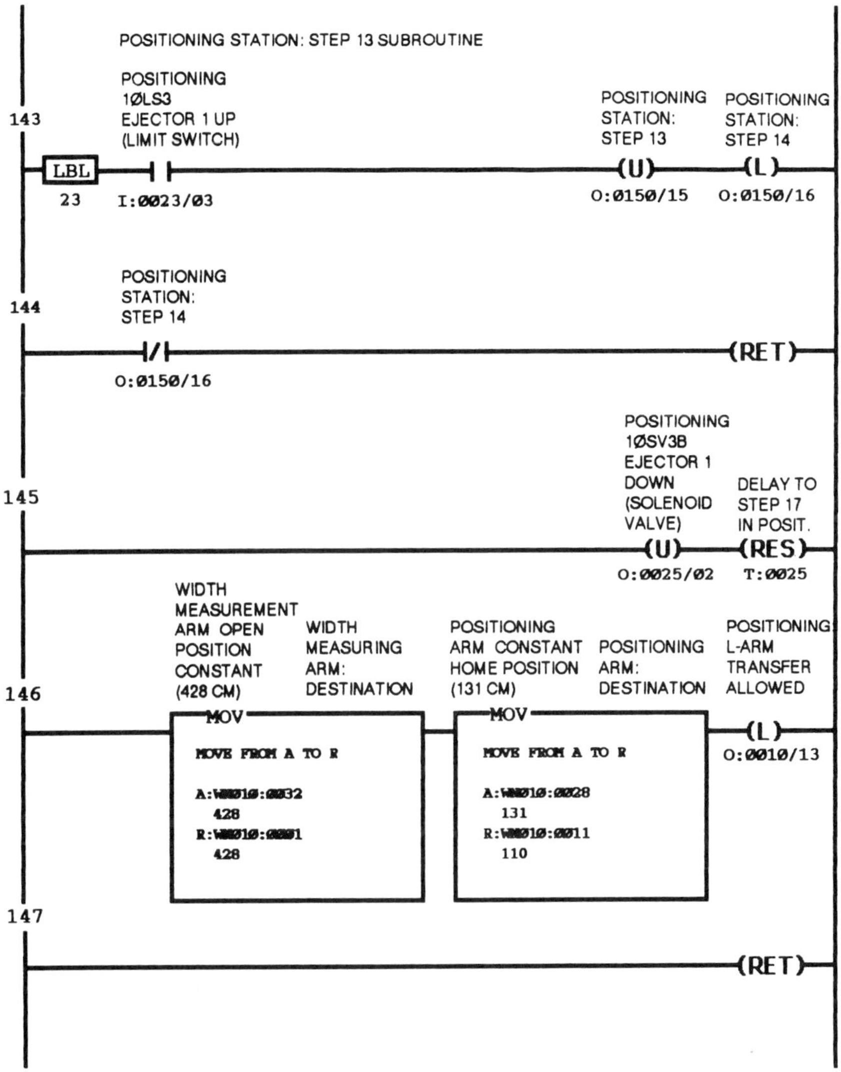

Figure 7. PLC ladder logic program for Step 13 subroutine.

A malfunction associated with solenoid valve 1ØSV3A causes ejector #1 not to go up, resulting in a dead state associated with the examine-on test of I:ØØ23/Ø3 on rung RS-143 of the Step 13 subroutine.

To incorporate the knowledge from this simulation into the Sequential-Functional expert system hierarchy, refer to Figure 8. The simulation

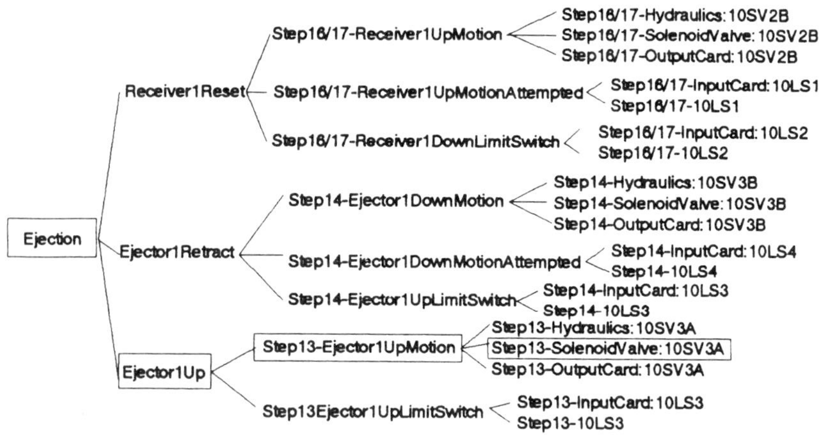

Figure 8. Ejection branch of the diagnostic hierarchy.

example is compiled in the "Step 13-Ejector1UpMotion" node, which tests whether ejector #1 has actually moved up. Visual information of the ejector position along with the status of bit address I:ØØ23/Ø3 on RS-143 is used to establish whether this is a potential malfunction, and the sub-node, "Step 13-SolenoidValve:1ØSV3A," asks the user to perform a field test to identify whether the solenoid valve is the malfunction. The parent node of "Step 13-Ejector1UpMotion," "EjectorUp," uses more general symptomatic information to establish whether the upward motion sub-function of "Ejection" has been executed.

6.7. *Database*

A common feature of expert systems for diagnosis is data-induced, goal-directed behavior. As such, expert systems are designed to elicit various forms of data to produce diagnostic hypotheses. Clearly shown with respect to the D-ArmBeginDownMotion specialist example given earlier, data in the form of bit address values, word address values, and visual data were requested. The values were then used in various tests to establish or respect to the D-ArmBeginDownMotion specialist example given earlier, data in the form of bit address values, word address values, and visual data a hypothesis (Bhatnagar *et al.*, 1986). As a result it is advantageous to have a single and separate database module that can perform the functions necessary to acquire the necessary data. The direct benefits are:

- Separate modules for diagnosis and data acquisition making programming commitments to their respective problems are more efficient and more easily maintained.

- A database module that can be connected to the plant data communication channels serves as a buffer between the plant environment and the expert system.

In WRAPITUP, this concept of an intelligent database for data acquisition stems from the idea that production and analysis of the diagnostically useful data is done more efficiently outside the expert system program. Unlike the diagnostic hierarchy, the structural organization of the intelligent database is centered around objects that represent data types that share the same kind of analysis methods. The database therefore responds to specific questions in the diagnosis module and answers these questions by making use of knowledge about the "sources" of the data. As shown in Figure 5, the "diagnostic request" for a data item is in the form of the DB-FETCH command.

Two sources of data are recognised according to the method needed to access the data. Observed data and field test data are obtained through direct user inquiry. Since no processing of the data is required, the query is made directly from knowledge groups. The PLC data, however, are obtained through communication with the PLC database. To accommodate the various aspects of acquiring and storing the data, an object-oriented database was constructed in which two classes of datatypes are recognized: BitAddressValues and WordAddressValues. The graphical representation of the database is shown in Figure 9. Associated with each of these data objects are methods for examining and validating the data. The actual structure of the data classes is shown in Figure 10. This figure shows a partial list of the bit addresses along with various properties. The properties of these bit addresses include their numerical value (in this figure all the values are NIL, meaning that data in the form of Os and 1s have not yet been received from the PLC) and pertinent documentation of the data addresses.

The database is also responsible for calling upon the PLC database for the necessary bit and word address values and performing limit checking on the data. When a diagnosis is initiated, the database automatically acquires the data values and notifies the user of any missing data. It is important to note that not all data available in the PLC database are useful

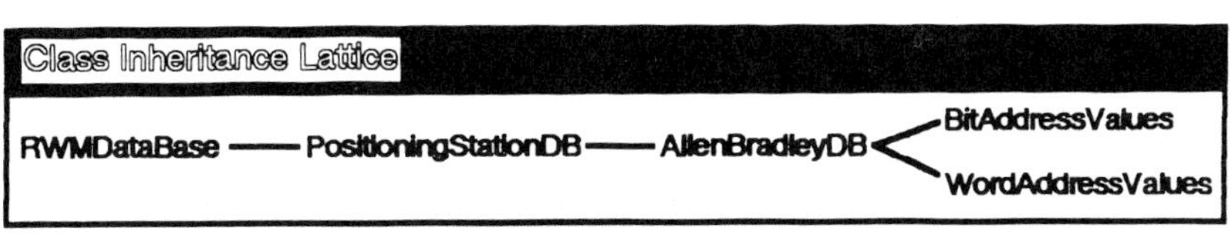

Figure 9. The database hierarchy for WRAPITUP.

```
[DEFCLASS BitAddressValues
(MetaClass Class Edited:                (* edited: "1-May-88 19:10"))
(Supers AllenBradleyDB)
(InstanceVariables (ivs ( I000:0001/07
   I000:0002/00 I000:0002/01 I000:0004/01 I000:0004/02
   I000:0004/03 I000:0004/06 I000:0004/07 I000:0004/11
   I000:0005/00 I000:0023/00 I000:0023/01 I000:0023/02
   I000:0023/03 I000:0023/04 I000:0023/05 I000:0023/06
   I000:0023/07 I000:0023/10 I000:0024/04 I000:0024/05
   I010:0001/00 I010:0001/01 I010:0004/00 I010:0004/01
   0000:0000/03 0000:0004/12 0000:0010/01 0000:0010/03
   0000:0010/13 0000:0010/17 0000:0011/02 0000:0011/03
   0000:0011/07 0000:0011/15 0000:0011/17 0000:0012/04
   0000:0024/15 0000:0025/00 0000:0025/01 0000:0025/02
   0000:0025/03 0000:0025/04 0000:0025/05 0000:0100/01
   0000:0101/16 0000:0102/10 0000:0110/10 0000:0120/00
   0000:0120/17 0000:0150/00 0000:0150/01 0000:0150/02
   0000:0150/03 0000:0150/04 0000:0150/05 0000:0150/06
   0000:0150/07 0000:0150/10 0000:0150/11 0000:0150/12
   0000:0150/13 0000:0150/14 0000:0150/15 0000:0150/16
   0000:0150/17 0000:0151/00 0000:0151/01 0000:0152/01
   0000:0170/01 0000:0170/15 0000:0170/16 0000:0170/17
   0000:0174/13 T000:0022/15 T000:0034/15 T000:0045/15))
 (I000:0001/07  NIL  question  "Please give the value of bit address I000:0001/07
(EQ 1 if block transfers to analog okay)"  doc  (* block transfers to analog okay))
 (I000:0002/00  NIL  question  "Please give the value of bit address I000:0002/00
(EQ 1 if wrap line is on)"  doc  (* wrap line is on))
 (I000:0002/01  NIL  question  "Please give the value of bit address I000:0002/01
(EQ 1 if control voltage okay)"  doc  (* control voltage on))
 (I000:0004/01  NIL  question  "Please give the value of bit address I000:0004/01
(EQ 1 if diameter and width okay)"  doc  (* diameter and width okay))
 (I000:0004/02  NIL  question  "Please give the value of bit address I000:0004/02
(EQ 1 if laser reading okay okay)"  doc  (* laser data checking is complete or manual I.D.
entry is complete))
 (I000:0004/03  NIL  question  "Please give the value of bit address I000:0004/03
(EQ 1 if scale reading okay)"  doc  (* scale reading okay. Move on to Step 9))
 (I000:0004/06  NIL  question  "Please give the value of bit address I000:0004/06
(EQ 1 if scale reading time out)"  doc  (* scale reading time out))
 (I000:0004/07  NIL  question  "Please give the value of bit address I000:0004/07
(EQ 1 if manual I.D. number is requested)"  doc  (* manual I.D. number is requested))
• • •
```

Figure 10. The data structure within the BitAddressValues database node.

for the diagnosis. In fact, the diagnosis requires a relatively small number of data addresses to interpret what has occurred in the PLC program. For WRAPITUP, only 15 word address values and 77 bit address values are potentially needed, while the PLC database contains thousands of data address values.

7. General Aspects of the Expert System: "WRAPITUP"

Figure 11 shows the overall schematic of the expert system WRAPITUP. When the positioning station of the RWM experiences a dead state failure,

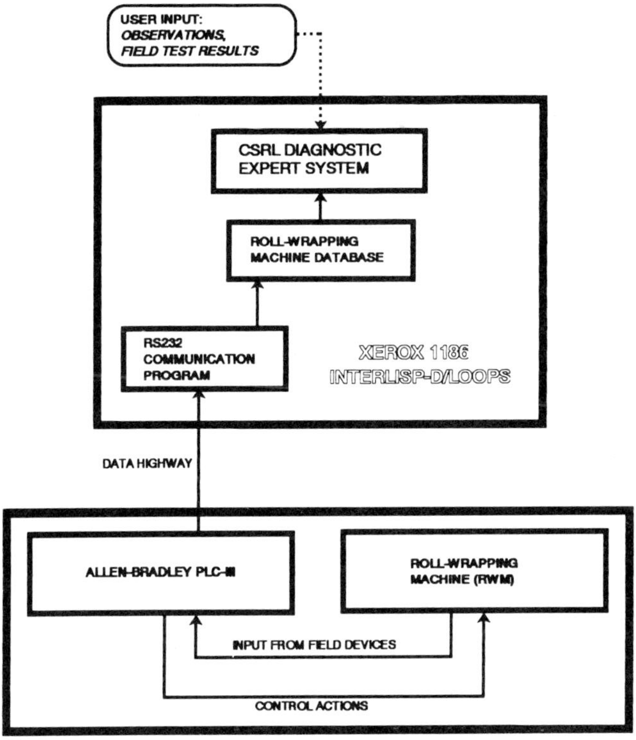

Figure 11. Schematic of the overall expert system.

WRAPITUP, as an advisory system, is invoked by the user. Once started, an RS232 communication program sends an inquiry to an Allen-Bradley (AB) PLC-III for data. The AB returns a series of 15 word address values and 77 bit address values which represent a "snapshot" of diagnostic data to be used by the expert system. The RS232 program checks for data communication errors between the PLC and the expert system[9]. Once the data is received, the expert system begins the actual diagnosis. Although the predominant source of diagnostic data comes directly from the Allen-Bradley, the expert system may also ask questions of the user during the diagnostic session. These questions usually involve supplying visual information, such as "Is the diameter-measuring arm all the way up?" The diagnostic session is complete when WRAPITUP has successfully found a

[9] The expert system computer in the current phase of the implementation is a Xerox 1186 running CSRL on top of a LOOPS/Interlisp-D environment.

malfunctioning hardware device responsible for the dead operating state. Listed below are some overall features of WRAPITUP.

- The expert system currently can diagnose a total of 113 different hardware malfunctions (both electrical and mechanical).
- The visual information requested by the system during a diagnostic session assumes a low level of expertise, i.e., the questions asked are unambiguous and require only Yes or No responses.
- On-line [10] testing has demonstrated that complete diagnostic sessions take only one or two minutes and that three or four questions are asked of the user.
- The system is robust in that virtually every test case has resulted in a correct diagnosis. These test cases include not only those involving intentionally-caused malfunctions generated for test purposes, but also some that occurred during the operation of the RWM. Several of these were previously unseen.
- For cases in which the system does not establish a specific hardware malfunction, the system can be expected to degrade gracefully. When such a diagnostic situation occurs (due to an unforeseen malfunction), the system does *correctly* focus into a very specific malfunction. In such cases, the system is easily augmented to incorporate these new diagnostic situations.

8. Discussion and Summary

8.1. Testing of the Expert System

WRAPITUP has been subjected to two phases of testing to evaluate its performance. Providing the most thorough evaluation of the expert system's robustness and ability to reach accurate diagnostic conclusions, the first phase of testing involving using the system to diagnose field device malfunctions that were intentionally caused. Approximately 50 diagnostic scenarios were run on the system, relating to 50 different malfunctioning hardware items. With some minor refinements in the knowledge, the system was able to identify the malfunctioning hardware item in every case.

In the second testing phase, which is currently progressing, the system is

[10] "On-line" means that the expert system is in direct data communication with the PLC. A snapshot of internal PLC data is requested and received when the user observes a dead state and invokes the expert system.

being used to diagnose actual dead state operating problems on the positioning station. The purpose of this testing phase is to evaluate both the accuracy of the decisions and speed in comparison to the expert diagnostician. In virtually all cases so far, the system consistently has been able to reach the correct diagnosis. Of primary note is that the expert system continues to demonstrate its robustness, even for several diagnostic scenarios previously unseen (some of these scenarios were new to the diagnostician as well). There have been two exceptions in which no specific hardware device was identified. The system, however, did degrade gracefully in that the lowest-level hypothesis established by the expert system was sufficiently detailed to help the diagnostician in identifying both the malfunctioning device and the problems with the expert system knowledge.

With respect to run-time diagnostic speed, the expert system has typically completed a diagnosis in less than two minutes and has required a maximum of 4 pieces of observed and/or field test information from the user. This is substantially faster than the expert who on the average required three hours (ranging from 1 hour to 12 hours) to reach the correct diagnostic conclusion. The result attests to the efficiency of compiled forms of knowledge, the effectiveness of levaraging the decompositional knowledge, and the speed of the computer.

8.2. Maintenance

Once the knowledge in the expert system is established, the ability to maintain and update this knowledge is essential for the viability of the system. To facilitate this, CSRL provides a user interface for modifying or adding onto the hierarchy and for editing the knowledge groups. The hypothesis nodes can be moved, added, deleted, and renamed by simple menu-driven commands attached to the hierarchy. Additional menu commands attached to each node allow the user to edit a knowledge group by adding or changing the desired hypothesis-testing expressions. The mechanism of modifying both the hierarchy and the knowledge groups in this manner is straightforward and easy to learn. The majority of the maintenance effort is appropriately spent on organizing the hierarchy and compiling the necessary hypothesis-testing knowledge.

Due to the robustness demonstrated by the expert system thus far, knowledge maintenance has involved adding only a small number of nodes to reflect information about new malfunctions and modifying knowledge groups to reflect minor changes in the ladder logic. From a maintenance viewpoint, the Sequential-Functional structure of the hierarchy has proven to be an effective framework for modularizing the knowledge for easy modification. Because the hypothesis-testing knowledge is local, changes to a particular knowledge group have no effect on the rest of the system.

8.3. Knowledge Engineering

With the recognition that the diagnosis was classificatory and that CSRL would provide the supporting programming framework, the knowledge engineering involved decomposing the problem into an appropriate Sequential-Functional hierarchy and determining the interpretive knowledge used to test each hypothesis. We found that the analysis of previous diagnostic cases was most useful for establishing a coarse version of the hierarchy. By considering a number of available case studies, an appropriate hierarchical structure could be roughed out, and for many nodes the testing knowledge could be reasonably delineated. It should also be noted that the modularity of the hierarchy was beneficial for focusing the knowledge acquisition process.

While the diagnostic case studies helped develop whole branches of the diagnostic hierarchy, they were not available for every potential malfunction scenario on the RWM. As a result, a procedure was needed to develop the rest of the diagnostic hierarchy for every potential malfunction without actually having to experience each malfunction. This was possible because the PLC program offers a complete behavioral simulation of the process functions. Using the PLC program, each hardware item was hypothetically analyzed to determine what would occur should this hardware item fail. The RWM functions that would be affected were enumerated, along with the symptomatic information used to establish these as potential malfunctions. The result was a very robust diagnostic system which has demonstrated the capability of resolving new malfunctions.

Acknowledgments

This research is supported by the Westvaco Corporation and is also based on work supported by the National Science Foundation under Grant No. CBT-8703746.

References

Allen-Bradley Bulletin 1775: PLC-3 Programmable Controller Programming Manual. Allen-Bradley Company (1983).

Bhatnagar, R., *et al.* An Intelligent Database for Process Plant Expert Systems. *Proceedings of the ISA*, Houston, TX, October 16–21 (1986).

Chandrasekaran, B. Towards a Taxonomy of Problem Solving Types. *AI Magazine*, 4 (1) pp. 9–17 (1983).

Chandrasekaran, B. Expert Systems: Matching Techniques to Tasks. In *Artificial Intelligence in Business*, (W. Reitman, ed.), Ablex Publishing, pp. 116–132 (1984).

Chandrasekaran, B. Generic Tasks in Knowledge-Based Reasoning: High-Level Building Blocks for Expert System Design. *IEEE Expert*, 1 (3) pp. 23–30 (1986).

Chandrasekaran, B. What Kind of Information Processing is Intelligence? A Perspective on AI Paradigms and a Proposal. To appear in *Source Book on the Foundations of AI,* Patridge and Wilks, ed.), Cambridge University Press (1987).

Chandrasekaran, B., and Mittal, S. On Deep Versus Compiled Approaches to Diagnostic Problem Solving. *Int. J. of Man Machine Studies*, 19, pp. 425–436 (1983a).

Chandrasekaran, B., and Mittal, S. Conceptual Representation of Medical Knowledge for Diagnosis by Computer: MDX and Related Systems. In *Advances in Computers* (M. Yovits, ed.), Academic Press, NY, pp. 217–293 (1983b).

Chandrasekaran, B., Smith, J. W., and Sticklen, J. Deep Models and their relation to Diagnosis. Laboratory of Artificial Intelligence Research, Department of Computer and Information Science, The Ohio State University, Columbus, OH (1987).

Chandrasekaran, B., and Punch, W. F. III. Hierarchical Classification, Its Usefulness for Diagnosis and Sensor Validation. Presented at *Second AIAA/NASA/USAF Symposium on Automation, Robotics and Advanced Computing,* (1987).

Chandrasekaran, B., Gomez, F., Mittal, S., and Smith, J. W. An Approach to Medical Diagnosis Based on Conceptual Structures. *Proc. Fifth Int. J. Conf. on AI*, pp. 134–142 (1979).

Davis, J. F. A Task-Oriented Framework Facilitates Expert System Development. In *Proc. Foundations on Computer Aided Process Operations*, pp. 695–700 (1987).

Davis, J. F., Punch, W. F. III, Shum, S. K. and Chandrasekaran, B. Application of Knowledge-Based Systems for the Diagnosis of Operating Problems. Presented at AIChE 1985 Annual Meeting, Chicago, IL (1985).

Davis, J. F., Shum, S. K., Chandrasekaran, B., and Punch, W. F. III. A Task-Oriented Approach to Malfunction Diagnosis in Complex Processing Plants. *NSF-AAAI Workshop on Artificial Intelligence in Process Engineering*, Columbia University, New York (1987).

Finch, F. E. and Kramer, M. A. Expert System Methodology for Process Malfunction Analysis. *10th World Congress on Automatic Control*, Munich, FRG, IFAC, (1987).

Jones, C. T., and Bryan, L. A. *Programmable Controllers: Concepts and Applications.* International Programmable Controls, Inc., Atlanta (1983).

Kramer, M. A. Malfunction Diagnosis Using Quantitative Models and Non-Boolean Reasoning in Expert Systems. *AIChEJ.*, 33, (1) pp. 130–140 (1987a).

Kramer, M. A. Expert Systems For Process Fault Diagnosis: A General Framework. *FOCAPO* (1987b).

Kramer, M. A., and Palowitch, B. L. Expert System and Knowledge-Based Approaches to Process Malfunction Diagnosis. AIChE Annual Meeting, Chicago, IL (1985).

Myers, D. R., Davis, J. F., and Hurley, C. E. Application of Artificial Intelligence to Malfunction Diagnosis of Sequential Operations Which Involve Programmable Logic Controllers. *Proceedings of AAAI Workshop on Artificial Intelligence in Process Engineering*, St. Paul, MN., AAAI, (1988).

Myers, D. R., Davis, J. F., and Hurley, C. E. Application of Artificial Intelligence to Malfunction Diagnosis of Sequential Operations which Involve Programmable Logic Controllers. *Proceedings from Spring AIChE Meeting*, Houston, AIChE (1989).

Otter, J. D. *Programmable Logic Controllers: Operation, Interfacing, and Programming.* Prentice Hall, Inc., Englewood Cliffs, New Jersey (1988).

Pessen, D., and Huebl, W. *The Design and Application of Programmable Sequence Controllers for Automation Systems.* Longman Group Limited, London (1979).

Ramesh, T. S., Shum, S. K., and Davis, J. F. A Structured Framework for Efficient Problem Solving in Diagnostic Expert Systems. *Computers in Chemical Eng.*, 12 (9/10), pp. 891–902 (1988).

Rich, S. H., and Venkatasubramanian, V. Model-Based Reasoning in Diagnostic Expert Systems for Chemical Process Plants. *Comput. Chem. Engng.*, 11 (2), pp. 111–122 (1987).

Shum, S. K., Davis, J. F. Punch, W. F. III, and Chandrasekaran, B. An Expert System Approach to Malfunction Diagnosis in Chemical Plant. *Comput. Chem. Engng.*, 12 (1) pp. 27–36 (1988).

4 Fault Detection and Diagnosis Using Artificial Neural Networks[1]

JOSIAH C. HOSKINS[2] AND
DAVID M. HIMMELBLAU

Department of Chemical Engineering
University of Texas at Austin
Austin, Texas

Abstract

Artificial Neural Networks (ANNs) have proven effective at solving problems in a wide variety of areas, such as image processing, speech recognition, and motor control. We describe here the application of ANNs to fault detection and diagnosis. Fault detection is a type of pattern recognition, and ANN classifiers appear to be a reasonable alternative to traditional classifiers.

ANNs consist of a number of simple neuron like processing elements locally interacting through a set of unidirectional weighted connections. Knowledge is internally represented by the values of the weights and the topology of the connections. Learning involves modifying the connection weights. These networks can learn and adapt themselves to inputs from the actual processes, thus allowing representation of complex engineering systems, which are difficult to model either with traditional physical engineering relations or knowledge–based expert systems.

Several factors motivate the use of artificial neural networks for fault diagnosis: their potential for high computation rates provided by massive parallelism, their capacity for a greater degree of robustness or fault

[1] Reprinted and adapted with permission from *Computers and Chemical Engineering*, Volume 12, J. C. Hoskins and D. M. Himmelblau, "Artificial Neural Network Models of Knowledge Representation in Chemical Engineering," Copyright 1988, Pergamon Press, PLC.

[2] Present address, Schlumberger Laboratory For Computer Science, Austin, Texas.

ISBN 0-12-480575-2

tolerance due to the distributed representation, and their ability to adapt and continue learning to improve performance. Furthermore, they can accommodate the uncertain, nonstationary, and nonlinear nature of typical chemical processes.

This chapter describes the features of neural networks that are desirable for knowledge representation for fault detection in chemical engineering processes. It explains the characteristics of neural networks, focusing on the nodes, the connections, and their learning capabilities. The chapter also describes a neural network design and simulation environment that can be used to study practical applications.

To demonstrate the feasibility of applying artificial neural networks to steady-state processes, we describe fault detection and diagnosis in two examples of simple chemical operations. The compactness and ease with which fault discrimination occurred in both cases emphasize the elegance as well as the pragmatics of the ANN approach. Applied to fault diagnosis, ANNs obviate the need for explicit feature extraction. Once the network was trained, it provided rapid classification since the propagation through the network is essentially a series of simple matrix operations that may or may not be executed in parallel. The results demonstrate the effectiveness of artificial neural networks for real-time on-line discrimination among faults.

1. Introduction

Artificial neural networks (ANNs) have proven effective in solving problems pertaining to pattern recognition such as speech recognition, image processing, and motor control. The application of ANNs to fault detection and diagnosis in engineering is relatively new (Hoskins and Himmelblau, 1988a), although fault detection and diagnosis is well known as a form of pattern classification. In the field of chemical engineering, ANNs appear to be particularly suitable for tasks requiring pattern recognition or continuous input-output control, especially for processes with uncertain models and data (Hoskins and Himmelblau, 1988b).

Several factors motivate the use of ANN models (Lippmann, 1987):

- They have the ability to use vast amounts of sensory information.
- They can respond with high speed to sensory inputs. Other techniques are limited by the time it takes to execute them in electronic hardware.
- They exhibit a greater degree of robustness and fault tolerance due to their distributed representation.
- They have general mapping capabilities, whereas traditional fault detection techniques involve detailed modeling of the system.

- The knowledge in an ANN is largely specified through learning, whereas most other techniques rely on a prespecified algorithm.

Many areas of chemical engineering, such as fault detection and diagnosis, process control (Hoskins and Himmelblau, 1988c), process design, and process simulation, can take advantage of these above properties to organize and detect features from unpredictable and/or imprecise process data.

The purpose of this chapter is to illustrate an ANN approach to knowledge representation in chemical engineering processes. The multilayer feed-forward analog perceptron model, a particular neural network model described in Section 2, is used in Section 4 to illustrate the feasibility of employing such networks for fault detection and diagnosis via two examples. These examples demonstrate the ability of neural network models to learn and adapt themselves to different statistical distributions of inputs involving nonlinear mappings and to generalize spontaneously, thus allowing the classification of similar inputs not used in training the network. In addition, we show that neural network models can exhibit the rule-following behavior of knowledge-based expert systems without containing any prior explicit representation of the rules.

2. Characteristics of Artificial Neural Networks

An ANN is a distributed information processing system composed of many simple computational elements (nodes) locally interactiang across very-low-bandwidth channels (connections) (Anderson and Rosenfeld, 1988). Most ANNs can be described as a directed graph (i.e., digraph) where the vertices are called *processing nodes* and the edges are called *connections*. Inputs and outputs of the network are simply additional connections originating or terminating in the external environment. The structure of these connections to the external environment is determined by the problem to which the network is applied. The architecture of the ANN (see Figure 1) is specified by the node characteristics, the network topology and the learning procedure(s). The next few subsections describe: 1) the multilayer feed-forward neural network topology composed of fully connected analog perceptrons, each having a sigmoid input–output response, and 2) the generalized delta rule learning heuristic[1] used in the experiments described in this chapter.

[1] The term "heuristic" is used instead of "algorithm" because the generalized delta rule does not guarantee a solution (see Romanycia and Pelletier, 1985).

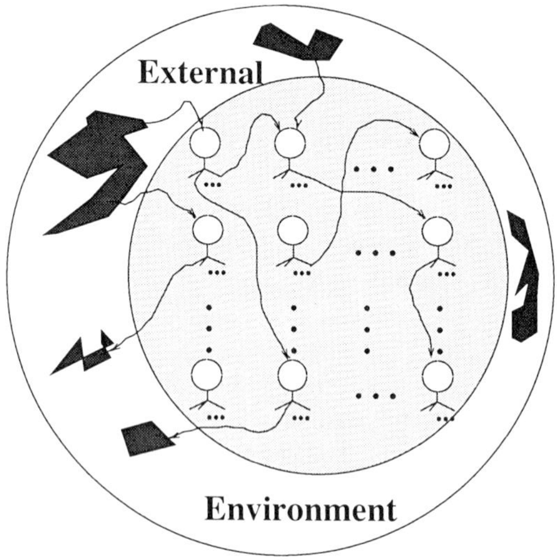

Figure 1. Artificial neural network.

2.1. Processing Nodes

Nodes in artificial neural networks are simple processors inspired by their biological counterparts. They can be described by a simple function that provides a mapping from n-dimensional space (inputs received via connections to other nodes or from the external environment) to one-dimensional space (a single output value, which becomes the input connections to other nodes or to the external environment). The network is inherently parallel, so that many nodes can carry out their computations simultaneously. Two features of a typical artificial neuron are illustrated in Figure 2: 1) the activity a (or state of the node) resulting from the inputs, and 2) the output, which is a function of the activity, $f(a)$.

Processing nodes contain an internal structure that consists of a number of functions and related data items. Typical functions include an activation function and an output function, while typical data items include input connection weights and local memory variables (e.g., activity value, output value, learning parameters, etc.). Figure 3 illustrates a processing node and its internal structure. All operations performed by a processing element are determined by its *transfer function*. The transfer function is usually composed of at least two functions: the *activation function*, which is a rule for combining the input connection signals as modulated by the connection weights to provide a state of activation, and the *output function*, which

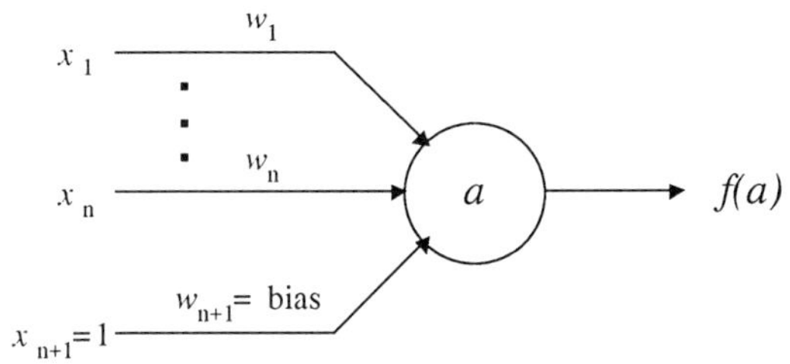

(a) processing node receiving adjustable weighted inputs

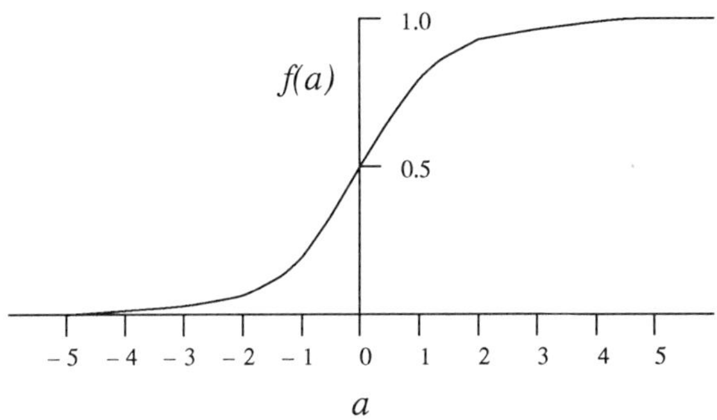

(b) output of a processing node as a function of its activity

Figure 2. Features of an artificial neuron. (a) Processing node receiving adjustable weighted inputs. (b) Output of a processing node as a function of its activity.

maps (using the local memory values) the current state of activation to an output signal.

2.1.1. Activation State

Each node must have a rule for combining the inputs from other nodes to provide a new state of activation. This activity reflects the intensity with which the input impinges on the node. The most common relation used to obtain the activity of node j is simply the weighted sum of the inputs given by the inner product

$$a_j(\mathbf{x}) = \mathbf{w}_j^{\mathrm{T}} \cdot \mathbf{x}, \tag{1}$$

where

$$\mathbf{x} = (x_1, x_2, \ldots, x_n, 1)^{\mathrm{T}}$$

is the measurement or input pattern vector, which reflects the state of the

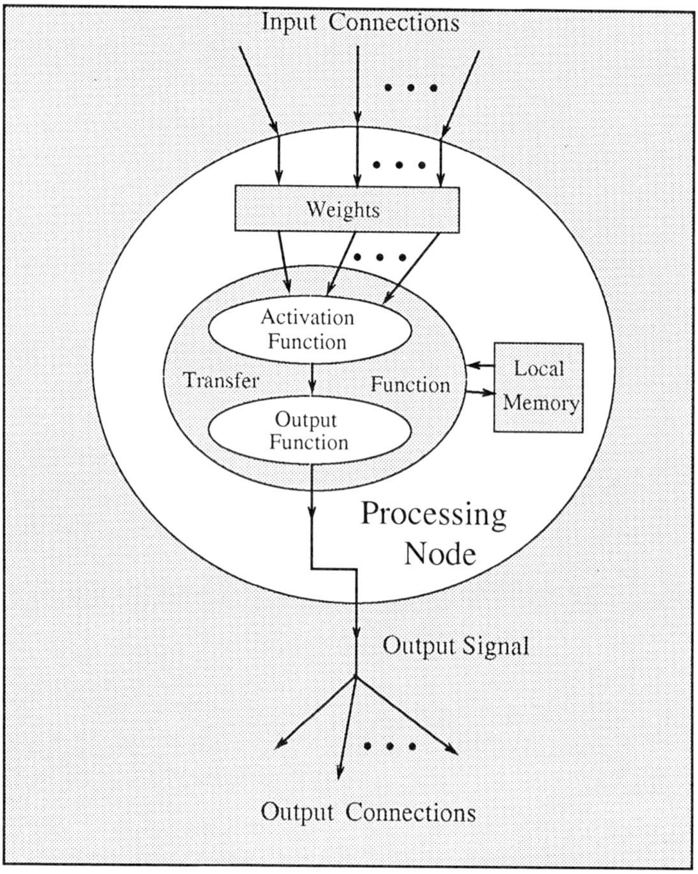

Figure 3. An artificial neuron.

process, and

$$\mathbf{w}_j^{\mathrm{T}} = (w_{1j}, w_{2j}, \ldots, w_{nj}, w_{n+1,j})$$

is the weight or gain vector. The input pattern vector is augmented with a constant input (here equal to unity), which is a standard technique for implementing the variable bias (Nilsson, 1965) that is augmented to the weight vector. However, since the same quantity is appended equally to all patterns, the basic geometrical properties of the pattern classes are not disturbed. The measurements in the pattern vector $\mathbf{x}$ and the weights in the weight vector w_j^{T} are real-valued and dimensionless. The activation values are dimensionless, unbounded real numbers. The state of activation $\mathbf{a}(t)$ over the set nodes captures what the system is representing at any time t.

Thus, processing in a neural network can be viewed as the evolution of a pattern of activity through time.

2.1.2. Output Function

The node must have an output function that maps the current state of activation to an output signal. Typically, the output function used is a threshold function for which no output is produced unless the node's activation exceeds a particular threshold value. This threshold forms the basis for discriminant (or decision) functions, which result in cutting the feature space $\mathbf{R}^m$ into regions corresponding to different classifications (also called decision regions). The output function for node j used in this work was a sigmoid logistic function given by

$$f_j(a_j) = \frac{1}{1 + e^{-a_j}}, \tag{2}$$

where a_j is the activity for node j; f_j is shown in Figure 2(b). The choice of Equation 2 to approximate a threshold logic unit embodies a number of concepts taken from the pattern-recognition literature. Specifically, the activity a_j can be viewed as a familiar linear discriminant function (Tou and Gonzalez, 1977) which, when transformed by f_j (a generalization of the *threshold logic unit*), acts as a nonlinear discriminant function (i.e., a nonlinear feature detector). More important, since f_j is differentiable and nondecreasing, it can be used in conjunction with the back propagation learning procedure to train multilayer nets, as described in Section 2.3.

2.2. Network Topology

As noted earlier, an ANN can be regarded as a collection of processing nodes and connections. However, almost all neural networks have considerable structure beyond this simple representation. For example, most ANN architectures group the processing nodes into disjoint subsets called *slabs* (sometimes called *layers*) in which all the processing nodes have essentially the same transfer function. Processing nodes can send connections to other processing nodes in the same slabs, as well as to processing nodes on other slabs. However, we employed only feed-forward connections. Figure 4 shows an ANN with three slabs.

Many ANN topologies have been proposed (Hopfield, 1982; Feldman and Ballard, 1982; Widrow and Sterns, 1985; Rumelhart and McClelland, 1986; Kohonen, 1988). Each differs in the number and character of the processing nodes, the connections, the training procedures, and whether the input/output values are continuous or discrete. The network topology

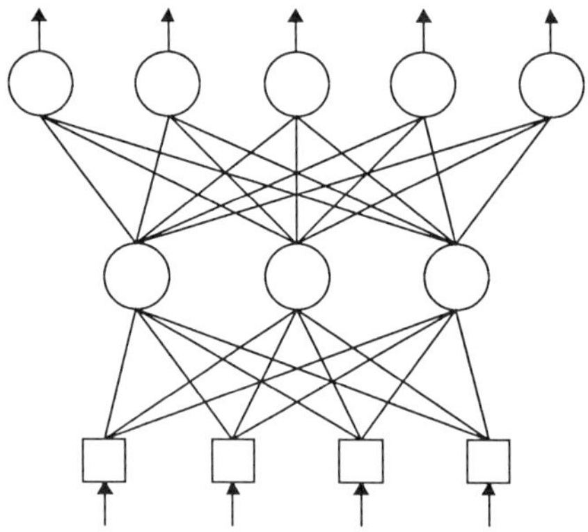

Figure 4. Two-layer neural network with a single hidden layer. Circles denote nodes. Arcs denote connections. The arrows denote connections to the outside world. The boxes represent preprocessing of the inputs.

illustrated in this chapter is specified by a multilayer, feed-forward neural network (see Figure 4). Formally, a multilayer ANN is an acyclic, unidirectional n-partite (i.e., n-layered for ANN) diagraph D. The set of vertices (i.e., nodes for ANN) are partitioned into n disjoint subsets V_1, $V_2, \ldots, V_n$ called *partite sets* such that every directed edge (i.e., connection in ANN) in D joins a vertex $v \in V_i$ to a vertex $u \in V_j$ for all $i \neq j$. A layer can be defined as a set of vertices in which no two vertices are adjacent (i.e., joined by a directed edge) (Chartrand and Lesniak, 1986).

Although the network described in this section can be extended to systems having many layers, it is restricted here to a system of two layers of processing nodes. The arcs in Figure 4 represent unidirectional feed-forward communication links connecting the nodes. Each node is fully connected to the preceding layer. Associated with each link is a signed real number, called a *weight*, that modulates the output passing through the connection. Weights can assume positive or negative real values representing an excitatory or inhibitory influence upon a node. The knowledge of the network is stored in the strengths of the interconnections along with the interconnection pattern. The calculations within a computational layer may take place in parallel; however, the computational layers themselves are asynchronous.

The input layer receives k signals from the outside world. These signals may undergo preprocessing as denoted by the boxes in Figure 4, but the inputs are not computational nodes of the same type as the circles in the neural network. The hidden layer receives input from an anterior layer and sends output to a subsequent layer, thus, it never communicates directly with the outside world. The purpose of the hidden nodes and their connections is to serve as a repository for the knowledge representation contained in the network. The outputs for the l hidden nodes are determined by first summing all the weighted inputs to the node and then passing this activity through the sigmoid functions $\mathbf{f}$ resulting in an l-vector $\mathbf{h}$ given by

$$\mathbf{h} = \mathbf{f}(\mathbf{W}_h^T \cdot \mathbf{x}), \tag{3}$$

where the nonlinear functions $\mathbf{f}$ are applied to each component of their arguments. $\mathbf{W}_h$ is the matrix containing the weight vectors for the nodes in the hidden layer. The vector $\mathbf{x}$ is the augmented input vector, which takes the biases into account. The output layer communicates m decision responses to the outside world. The values of the output layer responses form an m-vector $\mathbf{y}$ given by

$$\mathbf{y} = \mathbf{f}(\mathbf{W}_o^T \cdot \mathbf{h}), \tag{4}$$

where $\mathbf{W}_o$ is the matrix containing the weight vectors for the nodes in the output layer and $\mathbf{h}$ is the output vector of the hidden layer in Equation 3. Although the same nonlinear output function in Equation 2 was used to yield values for both the hidden and the output nodes of Equations 3 and 4, nodes in different layers can assume dissimilar roles. The task of the output nodes is to provide a decision response, and thus a nonlinear discriminant function that approximates a threshold function is a requirement.

2.3. Learning

Learning (or training) in ANNs involves adjusting the weights on the connections to achieve some performance criteria. ANN learning procedures can be partitioned into three categories based on the quality of information required during learning.

(1) *Error-correcting procedures.* These procedures are a form of supervised learning and, as such, demand detailed instructional information in the form of a teacher that specifies the desired output vector corresponding to an input vector.
(2) *Reinforcement procedures.* Reinforcement learning procedures, which are a form of weakly supervised learning, require less detailed, more selectional information (Edelman, 1978) with only a single scalar evaluation of the output, perhaps at infrequent intervals.
(3) *Unsupervised procedures.* These procedures construct internal models that capture regularities in their input vectors without receiving any additional information.

The difference in the information requirements of each learning procedure is germane to the difficult *credit assignment* problem (Cohen and Feigenbaum, 1982) of determining which actions or decisions deserve credit (or blame) for improvements in the overall learning system.

The most effective way to train ANNs for fault detection and diagnosis is supervised learning. A neural network can automatically store knowledge by learning from historical records that identify faults, and it has the characteristic of associative memory. Information about faults can be learned by training the network on a set of data that includes both the values of steady-state process variables for normal conditions and those for identified faulty conditions. If data cannot be found in past daily reports of process maintenance and data logs, the necessary data can be collected from a designated process quality control program designed to identify faults. One way of speeding up the learning is to start by using process simulators into which faults can be introduced. Because the network can filter out noise and classify faults, and because the information needed to train the network can be easily collected, ANNs can provide fully automatic, quantitative, stable, highly sensitive, and inexpensive diagnosis of faults, even in the presence of noise. Hoskins and Himmelblau (1988a) have demonstrated that artificial neural networks have considerable potential when applied to fault diagnosis of chemical processes.

Unsupervised learning procedures require neither teachers nor critics. These techniques employ some form of unsupervised learning, or clustering, often based on Hebb's hypothesis that repeated pairing of pre- and postsynaptic activity strengthens synaptic efficacy. While clustering is likely to play an important role in sophisticated problem-solving systems, it does not by itself provide the necessary means for a system to improve performance in tasks determined by factors external to the system, such as, for example, the task of controlling an environment having initially unknown dynamics. For this type of task, a learning system must form clusters that are useful in the system's interaction with its environment. Thus, it seems necessary to consider networks that learn under the influence of some sort of evaluative feedback, but this feedback does not need to be so informative as to include individualized instruction to each adaptive element.

Regardless of the method used to train a network, once the pattern of activation is established, the network can classify a new input pattern. The adaptive nature of ANNs allows the weights to be learned by experience, thus producing a self-organizing system. Much of the recent interest in ANNs can be attributed to new and more effective learning heuristics.

The following examples illustrate a robust learning heuristic for multilayered, feed-forward ANNs, called the generalized delta rule (GDR) or back-propagation learning rule (Rumelhart and McClelland, 1986; Parker, 1985; Le Cun, 1985; Werbos, 1974). Superficially, the GDR appears to be a straightforward, gradient-descent technique to minimize an objective function $J(w)$ that is differentiable with respect to each of the weights in the network. If the objective function is also a function of the output of the network, this implies that each f_j is differentiable with respect to node j's weights. However, the main contribution of Rummelhart and others was to provide a distributed, gradient-descent technique of optimization such that the gradient-descent expressions appear as discrete, weight-update rules that use only information locally available to each node and an error correction quantity that is back-propagated from nodes in the penultimate layer.

The objective function $J(w)$ to be minimized in GDR learning is the mean-squared error between the actual outputs from the output layer and the desired or target outputs for all the input patterns. This function is given by

$$J(w) = \sum_p E_p, \tag{5}$$

where E_p is

$$E_p = \frac{1}{2}\|\mathbf{y}_p - \mathbf{d}_p\|^2. \tag{6}$$

Given a training input vector, $\mathbf{x}_p$, $\mathbf{y}_p$ is the output vector generated by the forward propagation of the activation through the network, and $\mathbf{d}_p$ represents the desired output vector associated with $\mathbf{x}_p$.

The GDR learning procedure is purely deterministic. If at the start of the learning phase the nodes within a hidden layer begin with the same connectivity and symmetric weights, the learning procedure yields no change in the weights between the hidden layer and the output layer. In response to this dilemma, GDR learning begins by initializing all connection strengths (and biases) to small real values that are randomly selected between −0.5 and 0.5. A training pattern is presented to the input nodes, and the outputs of the input nodes are then propagated in a feed-forward fashion to produce $\mathbf{y}_p$. If each error (Equation 6) for the presented pattern is below a specified learning tolerance (0.01 in the example below), no weight changes occur. Otherwise, the weights are adjusted according to

$$\Delta_p w_{ii} = \eta \delta_{pj} y_{pj} + \alpha \Delta_{p-1} w_{ji}, \tag{7}$$

where $\Delta_p w_{ji}$ is the change in the weight from node i to node j after presentation p, η is a gain term usually called the learning rate, and α is a momentum term that smooths the effect of dramatic weight changes by adding a fraction of the weight change, $\Delta_{p-1} w_{ji}$, from the previous presentation, $p-1$. The error signal δ_{pj} is a measure of the distance from the activation level of node j to its desired level after presentation p.

The GDR provides two rules for calculating the error signal of a node. For an output node,

$$\delta_{pj} = (d_{pj} - y_{pj}) f'(a_{pj}), \tag{8}$$

where d_{pj} is the desired activation level for node j for input pattern p, and $f_j(a_{pj})$ is the derivative of the output function for node j with respect to activity generated by the input pattern p.

The major contribution of the GDR lies in its formula for computing the delta values for the hidden nodes. Before the GDR, there was no learning procedure for multilayer, feed-forward networks. Thus, research was limited to single-layer networks able to discriminate only between classes of linearly separable sets (for a detailed description of single-layer perceptrons and their limitations, see Minsky & Pappert 1969). For an arbitrary node j in a hidden layer, the rule for calculating the error signal is

$$\delta_{pj} = f_j'(a_{pj}) \sum_k \delta_{pk} w_{kj}, \tag{9}$$

where the summation is over all the k nodes to which the node j sends output.

As the name *back propagation* suggests, the basic idea behind this computation of error signals for the hidden nodes is to propagate errors back through the system that are based on observed discrepancies between the values of output nodes and the expected output for a training pattern. The error signal focuses on adjusting connection strengths that are most responsible for the output error. Thus, after performing the differentiation, the error signal becomes

$$\delta_{pj} = \begin{cases} (d_{pj} - y_{pj}) y_{pj} (1 - y_{pj}) & \text{if } j \in L \\ y_{pj}(1 - y_{pj}) \sum_k \delta_{pk} w_{kj} & \text{otherwise,} \end{cases} \tag{10}$$

where L is the set of indices for the output nodes.

Learning involves two phases (see Figure 5). First, the inputs are propagated in a feed-forward fashion through the network to produce output values that are compared to the desired outputs, resulting in an error signal

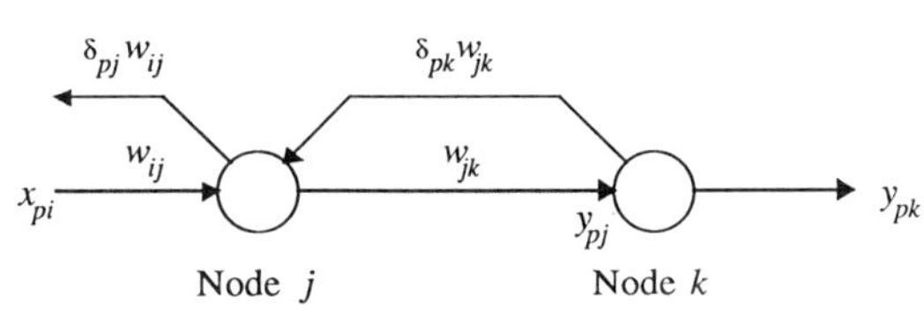

Figure 5. Back propagation of the error signal.

for each of the output nodes. Second, the error is propagated backward through the network using the following procedure. The deltas (error signals) for the output layer are calculated first, and these deltas are used recursively to calculate the deltas for the adjacent hidden layer using equation 10. These deltas can then be used to update all the weights in the network.

A form of training that represents unsupervised learning is Kohonen's (1988) self-organizational mapping. The map is generated by establishing a correspondence between inputs and neurons in a network such that the topological neighborhood relationships among the inputs are reflected as faithfully as possible in the arrangement of the corresponding neurons in the net. The correspondence is obtained iteratively by a sequence of training steps with the weights of the "winner" adjusted by

$$\Delta w_{ji} = \eta(t)[x(t) - w_{ji}(t)], \tag{11}$$

where $\eta(t)$ is a gain, $x(t)$ is the input, and the weights $w(t)$ are functions of time. Both the gain and the size of the neighborhood around the winner are slowly decreased with time. However, learning in Kohonon self-organization is slow and may depend on the order of presentation of inputs. Furthermore, it requires normalized inputs, proper initialization of the weights, and the network needs to be trained on data that is statistically meaningful relative to the total input the net will experience in practice.

2.4. *Recognition of Patterns*

Once the network is trained, any new pattern vector presented at the input nodes is immediately classified in real time on-line. Training develops basins of attraction around equilibrium points, and new patterns are classified according to these basins. Hopefully training does not develop spurious equilibrium points so that input patterns are misclassified.

3. ZNL Architecture

ZNL (Zone Node Layer) is a neural network design and simulation environment (see Figure 6) that we built to experiment with connectionist networks. ZNL is implemented in the C programming language using LEX, a lexical analyzer, and YACC, which converts a BNF specification of a language and semantic actions written in C into a compiler for that language, for parsing. ZNL consists of three parts: (1) the ZNL language, (2) the ZNL interactive simulator, and (3) the ZNL build module. The ZNL environment is meant to be as general-purpose as possible, and no attempt is made to force a particular neural network model on the user.

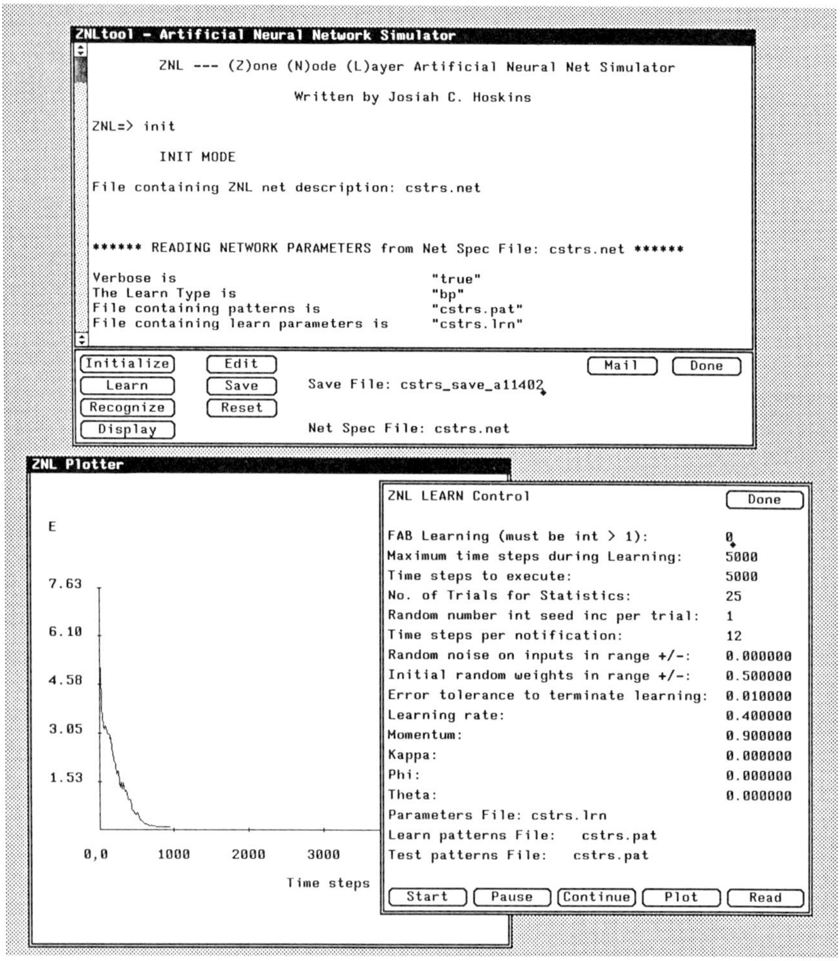

Figure 6. ZNL artificial neural network simulator.

Z(NL)^2, the ZNL Network Language, provides a textual representation of neural network architectures. The purpose of this endeavor is two-fold. First, the Z(NL)^2 language allows you to specify network architectures unambiguously, thus increasing the potential for researchers to formulate and communicate their architectures without becoming immeshed in the details of the implementation of the network. Second, a compiler or interpreter (such as the ZNL Interactive Simulator) can be built for the computational environment of choice, thus allowing for portability of network architectures to very different computational environments. The Z(NL)^2 language also allows the state of the network to be stored.

ZNLIS, the ZNL Interactive Simulator, compiles (or interprets) $Z(NL)^2$ source code into an executable network simulator. A ZNLIS session typically consists of reading in a $Z(NL)^2$ network specification, a training session in which the network learns from experience, and recognition in which new patterns presented to the network can be evaluated.

ZNLBM, the ZNL Build Module, allows a user to replace key functions in ZNLIS, thus encouraging experimentation with different neural net models and learning algorithms.

4. Fault Detection and Diagnosis Examples

Because modern chemical plants are extremely complex, they are susceptible to equipment malfunction and operator error. The complexity hampers the operator's ability to diagnose and eliminate potential process upsets or equipment failures before they can occur (Himmelblau, 1978). Hence, a continuing question in chemical engineering is how to use the process state vector to make or aid decisions about possible action or control at each time increment. Current techniques rely on expert systems, modeling using classical techniques in the time or frequency domains, and statistical analysis. Neural network classifiers can learn and adapt themselves to different statistical distributions of inputs involving arbitrary nonlinear mappings (see Kolmogorov, 1956 for analysis). These classifiers use only a restricted set of parameters while making weaker assumptions about the shapes of underlying distributions than traditional statistical classifiers.

This section presents two examples of how an ANN approach to knowledge representation is applied to fault detection and diagnosis:

(1) *The CSTR example* illustrates the diagnosis of faults for a simple system composed of three continuous-stirred-tank reactors.

(2) *The chemical reactor example* illustrates the application of ANNs to fault diagnosis in a chemical reaction process in which heptane is converted catalytically to toluene.

4.1. Example 1: Fault Diagnosis for CSTRs in Series

The CSTR example illustrates the diagnosis of faults for a simple system composed of three continuous-stirred-tank reactors in series. Each piece of equipment is operated isothermally and with constant fluid volume (i.e., no fluid accumulation is permitted). The flow sheet for the process is illustrated in Figure 7. The state variables used as fault indicators are the flow rate, the temperature, and the concentration of components A and B in streams 1 and 4. All six state variables are monitored, and their sensors are assumed to function properly (these constraints can be relaxed). All readings are taken at steady state.

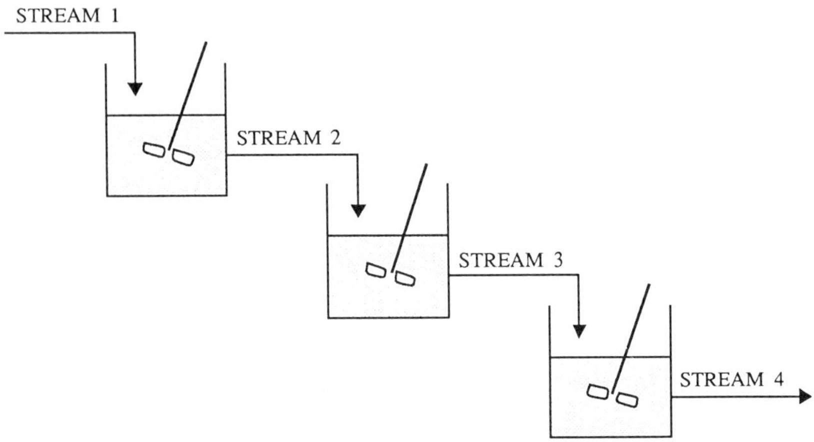

Figure 7. Three continuous stirred tank reactors in series.

Table 1. List of selected faults.

A	Inlet Concentration of Component A	Low
B	Inlet Concentration of Component A	High
C	Inlet Flow Rate	Low
D	Inlet Flow Rate	High
E	Temperature	Low
F	Temperature	High

Table 2. Sensor measurement patterns for six selected faults.
FR = flowrate ($ft^3 - min$), *T* = temperature (^0F), *C* = concentration (lb–moles/ft^3); superscript denotes stream; subscript denotes component

	Measurements					
Faults	FR	T	C_A^1	C_B^1	C_A^4	C_A^4
A	18.0	190	0.3	3.18	0.2275	3.252
A	18.0	190	0.6	2.88	0.3755	3.104
B	18.0	190	1.3	2.18	0.5990	2.881
B	18.0	190	1.6	1.88	0.6681	2.812
C	13.0	190	1.0	2.48	0.4475	3.033
C	15.0	190	1.0	2.48	0.4777	3.002
D	22.0	190	1.0	2.48	0.5600	2.920
D	26.0	190	1.0	2.48	0.5958	2.884
E	18.0	150	1.0	2.48	0.8960	2.584
F	18.0	210	1.0	2.48	0.3102	3.170
F	18.0	230	1.0	2.48	0.1703	3.310

Table 1 lists six selected faults, labeled A through F, each affecting the operation of this process. The possible faults involve the system flow rate, the temperature, and the inlet concentration of component A. Table 2 shows the sensor measurement patterns of the values of the six state variables associated with the six selected faults used for training the network.

4.1.1. Training the Network

Figure 8 shows the network architecture used for fault detection and diagnosis. It consists of six inputs corresponding to the six state variables of

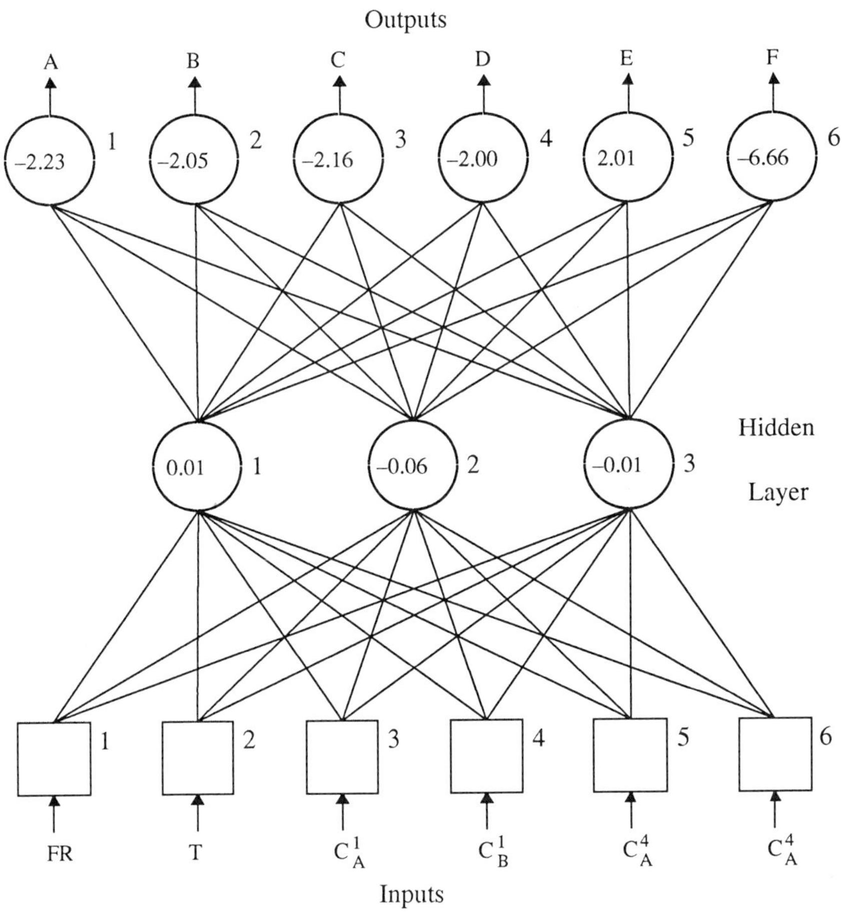

Figure 8. Trained network for the example process. The numbers in the circles represent the biases of the nodes. For values of the weights associated with the arcs, see Table 3.

the system, three hidden nodes, and six output nodes corresponding to the six respective process faults listed in Table 1. All nodes in each layer are fully connected to the nodes in an adjacent layer, resulting in 45 connection links, nine of which are bias or threshold links. Threshold values can be learned by assuming that the augmented weight is equivalent to a weight from a node that is always turned on to give an output of unity (called a truth node).

Twelve measurement patterns were used to train the network: the 11 measurement patterns listed in Table 2, plus one measurement pattern for a normally operating system. These measurement patterns were obtained from a digital computer simulation program designed to model the dynamics of an arbitrary combination of reactors and/or vessels. A second-order reaction was assumed to occur (2A $\rightarrow$ 2B) between components A and B in each tank with a frequency factor of 5.0×10^{14} (lbmole)/(ft^3)(min) and an activation energy of 4.47×10^4 (Btu/lbmole). Because the values of the sensor readings in the input patterns to the network are factors in the equations that update the values of the weights in the learning procedure, measurements with larger magnitudes exert a greater influence on learning. To remove this bias, the simulated measurement data were scaled (as indicated by the preprocessing boxes depicted in Figure 8) so that the inputs to the network varied continuously over the range of −1.0 to 1.0.

Before commencing the simulation for this example, we wrote a network specification file in $Z(NL)^2$. We started the simulation session by issuing the ZNLIS command to initialize the network. Once the network specification was compiled, the next step was to start the learning mode. The twelve patterns were presented to the network, and a learning criterion of 0.01 for each pattern error E_p was set to terminate the learning process. During training, the target values used for the output nodes were set to 0.1 and 0.9 rather than 0 and 1. If the values 0 and 1 are used as target values, the logistic activation function tends to generate infinitely large weight values, which can produce truncation errors causing the weights to become frozen (i.e., the value of the term $y_{pj}(1 - y_{pj})$ in equation 10 approaches zero). Values of 0.25–0.9 for η (the learning rate) and 0.99 for α (the momentum) were used in the learning phase.

The number of time steps (i.e., the number of presentations of each measurement pattern) required to train the network is a function of the number of hidden nodes present in the hidden layer. Figure 9 shows the relationship between a minimum learning rate and the number of hidden nodes. The weights and biases for a trained network having three hidden nodes and using a learning rate of 0.25 are shown in Table 3. To determine whether the network had indeed classified the input patterns, the same set

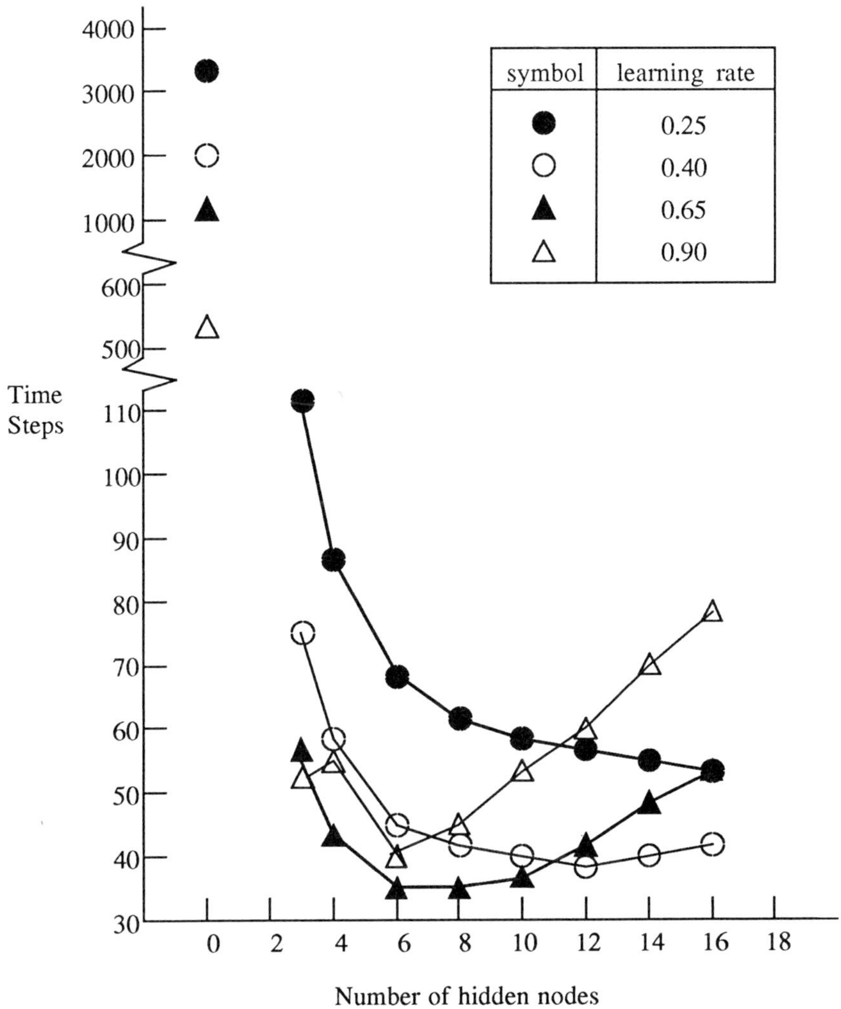

Figure 9. Number of time steps versus number of hidden nodes at different learning rates.

of patterns was passed through the network in the ZNLIS recognition mode. Recognition of a learned pattern took only a small fraction of a second on a Sun Microsystems 3/160. All patterns learned were quickly and easily recognized. Therefore, the network was able to rapidly learn the correct association between the process faults and the measurement patterns rapidly and autonomously using only stimuli from its external environment.

Table 3. Weights for the trained network in Figure 8.
Output letters correspond to the faults in Table 1.

	Hidden nodes			
Inputs	Node 1	Node 2	Node 3	Bias
FR	−0.21	5.70	−4.79	
T	3.94	1.83	−0.53	
C_A^1	3.83	−1.43	−1.57	
C_B^1	−4.02	1.11	1.31	
C_A^4	−0.65	−2.24	−2.36	
C_A^4	0.58	1.91	2.97	
Bias	0.01	−0.06	−0.01	
Outputs				
A	−4.67	1.99	2.71	−2.23
B	5.12	−2.28	−2.99	−2.05
C	0.16	−5.18	5.08	−2.16
D	−0.28	4.62	−4.65	−2.00
E	−4.09	−2.57	−1.64	2.01
F	4.46	2.77	1.76	−6.66

4.1.2. Analysis of the Use of ANN Classifiers

An ANN classifier operates similarly to a classical decision theory classifier. The recognition problem to be solved by the ANN is to generate the decision boundaries separating the set of M different pattern classes, denoted by $\omega_1, \omega_2, \ldots, \omega_M$, based on the observed measurement vectors. The decision boundaries associated with each ω_i can be defined by the discriminant functions, $g_1(\mathbf{x}), g_2(\mathbf{x}), \ldots, g_M(\mathbf{x})$. These functions are scalar and single-valued functions of the input pattern $\mathbf{x}$. The rule for classification is

if

$$g_i(\mathbf{x}) > g_j(\mathbf{x}) \text{ for all } j \neq i, \tag{12}$$

then

$$\mathbf{x} \in \omega_i$$

In other words, if the ith decision function $g_i(\mathbf{x})$ has the largest value for a pattern $\mathbf{x}$, then $\mathbf{x} \in \omega_i$(the pattern $\mathbf{x}$ is in the class ω_i).

Learning Linearly Separable Decision Regions. Early ANNs used linear discriminant functions. Inputs were passed through some type of hard limiting nonlinearity with outputs of -1 or $+1$, or threshold logic with outputs of 0 or 1. The networks were limited to a single-layer network architecture due to the absence of multilayer learning rules. Geometric hyperplane analysis (Burr, 1986) shows that a single-layer neural network is limited to hyperplane (linearly separable) decision regions. Thus, a single-layer ANN must rely on the coding provided by the outside world, which allows the network to map only similar inputs to similar outputs. The learning results shown in Figure 9 indicate that the fault detection and diagnosis can be accomplished with no hidden layer, which in turn suggests that the decision regions are linearly separable. However, it seems to be advantageous, even in the linearly separable case, to use a hidden layer, because the rate of learning is greatly enhanced. The number of time steps (presentations of each pattern) for the network with no hidden layer varied from 540 to over 3000 for the range of η investigated. These investigations produced no solutions for the network containing one hidden layer with only one or two hidden nodes for trials consisting of 100,000+ time steps (the E_p's were seen to oscillate somewhat randomly).

Figure 9 suggests that there are a number of factors affecting the optimum number of hidden nodes. The number of hidden nodes must be large enough to form a decision region that is sufficiently complex for the given problem, an arrangement not possible for just one or two hidden nodes. The convergence rate is seen to depend not only on the number of hidden nodes, but also on the learning rate, η. Increasing the number of hidden nodes results in decreasing the convergence rate until a minimum convergence rate is detected. A further increase causes a degradation of the convergence rate. This behavior was observed for all of the learning rates investigated. The minimum convergence rate occurred at a different number of hidden nodes for different learning rates. The lower learning rates (0.25 and 0.40) generated better performance over a large range, while a learning rate of 0.65 yielded the best convergence rate. Higher learning rates greatly restricted the range of the number of hidden nodes with respect to rapid learning. Lower learning rates allowed larger ranges but generated longer training times.

To illustrate the internal working of the ANN, Figure 10 shows the patterns of activity generated during classification. The ANN classifier has one output node corresponding to each class (fault to be identified). For a pattern of inputs that represented a normally operating system, the network revealed activity levels of near zero for the hidden nodes, which in turn generated strongly negative activity levels for all the output nodes. Note in Figure 8 that the biases for the hidden nodes were all approximately zero, so that the activity from the inputs was passed through unbiased. Figure 2b shows that for a strongly negative activation value the output of the node is near zero. Hence, the network in effect discovered that the inputs for a normally operating system cause all the fault classes to be rejected, because none of the ouput nodes corresponding to a fault was turned on.

Figure 10 shows the relative activities of the hidden layer and output layer nodes for all six fault patterns. For example, the pattern of activity produced for the seventh measurement pattern from Table 2 corresponds to fault class D in Figure 10. The hidden nodes had near zero activity for node 1, strongly positive activity for node 2, and strongly negative activity

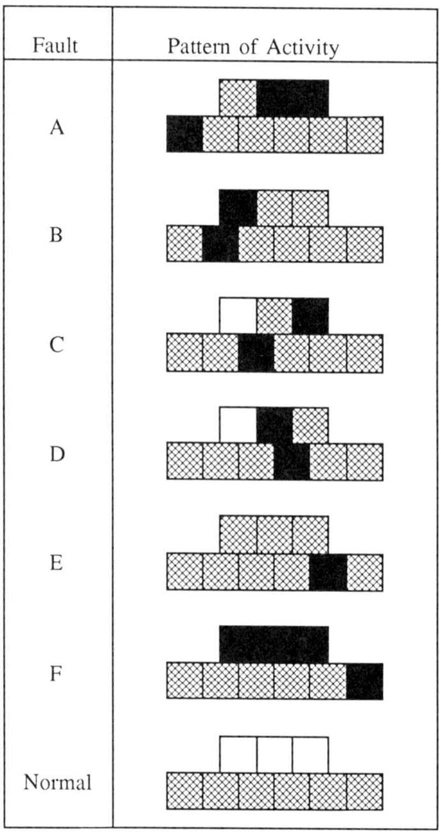

Figure 10. Patterns of activity for the six faults in Table 1 and the normal operating state. The three boxes in the top row of each pattern correspond to hidden nodes, and in the bottom row to output nodes.
■ significant positive activity ▩ significant negative activity
□ near zero activity

for node 3. The outputs for the hidden nodes were 0.39, 0.99, and 0.00, respectively, and the corresponding weights to output node 4 were −0.28, 4.62, and −4.65. Note that the bias for output node 4 was −2.00, which means that node 4 was always turned off unless there was enough activation from the hidden nodes to overcome the bias. In this case, the activation from the hidden notes was 4.46, which was strong enough to overpower the bias and turn on the fault class D output node. However, not only must the correct fault class be activated, but the other classes must be rejected. An interesting fault class that must be rejected is E (output node 5). For E, the bias of the node was positive, which implies a naturally turned-on state. The weights associated with the links from the hidden nodes were −4.09, −2.57, and −1.64, which resulted in an activation of −4.14, enough to swamp the 2.01 bias and turn off the node. A similar analysis holds for the other fault classes rejected for this case.

Effect of Sensor Noise on Discrimination During Learning. One of the handicaps that fault detection and diagnosis must overcome is imprecise sensor measurements. False alarms caused by large random values or outliers in measured data can lead to the breakdown of any fault detection technique. In an attempt to determine whether ANNs are suitably robust candidates, we added random noise to the inputs during the learning phase:

$$\mathbf{x}' = \mathbf{x} + \mathbf{n}, \tag{13}$$

where $\mathbf{n}$ was a noise vector in which each element contained a random real value in the range of 0–10% of the input, and $\mathbf{x}'$ was the input vector with the added noise. Noisy data require more complex decision regions. The learning results are shown in Figure 11. No solution was found for the single layer with no hidden nodes, which implies that the decision regions are not linearly separable for the example system with noisy inputs. By using a nonlinear discriminant function (equation 2), a multilayer feedforward ANN, and the GDR learning procedure, the network could properly classify inputs. Furthermore, geometric analysis for a multilayer network (Lippmann, 1987) shows that a two-layer perceptron is sufficient to form convex decision regions, and three layers are sufficient to form any desired decision region. The plots in Figure 11 exhibit the same general trends as in the linearly separable case shown in Figure 9. Note, however, that the range of the number of hidden nodes for low convergence rates becomes smaller, especially for the high learning rates.

For complex decision regions it may be useful for the ANN to approximate a logical (rule-like) decision by providing a decision score invariant for any input vector in a given class. Thus, the discriminant function for

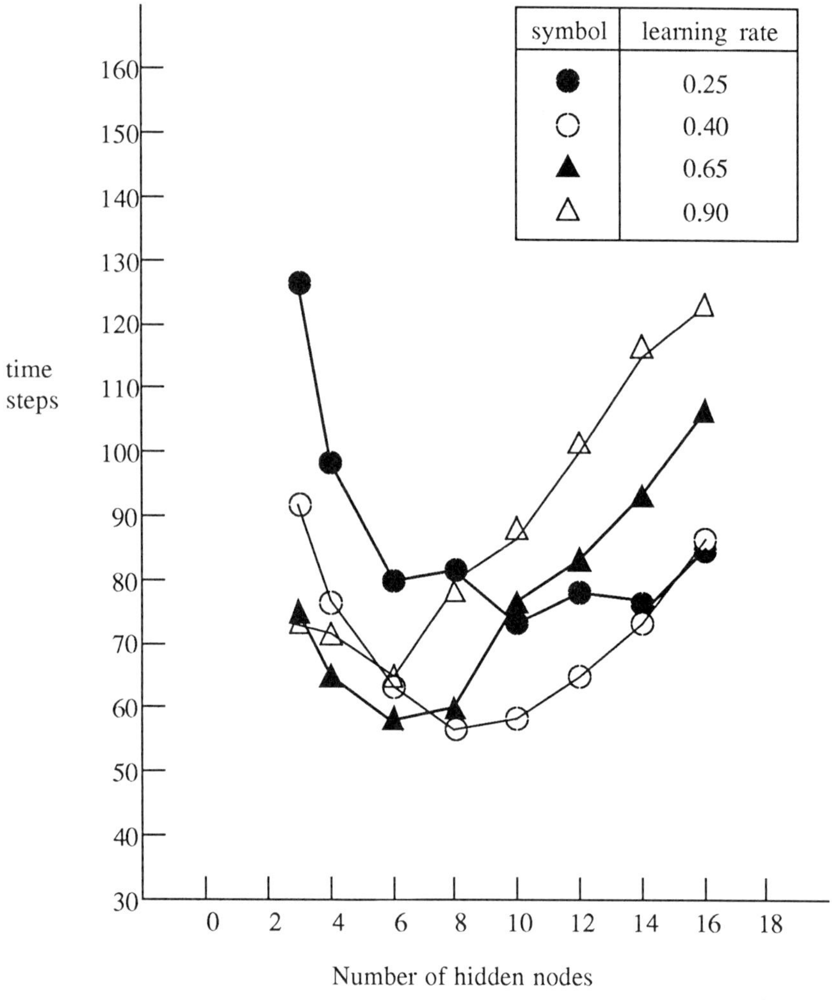

Figure 11. Number of time steps versus number of hidden nodes at different learning rates with imprecise inputs.

each output node provides a measure, scaled from 0 to 1, of how much the corresponding class associated with that node.

The use of ANNs in solving chemical engineering problems is appealing because of the capacity of ANNs to learn by recognizing patterns similar to those in the learning set. While learning, ANNs can capture high-order constraints between data, yet stress the discrimination power between the different classes. ANN techniques are thus potentially more effective than many approaches in pattern recognition (Tou and Gonzalez, 1977) that use a separate model for each class and usually make assumptions about the distribution of their pattern vectors.

4.1.3. Rule-Following Behavior

While the results discussed in the previous section demonstrate the excellent performance of the ANN in classifying the given set of measurement patterns, they provide little information about the internal representation of the network and its relation to the system under study. Expert systems use rules to provide a formal way of representing knowledge, especially when the domain knowledge results from empirical associations developed through experience in solving problems (Waterman, 1985). In contrast, ANNs use abstract internal representations, which can exhibit rule-following behavior without containing explicit representations of the rules (Derthick and Plaut, 1986). ANN rely on inductive inference [Holland *et al.*, 1986] or generalization to recognize invariances, capture high-order correlations, form complex, high-dimensional inputs, and generate relationships between the inputs and outputs from a set of examples.

The inductive inference problem (Angluin and Smith, 1983) is concerned with the study of how knowledge can be represented by inference rules learned from a set of examples. The ANN approach to this problem is to generate an input-to-output mapping based on a training set. The encoded internal representation captures the similarity among inputs that results in generalizations of the examples available in the learning process. This ability to generalize gives the network the potential to map inputs not in the training set. Thus, if an ANN can be trained by a wide range of representative system behaviors, the network should be able to generalize to a larger range of circumstances.

To demonstrate the generalization capabilities of the ANN, the percent misclassification was examined for this example with the three hidden nodes and the learning rate parameter set to 0.40. To test the ability of the network to recognize new input patterns, 35 measurement patterns not used in the training process were presented to the trained network. These new measurement patterns were chosen to be representative of the measurement space for the six faults and the normal operating system. Figure 12 shows the percent of correct classifications as a function of the number of input patterns (with added noise) used to train the network. Perfect generalization occurred when only two measurement patterns in the training set had been used for each fault group (i.e., 12 input patterns total in the training set, two of which were used for the normal operating system). Increasing the number of measurement patterns maintained the

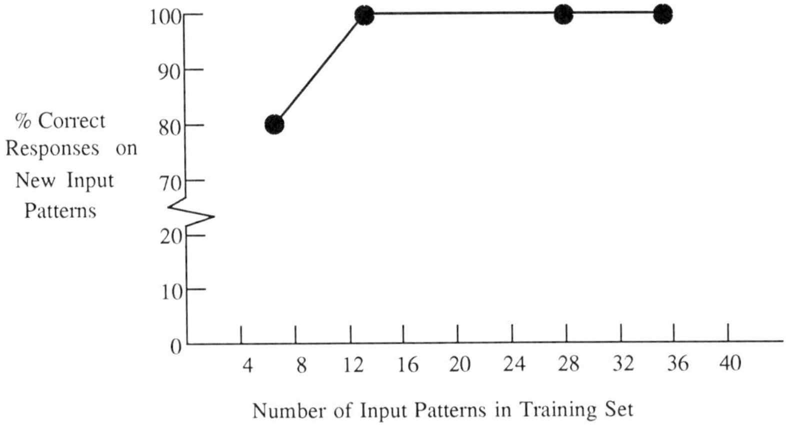

Figure 12. Generalization capacity versus training set size.

same excellent performance level. However, some failure to generalize efficiently was observed for the training set containing only seven training patterns because the training set was so restricted that it was not representative of the general mapping. This result was not surprising since the seven-input pattern training set included no example patterns that were representative of the crossover between the normal and faulty region. There is little guidance in the literature regarding the choice of representative sets of examples for training with continuous inputs because nearly all previous studies have investigated binary inputs. Most studies also used a closed set of possible inputs. More studies are needed to direct the development of heuristics that assist the choice of representative training sets.

In the examples presented here, inference rules are not explicitly stated in the model. However, generalizations about the interdependencies between the state variables of the process emerge from a representative set of training patterns. The excellent generalization exhibited by the network shows that the structure extracted from the training examples agrees with the structure attributed to the domain. These generalizations could, of course, be coded in the rules of an expert system or even in an engineering mathematical model. However, if changes in the process occur, the rules or the model itself must be changed to compensate, whereas in a neural network model the architecture need not change to adapt to deviations in the process: the network simply acquires more knowledge from experience.

4.2. Example 2: Fault Diagnosis in a Chemical Reactor

This section describes how to apply ANNs effectively to fault diagnosis in a chemical reactor. The network architecture used is composed of a two-

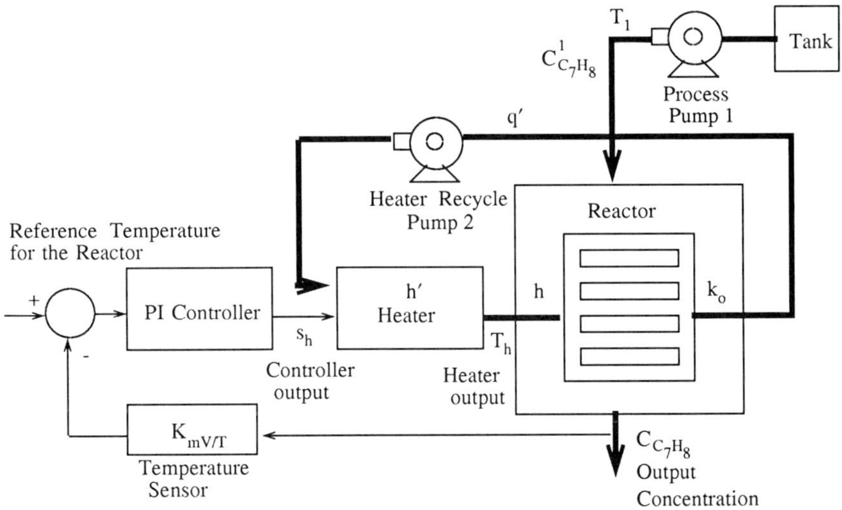

Figure 13. Process to be diagnosed.

state multilayer neural network, and it is used to discriminate among the causes of faults as well as to estimate the degree of deterioration represented by the faults in the presence of noisy measurements.

4.2.1. Simulated Faults in the Process

This second example illustrates a process in which heptane is converted catalytically to toluene in a PI-feedback-controlled chemical reactor that is operated near its steady-state operating conditions. No catastrophic and/or abrupt event occurs because any process fault is in the incipient stage and the effect of any fault is strongly regulated and alleviated by the controller. Thus, no change in reactor temperature occurs as a result of a fault. Figure 13 shows the process. Heptane stored in the tank is fed to the reactor through process pump 1. The catalytic reaction occurring in the reactor is

$$C_7H_{16} \rightarrow C_7H_8 + 4H_2.$$

The reaction rate is controlled by the temperature in the reactor. Steam supplied to the heat exchanger in the reactor is recycled to the inlet of the heater via recycle pump 2. The outlet valve in the heater is controlled by the PI controller. The controlled variable is the reaction temperature, and the controller regulates the temperature to the specified value. The heater control comes from an electronic circuit or a process computer that is always in a nonfaulty operating mode. We used lumped models of the reactor, heater, and controller to simulate various faults.

To illustrate the use of an ANN in diagnosing faults, let us categorize five possible causes of faults, plus a no-fault (normal) condition, as follows:

Fault 1 Deterioration of catalyst performance due to physical and/or chemical deterioration of the catalyst leading to a decrease in the frequency factor for the catalyst, k_o, in the model.

Fault 2 Fouling of the heat exchanger surface in the reactor leading to a decrease of overall heat transfer coefficient, h, in the model.

Fault 3 Fouling of the heat exchanger surface in the heater leading to a decrease of overall heat transfer coefficient, h', in the model.

Fault 4 Partial plugging of the pipeline connected to pump 2 leading to a decrease in the volumetric flow rate, q, in the model.

Fault 5 Partial plugging of the pipeline connected to pump 2 leading to a decrease in the volumetric flow rate, q', in the model.

The ANN was used to discriminate among the above five faults from three measurements: the outlet concentration of $C_{C_7H_8}$, the outlet temperature T_h, and the controller output signal s_h. Table 4 qualitatively summarizes the relation of each input vector to a particular fault. For this example, five levels of each cause of a fault are illustrated, based on the data in Table 5.

4.2.2. Training the Network

Figure 14 shows the specific configuration of the network employed in this example. Each subnetwork is comprised of three layers.

- An input layer consisting of three input nodes corresponding to the three measured variables.
- A hidden layer with four nodes.

Table 4. Qualitative knowledge that relates the causes of faults and measurable variables when process is in quasi-steady state.

Fault		Change in Variables due to Faults		
No.	Cause	s_h	T_h	$C_{C_7H_8}$
1	Decrease in k_0	Decreased	Decreased	Decreased
2	Decrease in h	Increased	Increased	Unchanged
3	Decrease in h'	Increased	Unchanged	Unchanged
4	Decrease in q	Decreased	Decreased	Increased
5	Decrease in q'	Decreased	Unchanged	Unchanged

Table 5. Relation between the causes of the faults and the measurable process variables when the process is in the quasi-steady state. Values shown are normalized by the value of the variables in the normal steady-state condition.

Fault No.	Level	Cause	Change in Variables Due to Faults		
1	1	$0.95k_0$	0.99	0.99	0.98
	2	$0.90k_0$	0.98	0.99	0.95
	3	$0.85k_0$	0.97	0.99	0.92
	4	$0.80k_0$	0.96	0.99	0.89
	5	$0.75k_0$	0.95	0.99	0.86
2	1	$0.95h$	1.04	1.01	1.00
	2	$0.90h$	1.07	1.02	1.00
	3	$0.85j$	1.12	1.03	1.00
	4	$0.80h$	1.17	1.04	1.00
	5	$0.75h$	1.22	1.06	1.00
3	1	$0.95h'$	1.05	1.00	1.00
	2	$0.90h'$	1.11	1.00	1.00
	3	$0.85h'$	1.18	1.00	1.00
	4	$0.80h'$	1.25	1.00	1.00
	5	$0.75h'$	1.33	1.00	1.00
4	1	$0.95q$	0.98	0.99	1.02
	2	$0.90q$	0.95	0.99	1.05
	3	$0.85q$	0.92	0.98	1.07
	4	$0.80q$	0.90	0.97	1.11
	5	$0.75q$	0.87	0.97	1.14
5	1	$0.95q'$	0.95	1.00	1.00
	2	$0.90q'$	0.90	1.00	1.00
	3	$0.85q'$	0.85	1.00	1.00
	4	$0.80q'$	0.80	1.00	1.00
	5	$0.75q'$	0.75	1.00	1.00

- An output layer consisting of five output nodes corresponding to the five causes of faults.

The number of nodes in the hidden layer, four, was selected as optimal for minimizing the number of presentations of the input vectors during training. Figure 15 shows the relation between the number of nodes in the hidden layer and the number of presentations needed to satisfy the convergence condition that each output error be less than 0.1. Unscaled values

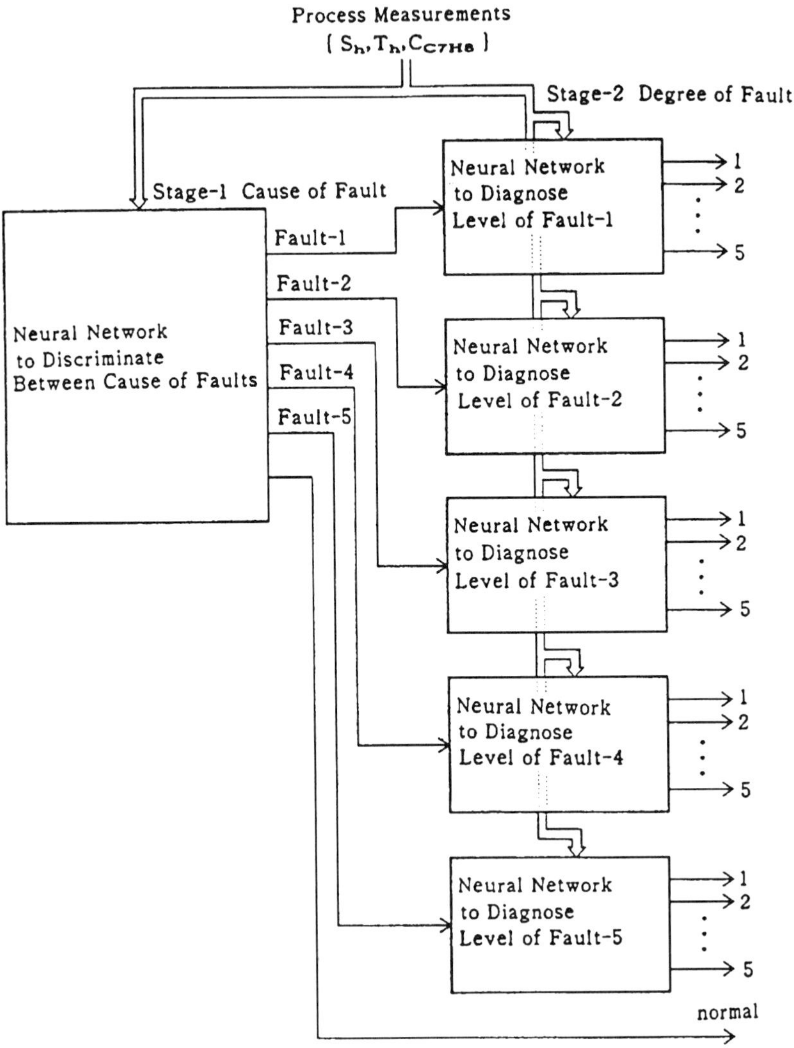

Figure 14. Two stage ANN fault diagnosis of a chemical process.

of the sensor measurements calculated from the data in Table 5 were used to train the network. The values of the outputs from the output nodes were trained to be composed of 0s and 1s in the same way as described for the CSTR in series example. The subnetwork was trained to discriminate by using the data of Level 1 for the five faults in Table 5 with learning rate $\eta = 0.1$ and momentum $\alpha = 0.9$.

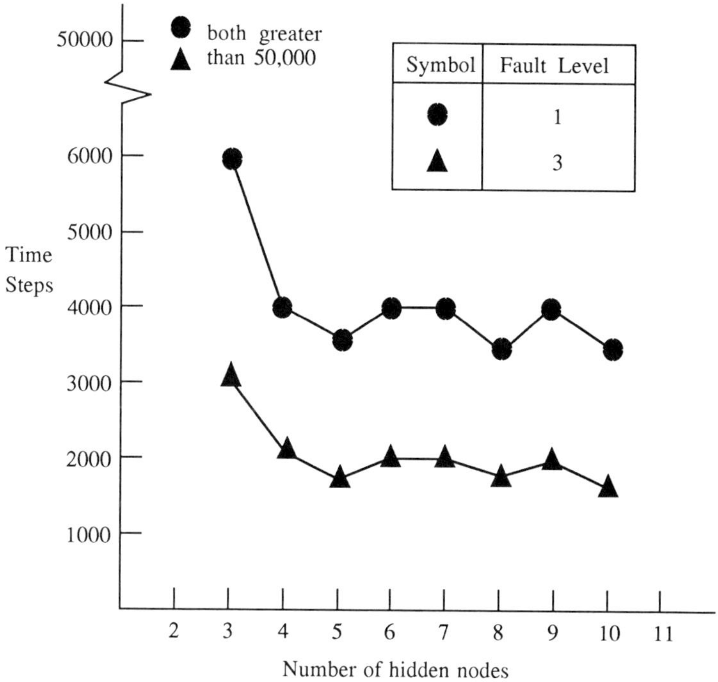

Figure 15. Number of time steps versus the number of hidden nodes for date of Levels 1 and 3 in Table 5.

The initial thresholds w_{n+1} were fixed values taken randomly from a uniform distribution in the range [0,1]. The initial guesses for the weights w_{ji} for the back-propagation procedure were determined suboptimally by a Monte-Carlo method.

Sensor measurements were sampled over a simulated 10-hour period with a sampling interval of 0.01 hour. The input nodes effectively scaled the inputs. Measurement noise was added to the deterministic values of the three sensor variables. The noise was normal random noise that had zero mean but had the following relative (and absolute) non-zero standard deviations:

Outlet concentration of $C_{C_7H_8}$: 1.0% (5.2 gmol/m^3)

Heater outlet temperature: 0.5% (4.5 K)

Controller output signal: 1.0% (2.2 mV)

The three noisy measurements were introduced to the trained network. If an output node had an activity level greater than 0.8, it was classified as

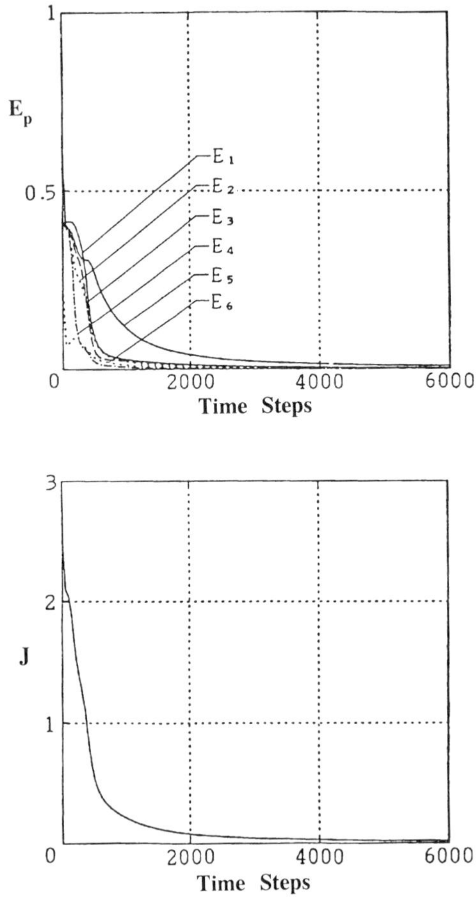

Figure 16. Convergence of the objective function when discriminating among faults.

having identified the fault; if the activity level was less than 0.2, it was classified as normal.

Figure 16 shows how the error for each output node and the total error decreased as the number of presentations of the error vector increased for Stage 1 of the network. Recently, Hoskins (1989) introduced a new heuristic, called *focused-attention back-propagation* (FAB) learning, that enhances the back-propagation procedure by focusing attention on the exemplar patterns that are most difficult to learn. Preliminary experiments with FAB show a speed-up of 1.5 to 10 times faster than back-propagation for certain continuous mappings; thus, we see a potential for reducing the learning times reported in Figure 16.

Table 6. Average activation levels of each output node of the network for five different levels (Level 1 to Level 5) of the five faults (Fault 1 to Fault 5). The network was trained using knowledge of the faults in Level 2 in Table 5. The inputs included noise.

Fault		Average Activation Level of Output Nodes				
No.	Level	Node 1	Node 2	Node 3	Node 4	Node 5
1	1	0.89	0.10	0.00	0.00	0.01
	2	0.93	0.06	0.00	0.00	0.02
	3	0.99	0.03	0.00	0.00	0.03
	4	0.99	0.01	0.00	0.00	0.05
	5	1.00	0.00	0.00	0.00	0.06
2	1	0.02	0.87	0.01	0.00	0.00
	2	0.05	0.90	0.03	0.00	0.00
	3	0.12	0.92	0.20	0.00	0.00
	4	0.33	0.92	0.49	0.00	0.00
	5	0.73	0.95	0.59	0.00	0.00
3	1	0.00	0.12	0.93	0.01	0.00
	2	0.00	0.07	0.97	0.02	0.00
	3	0.00	0.00	1.00	0.04	0.00
	4	0.00	0.00	1.00	0.06	0.00
	5	0.00	0.00	1.00	0.07	0.00
4	1	0.00	0.00	0.00	0.93	0.01
	2	0.00	0.00	0.00	0.97	0.02
	3	0.00	0.00	0.00	1.00	0.02
	4	0.00	0.00	0.00	1.00	0.02
	5	0.00	0.00	0.00	1.00	0.02
5	1	0.02	0.00	0.00	0.01	0.93
	2	0.04	0.00	0.00	0.03	0.97
	3	0.06	0.00	0.00	0.05	0.99
	4	0.08	0.00	0.00	0.05	0.99
	5	0.09	0.00	0.00	0.05	1.00

4.2.3. Classification Results

Table 6 shows the average of 1000 values of the activity levels of the five output nodes of the subnetworks for each fault (Fault 1 to Fault 5) at each deterioration level (Level 1 to Level 5) for Stage 2 of the network. For Fault 1, with any level of deterioration, when the measurements were fed to the network, Node 1 always had the highest activity level among the five output nodes.

Figure 17 shows the classification of outcomes of the five nodes by activity level when Fault 1 occurred. The output of Node 1 had high activity levels

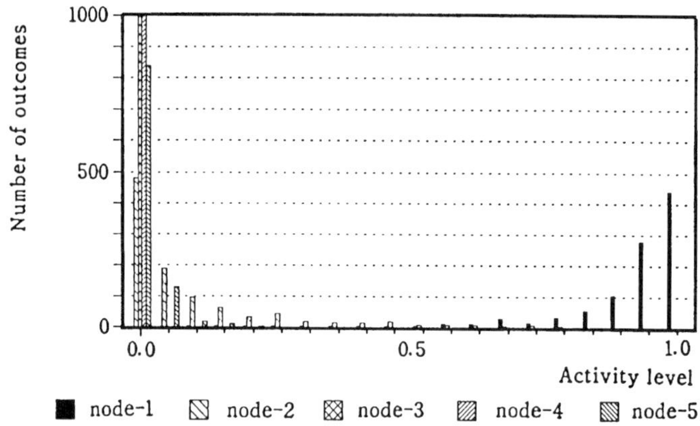

Figure 17. Classification of the output nodes when Fault 1 occurred for 1000 decisions (a total of 5000 outputs).

and fired, whereas the other nodes had low activity levels and thus did not fire. Similar results were obtained for the other respective faults. In the case of no fault, the outputs of the five nodes all had low activity levels and no node fired. Thus, even though the network was trained by knowledge of faults at Level 2, it could discriminate the causes of faults for any level of deterioration.

Table 7 and Table 8 show the classification matrices for Stage 1 when the faults at Level 2 and Level 4 occurred, respectively. These tables demonstrate that the network correctly diagnosed the cause of the faults with

Table 7. Classification matrix for the diagnosis of the causes of faults at Level 2 for the network trained using data of Level 2. Diagonal elements in bold type in the Table represent the percent diagnosed correctly. The process measurements included noise.

	Diagnosed					
Introduced	Fault 1	Fault 2	Fault 3	Fault 4	Fault 5	Normal
Fault 1	**88.2%**	0.5%	0.0%	0.0%	0.0%	0.3%
Fault 2	2.0%	**73.2%**	0.0%	0.0%	0.0%	0.0%
Fault 3	0.0%	0.3%	**94.2%**	0.0%	0.0%	0.0%
Fault 4	0.0%	0.0%	0.0%	**94.6%**	0.0%	0.0%
Fault 5	0.0%	0.0%	0.0%	0.0%	**99.5%**	0.0%
Normal	0.4%	0.0%	0.0%	0.0%	0.0%	**85.4%**

Table 8. Classification matrix for the diagnosis of the causes of faults at Level 4 for the network trained using data of Level 4.
Diagonal elements in bold type in the Table represent the percent diagnosed correctly. The process measurements included noise.

	Diagnosed					
Introduced	Fault 1	Fault 2	Fault 3	Fault 4	Fault 5	Normal
Fault 1	**100.0%**	0.0%	0.0%	0.0%	0.0%	0.0%
Fault 2	14.8%	**79.1%**	12.1%	0.0%	0.0%	0.0%
Fault 3	0.0%	0.0%	**100.0%**	0.0%	0.0%	0.0%
Fault 4	0.0%	0.0%	0.0%	**100.0%**	0.0%	0.0%
Fault 5	0.0%	0.0%	0.0%	0.0%	**100.0%**	0.0%
Normal	0.2%	0.0%	0.0%	0.0%	0.0%	**85.4%**

reasonable robustness. If instead of using an arbitrary activity level for classification (such as 0.8) to ascertain the most likely candidate for a fault, the ratio of the two highest activities is used to discriminate, the classification becomes more certain. Diagnostic results of faults at Level 2 and Level 4 via the ANN trained by the data at Level 5 were not as satisfactory as those obtained via the ANN trained by the data at Level 2.

Table 9 shows a classification matrix based on 1000 decisions for the diagnosis level of Fault 5 by Stage 2 of the network. The matrix shows that

Table 9. Classification matrix for diagnosing level the of deterioration of Fault 5.
The network was trained using data about Fault 5 in Table 5. Diagonal elements in bold type in the Table represent the % diagnosed correctly. The process measurements included noise.

	Diagnosed				
Introduced	Level 1	Level 2	Level 3	Level 4	Level 5
Level 1	**90.6%**	0.4%	0.0%	0.0%	0.0%
Level 2	0.0%	**98.1%**	0.0%	0.0%	0.0%
Level 3	0.0%	0.8%	**80.1%**	7.1%	0.0%
Level 4	0.0%	0.0%	1.8%	**81.9%**	3.0%
Level 5	0.0%	0.0%	0.0%	0.0%	**99.9%**

the second stage of diagnosis clearly discriminated among the levels of deterioration of the fault. These results demonstrate that a two-stage multilayer ANN can diagnose incipient faults from the three noisy steady state process measurements.

The advantage of the two-state network is that if knowledge about the cause of a new fault becomes available, only the network in the first stage has to be retrained for diagnosis to occur. When new information about the level of deterioration of a certain fault is received, only the one network corresponding to that fault has to be trained to accommodate the new knowledge.

5. Conclusion

Artificial neural networks possess a number of properties that make them attractive for knowledge representation in chemical engineering. They have the capacity to learn and store information about process faults via associated memory, and thus have an associated diagnostic ability with respect to faults that occur in a process. Knowledge of the faults to be learned by the network evolves from sets of data, namely values of steady-state process variables collected under normal operating conditions and those collected under faulty conditions, in conjuction with information about the degree of the faults and their causes. Such data can in principle be collected from organized perturbations of the process and/or historical records of data and observed associated faults, as well as from process simulators.

In this study the results of ANN simulation experiments applied to two examples of fault detection and diagnosis demonstrate many of the unique features of the ANN approach. The multilayer, feed-forward, analog-perceptron neural network can learn useful mappings that provide the correct association between fault classes and measurement patterns autonomously. The generalized delta learning rule applied to the multilayer network architecture is able to map the complex decision regions caused by adding sensor noise to deterministic measurement patterns. Rule-following behavior occurs without any explicit representation of rules due to spontaneous generalization, thereby allowing the network to classify similar input patterns not used to train the network. In conclusion, we have demonstrated, via two simple chemical engineering processes, how an artificial neural network (ANN) can learn and discriminate successfully among faults.

References

Anderson, J. A., and Rosenfeld, E. *Neurocomputing: Foundations of Research*. MIT Press, Cambridge (1988).

Angluin, D., and Smith, C. H. Inductive Inference: Theory and Methods. *ACM Computing Surveys*, 15, p. 237 (1983).

Burr, D. J., A Neural Network Digit Recognizer. *Proceedings IEEE International Conference on Systems, Man, and Cybernetics*, Atlanta (1986).

Chartrand, G., and Lesniak, L. *Graphs and Digraphs*. Wadsworth, Monterey (1986).

Cohen, P. R., and Feigenbaum, E. A. *The Handbook of Artificial Intelligence, Vol. 3.*, Kaufmann, Los Altos (1982).

Derthick, M., and Plaut, D. C. Is Distributed Connectionism Compatible with the Physical Symbol System Hypothesis? *Proceedings of the 8th Annual Conference of the Cognitive Science Society* (1986).

Edelman, G. M., Group Selection and Phasic Reentrant Signaling: A theory of higher brain function. In *The Mindful Brain: Cortical Organization and the Group-Selective Theory of Higher Brain Function*, MIT Press, Cambridge (1978).

Feldman, J. A., and Ballard, D. H. Connectionist Models and Their Properties, *Cognitive Science*, 6, pp. 205–254 (1982).

Himmelblau, D. M. *Fault Detection and Diagnosis in Chemical and Petrochemical Processes*. Elseviern, Amsterdam (1978).

Holland, J. H., Holyoak, K. J., Nisbett, R. E., and Thagard, P. R. *Induction*. MIT Press, Cambridge (1986).

Hopfield, J. J. Neural Networks and Physical Systems with Emergent Collective Computational Abilities. *Proceedings of the National Academy of Science*, 79, pp. 2554–2558 (1982).

Hoskins, J. C. *Speeding Up Artificial Neural Networks in the "Real" World*. MCC Technical Report Number STP-049-89, Austin (1989).

Hoskins, J. C., and Himmelblau, D. M. Artificial Neural Network Models of Knowledge Representation in Chemical Engineering. Computers & Chemical Engineering, 12 (9/10), pp. 881–890 (1988a).

Hoskins, J. C., and Himmelblau, D. M. Artificial Neural Network Models for Adaptive Chemical Process Control. Presented at the *AAAI-89 Workshop on AI in Process Engineering*, St. Paul, MN (1988b).

Hoskins, J. C., and Himmelblau, D. M. Automatic Chemical Process Control Using Reinforcement Learning in Artificial Neural Networks. Presented at *The First Annual International Neural Network Society Meeting*, Boston, MA (1988c).

Kohonen, T. *Self-Organization and Associative Memory*, Springer-Verlag, Berlin (1988).

Kolmogorov, A. N. On the Representation of Continuous Functions of Several Variables by Superpositions of Continuous Functions of a Smaller Number of Variables. *Doklady Akademii Nauk SSSR*, 108, pp. 179–182 (1956).

Le Cun, Y. Une procedure d'apprentissage pour reseau a seuil assymetriqur [A learning procedure for assymetric threshold network]. *Proceedings of Cognitiva*, 85, pp. 599–604 (1985).

Lippmann, R. P. An Introduction to Computing with Neural Nets. *IEEE ASSP Magazine*, 4, pp. 4–22 (1987).

Minsky, M., and Papert, S. *Perceptrons: An introduction to computational geometry*. MIT Press, Cambridge (1969).

Nilsson, N. J. *Learning Machines*. McGraw-Hill, New York (1965).
Parker, D. B. Learning-logic (TR-47). Center for Computational Research in Economics and Management Science, MIT (1985).
Romanycia, M. H., and Pelletier, F. J. What is a heuristic? *Comput. Intell.*, 1, pp. 47–58 (1985).
Rumelhart, D. E., and McClelland, J. L. (Eds.). *Parallel Distributed Processing: Explorations in the Microstructure of Cognition*. The MIT Press, Cambridge (1986).
Tou, J. T., and Gonzalez, R. C. *Pattern Recognition Principles*. Addison-Wesley, Reading (1977).
Waterman, D. *A Guide to Expert Systems*. Addison-Wesley, Reading (1985).
Werbos, P. J. Beyond Regression: New Tools For Prediction and Analysis in the Behavioral Sciences. PhD thesis in applied mathematics, Harvard University (1974).
Widrow, B., and Sterns, S. D. *Adaptive Signal Processing*. Prentice-Hall, Englewood Cliffs (1985).

5 A Modular Approach to Multiple Faults Diagnosis

EDUARDO MORALES[1] AND
HECTOR GARCIA

Departamento de Simulación
Instituto de Investigaciones Eléctricas
Cuernavaca, Morelos
México

Abstract

A model-based approach for fault diagnosis has recently received considerable attention. The applicability of these systems is severely restricted by the computational complexity involved in the process which depends, among other things, on the number of possible faults to consider, the number of components involved in the process, and the connectivity between these components. MFD, a model-based multiple-faults diagnosis system based on the system of de Kleer and Williams (1987), is presented. A modification on the basic constraint propagation method is introduced, which simplifies the analysis by dividing the propagation reasoning into groups of components. By effectively reasoning with groups of constraints, this modular strategy improves efficiency, reduces the computational overhead and augments applicability, increasing feasibility for model-based multiple faults diagnosis systems even in complex domains.

1. Introduction

Trying to establish the differences between a model of a system and the system itself has been a major concern of engineers and scientists. The diagnostic process, used for troubleshooting a system or as a refinement of a model, has received considerable attention, which is reflected in the

[1] Present address, the Turing Institute, Glasgow.

ISBN 0-12-480575-2

number of diagnostic systems described in the literature. When troubleshooting a system, a diagnostician compares the measurements taken on the system with the values he expects the system to have. Whenever a discrepancy is found the diagnostician hypothesizes the causes based on the behavior shown by the system and usually does further experiments to narrow down the hypotheses. Every new experiment is used to confirm or refute competing hypotheses.

Broadly speaking, the diagnostic process involves two phases: a failure identification phase where the differences between the observed and the expected behavior are recognized, and a failure location phase where usually a set of tests are executed to identify the faulty component(s). In the first phase several competing hypotheses are normally generated, each one including possibly more than one component. In the second phase, the diagnostician uses the observed behavior, the underlying causality of the system, and his or her experiential knowledge to narrow down the hypotheses and to suggest new experiments until a single consistent hypothesis is reached. In a sense the diagnostician is assigning credit or blame to the components involved in order to explain the observed behavior. In complex systems, the diagnostician tends to work by regions or divisions of the system that include several components. These divisions are based on artificial or physical boundaries that the diagnostician recognizes and uses to discriminate large numbers of components. These divisions help the diagnostician to focus the search and to simplify the analysis. New experiments can shift the attention from one region to another, but the analysis is usually restricted to one region at a time. The experiments are used to gather enough evidence to discriminate among the rest of the regions and concentrate on a particular one (which can itself be subdivided into regions).

A diagnosis is represented as a set of components (the winning hypothesis) such that if every component mentioned in the set fails and every component not mentioned functions correctly then the observed behavior is consistent with the system description. This can be rephrased as assuming a set of components to be faulty in order to explain the observed behavior.

Diagnosis has been the largest application domain for expert systems. Several strategies and capabilities have evolved rapidly over the last few years. Many of the earlier diagnostic systems used what is called *experiential* or *shallow* knowledge expressed as a set of production rules. More recently, a model-based approach has been preferred over or combined with traditional expert systems to gain robustness and overcome some previous limitations.

In this chapter we present a rationale for preferring a model-based over an experiential-based approach in diagnosis for Process Engineering.

MFD, a multiple faults diagnosis system based on work of DeKleer and Williams (1987) is introduced. The computational burden in this kind of system becomes intractable above a certain number of components and faults involved. To maintain the effectiveness of the process a new constraint propagation process is introduced to divide the process and propagate through regions reducing considerably the complexity. This group propagation technique aims to capture the diagnostic process by regions followed by a diagnostician, reducing the computational cost of the process and improving its applicability. Finally, conclusions from this work are drawn and future directions are indicated.

2. Shallow versus Deep Knowledge

Expert systems use production rules to provide the mapping between the observed behavior or symptoms and the underlying causes. These systems rely on the experience of a domain expert and on his ability to express his knowledge as a set of production rules. The so-called elicitation process, or process by which the domain knowledge is "extracted" from the expert, is normally very slow and susceptible to errors. The production rules are manually entered to cover all the possible domain failures that the expert has pondered. With this procedure it is never clear when a system is complete, even after a number of interactions, and it cannot diagnose unforeseen faults. A knowledge validation process might turn out to be quite difficult, especially in large knowledge bases or when tested in a dynamic environment. Additionally, these systems are very carefully handcrafted to a specific problem domain, narrowing their applicability.

In reaction to these problems, a more theoretical model-based approach began to emerge in the literature. This model-based or "deep" knowledge approach was initially incorporated into the already built expert system, in an attempt to obtain a graceful degradation in the systems, especially in the borderline of their knowledge or where new situations arose. An alternative approach used inductive algorithms to generate the required rules straight from examples of failures of the system, avoiding completely the problem of forcing the expert to provide the rules (see Mowforth (1986) for some examples). More recently a combination of a model-based approach with learning techniques, pioneered by Mozetic *et al.* (1988), has been used to generate examples of failures from a theoretical model of the system and then apply an inductive algorithm to synthesize a set of diagnostic rules. In all these approaches the knowledge validation process is carried out over the model of the system, which the domain expert is more familiar with.

3. The Model-Based Approach

There has recently been great interest in systems based on "deep" knowledge using either qualitative or quantitative reasoning. In contrast to the experiential approach, where sufficient knowledge for performing a task is given without any underlying representation of the causal mechanism, the model-based approach captures the underlying causal structures from which all the inferences are made. The model-based approach is suitable to process engineering problem, where a model of the system normally exists and which can be more accurately validated by an expert. We describe a technique called *constraint propagation* (Sussman and Steele, 1980), which can be equally applied to qualitative or quantitative reasoning, although in general any kind of theorem prover could be used.

Constraint propagation operates on state variables, values and constraints. A constraint gives the conditions that the state variable must satisfy. Given some initial values, constraint propagation assigns each state variable a value that satisfies the associated constraints. The state variable can take numerical values or trend values such as increasing, steady or decreasing.

For example we might have the state variables: `F`, `m` and `a`, and a constraint to express `F = m * a`. Given the initial numerical values of `F = 40` and `m = 4`, we can use a numerical constraint to calculate `a = 10`. Similarly, with the trend values of `a = inc` (for increasing) and `m = std` (for steady), we can use a qualitative constraint to derive `F = inc`. Note that all the values cannot always be derived by the system (e.g., in the example above, if we only know that `a = 9.8`, we will not be able to derive `m` and `F`. Similarly, knowing `m = inc` and `a = dec` is not enough to derive `F`).

Constraints can be viewed as conduits among which values can be propagated. A system can be modeled by a set of constraints arranged in a

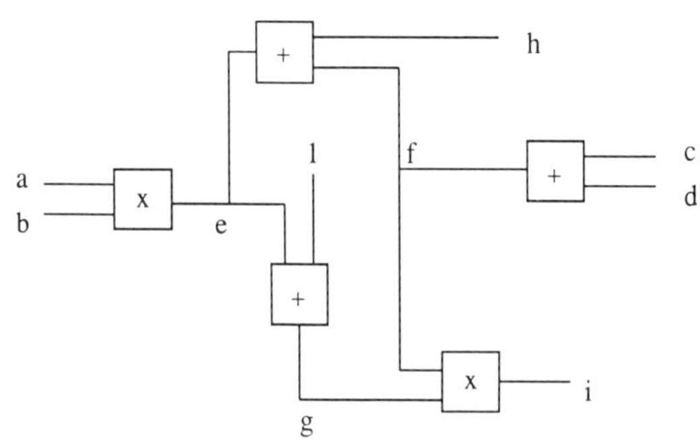

Figure 1. Constraint net.

constraint net where known values can be propagated to predict the behavior of the system. Each new state variable value can be propagated through the constraints for a better understanding of the system behavior. The propagation process may stop, usually for lack of information, although it is not always necessary to know all the state variable values in order to reach a diagnostic conclusion. Without a concrete diagnostic result, new measurements have to be taken. These new measurements might not be immediately available, and in general it might be necessary to consider a cost analysis, the usefulness of the new measure in terms of the amount of information that can be discriminated and even its realibility.

We can illustrate the above ideas with a simple example. Suppose we have the following model.

$$\mathtt{i} = ((\mathtt{a} * \mathtt{b}) + 1)(\mathtt{c} + \mathtt{d}) \tag{1}$$

$$\mathtt{h} = (\mathtt{a} * \mathtt{b}) - (\mathtt{c} + \mathtt{d}). \tag{2}$$

Its constraint net is shown in Figure 1.

Suppose that we have the following initial numerical values:

$$\mathtt{a} = 1$$

$$\mathtt{b} = 2$$

$$\mathtt{c} = 2$$

$$\mathtt{d} = 3.$$

With these numerical values, we can propagate through the constraints to obtain

$$\mathtt{e} = 2$$

$$\mathtt{g} = 3$$

$$\mathtt{f} = 5$$

$$\mathtt{i} = 15$$

$$\mathtt{h} = -3.$$

Similarly we can use the same constraint net, but now with qualitative values to qualitatively predict the behavior of the system. If we have the trend values

$$\mathtt{a} = \mathtt{std}$$

$$\mathtt{b} = \mathtt{dec}$$

$$\mathtt{c} = \mathtt{std}$$

$$\mathtt{d} = \mathtt{inc},$$

(assuming 1 = std) we obtain

```
e = dec
g = dec
f = inc
h = dec
i = ?.
```

Even if we cannot be sure of the trend for `i`, this might not be necessary for a diagnosis.

Suppose now that we only know the values for `a`, `b` and `c` as given above. With this information, we could still infer the values for `e` and `g` (as above), but will not be able to infer the values for `d`, `f`, `h` and `i`. If we are interested in knowing the value of `h` but `h` is too expensive to measure, we have to decide on some other variable to measure, in this case one of `f`, `d` or `i`. With the values of `a`, `b`, `c`, `e` and `g` as given above, we may decide on `i`. If we use numerical values and `i = 12`, `h` would be `h = -7` with `f = 9` and `d = 7`. Similarly, with qualitative values if `i = std`, `h` would be `h = dec` with `f = inc` and `d = inc`.

If we assume that all the measurements have equal costs, then the criterion for choice of measurements is to minimize the number of measurements. In the example above, if we were interested in knowing the value of `h`, we should measure `h` directly. This oversimplification is far from being typical, and more careful analysis on the information content of each measure has to be done to reduce the number of measurements (de Kleer and Williams, 1987). In general, some measurements are more expensive than others and some have a different reliability, for which a different analysis is required (Gallanti *et al.*, 1986).

Most diagnostic systems inputs are taken from real data coming from or measured in the environment. In a qualitative reasoning approach, the first step is to translate the numerical data into qualitative values. By abandoning the precision of numerical analysis, the qualitative approach has the advantages of every abstraction: less precise information is required in order to obtain a result or conclusion, and in general it is computationally less expensive and consequently faster. Its main disadvantage is that it provides less accurate and sometimes ambiguous results. However, in many applications a qualitative result may be all that is required, and even when precise results are needed, a qualitative analysis can provide a framework to guide further analysis.

We can use a qualitative reasoning approach either in a steady-state (time independent) or a time-based (time dependent) system. We can assume a steady-state environment in systems without dynamic components or whose description can be assumed static while the diagnostic process is taking place.

When time plays a crucial role in the process, we need to simulate the system over time to analyze its behavior (Kuipers, 1986). Given a set of qualitative constraints describing the structural relationships, the possible behavior of the system in a particular time interval can be derived from an intial qualitative state. Time can be represented as a totally ordered set of distinguished points that can be generated in the qualitative simulation process. The qualitative simulation proceeds by determining all the consistent qualitative values permitted to each parameter at each time. The simulation may branch into several possible states at each time, and some filtering criteria are usually applied to eliminate spurious states and to recognize cyclic behavior (see Kuipers (1986) for more details).

A recent approach to fault diagnosis has used a combination of constraint propagation techniques of a qualitative model in steady-state with an inductive learning algorithm. This approach was first pioneered by Bratko and his group (Bratko *et al.*, 1985; Mozetic *et al.*, 1988) to predict multiple cardiovascular disorders, and has recently been applied by Pearce (1988) to single faults of a satellite electrical system. The qualitative model is used to predict the behavior of the system for every possible component failure. These component-system behavior relations are used as input examples to an inductive algorithm such as (Michalski *et al.*, 1969) or ID3 (Quinlan, 1986) to induce a rule or a decision tree, respectively, for each failure. Simulation of faults can be achieved by changing the state of the components and firing the constraint rules until the system reaches a stable state. By considering all the possible failures of the system, we can construct a complete example set covering all the possibilities, thus generating a complete set of rules (provided the system behaves differently for each failure). Special domain rules can be used to reduce the exponential explosion when considering multiple faults (Bratko *et al.*, 1985; Mozetic *et al.*, 1988).

In general, a quantitative approach is preferred over qualitative reasoning when accurate results are needed (e.g., digital circuits). We now describe a multiple fault diagnosis system using constraint propagation with numerical values in a steady-state system that is especially suitable for digital circuits.

4. Multiple Faults Diagnosis

In the following subsections, a multiple-fault diagnostic system is discussed. It is the basis on which the system proposed in Section 5 is built. Therefore, although it is not the major purpose of this chapter, sufficient detail is given to the underlying mechanism of these kind of multiple faults diagnosis systems.

4.1. The MFD System

MFD (Morales and García-Mendoza, 1988a) is a multiple-faults diagnosis system based on physical models. It is based on the theory given by Reiter (1987) and follows the system of de Kleer and Williams (1987).

This diagnosis technique is based on the following premises:

(1) Taking a measurement does not affect the system behavior.
(2) The system is in stationary state, i.e., its state is independent of time.

MFD takes as inputs

- The model of the physical system in the form of a set of equations.
- A set of initial values.
- A set of measurements.

4.2. General Description of the Process

The equations are transformed into a constraint net and the initial values are propagated through it in order to predict the behavior that the system should show given those values, i.e., assuming the correct functioning of all the components. The basic constraint propagation method is extended by storing the constraints or components on which each propagated value depends in a dependency list. When a new measurement is taken, it is compared with the predicted value(s). If there is an inconsistency, the measurement is propagated through the net, generating new associated values with new dependency lists. Once an inconsistency is identified, it is used to generate hypotheses of groups of components that may be faulty and to restrict the propagation process. In the following subsections, we describe how these processes (taking measurements, propagating values, storing dependencies, identifying inconsistencies and generating hypotheses) interact until the diagnosis is completed.

4.3. Propagation of Values

The first step in MFD is to transform the set of equations into a constraint net. Each equation of the model represents one or more constraints given by the functional operators involved in the equation and is interconnected to other constraints by their common variables. The initial values are propagated through the net, generating an initial set of possible inferences (predictions). These inferences are compared against subsequent measurements to find the faulty component(s). All the predicted values for each variable have an associated set of dependencies (constraints) which contains all the components that were involved in order to obtain the value (de Kleer, 1986a). Note that the set of initial values, as well as any subsequent measurement, have an empty dependency list ([]). This means that the value is given and does not depend on the correct functioning of any of the system components. Every time a new measurement is taken, it is propagated through the net, and all the constraints involved in the new generated values are stored. Each propagation is done in an specific environment, where an environment corresponds to all the inferences made at a particular propagation. All the predicted values of each propagation have the same environment identification number, i.e., environment 0 corresponds to the initial propagation, environment 1 corresponds to the propagation performed with the first measurement, and so on.

In order to generate new values, we only consider values of the current environment and at most one value of a different environment, to prevent the generation of the same values with each propagation. We are assuming that all the constraints have at most three associated variables. It is clear that the given values (initial or measured) delimit the propagation process.

For example, if we express the equation `H = F * G` as the constraint `mult8 (F, G, H)` and we know the values for `F` and `G`, we can derive a value for `H`. This generated value will inherit the dependency lists (expressed in PROLOG list notation) of `F` and `G` plus the constraint involved in the propagation, in this case `mult8`. If

```
F = 4, [mult1, add6], 2.                    (1)
G = 2, [mult2, add3], 2.                    (2)
```

where (1) represents that variable `F` with a value of 4 depends on `mult1` and `add6` and was generated in the environment number 2, and (2) represents that `G` has a value of 2 in environment 2 and depends on `mult2` and `add3`. We can infer the value of `H`:

```
H = 8, [mult1, add6, mult2, add3, mult8], 2.   (3)
```

Note that `mult8` is added into the dependency together with the dependencies of `F` and `G`. This means that the value `H = 8` depends on the correct functioning of the constraints for `F`, `G` and the constraint from which the value was obtained.

If a conflict (difference) is found between a given (measured) value and a predicted (propagated) value, the dependencies of the propagated value represent a set (conflict set) of possibly faulty components. This conflict set is minimal if no proper subset of it is also a conflict set. Representing the information in terms of minimal conflict sets (MCS) saves memory space when representing possible conflicts. The constraint propagation process is guided by the minimum number of components in the dependency lists, thus producing minimal conflict sets. The MCSs also prevent further constraint propagation on any superset of a MCS. A conflict set gives us a set of components with at least one faulty component. Any further propagation will not provide us with any further information, as any superset of a conflict set is also a conflict set and will involve the same faulty component(s).

Going back to our example, if in the future we measure H = 7, the MCSs would be those sets of dependencies that support H ≠ 7. In this case there is just one predicted H with a different value and the MCS will be [mult1, add6, mult2, add3, mult8]. The only thing we can be sure within this MCS, is that at least one of its components is faulty.

We can have more than one predicted value for a variable. Suppose that we have

```
add8(A,B,C)   (meaning that C = A + B)
        A = 4, [mult1], 1
        A = 6, [mult2, mult4], 2
        B = 3, [add3], 2.
```

From this information we can infer C to be

C = 9, [mult2, mult4, add3, add8], 2 (from A = 6 and B = 3 both generated in environment 2)

C = 7, [mult1, add3, add8], 2(from A = 4 in environment 1 and B = 3 in environment 2).

If we measure C, there are three possibilities:

(1) If C = 9, the new MCS would be [mult1, add3, add8], the dependency list of C = 7, since this is in conflict with the actual value of C.
(2) If C = 7, the new MCS would be [mult2, mult4, add3, add8], the dependencies of C = 9.
(3) If C ≠ 9 and C ≠ 7, the new MCSs would be both the one in (1) and the one in (2) (i.e., [[mult1, add3, add8], [mult2, mult4, add3, add8]]).

The constraint propagation is guided by the connectivity of the constraints and the MCSs, ensuring that the conflicts will be the ones with the least possible number of components (i.e. minimal). The propagation process stops when the possibilities of generating new values are exhausted, when there is not enough information to continue the propagation, and/or when a dependency list of a derived function is a superset of a minimal conflict set.

In our implementation, when a component (constraint) is found to work correctly (i.e., when all of its I/O are measured and correspond to those values that were predicted), this constraint is eliminated from all the dependencies, from the conflict list and from the candidates search space (defined below), reducing significantly the complexity of the analysis.

4.4. Minimal Candidates

The possible diagnosis space, or *candidate space* is represented by the set of diagnoses (where the components not mentioned are guaranteed to work correctly and all those mentioned are possibly faulty) that are consistent with the observations. This set of candidates is decremented during the diagnosis process. So while the MCSs are incremented monotonically, the candidate space is decremented in the same way. Due to the fact that the size of the candidate space is an exponential function of the number of components or constraints (2^n) and that in a complex system, n could be over 50, it is important to represent and to manipulate the candidates in a minimal manner. A minimal candidate set or minimal diagnosis set (MDS) is one none of whose subsets is in the candidate space. In this way, despite the fact that the number of possible diagnoses grows exponentially with the number of components, in most cases the observations of the behavior of the system can be explained by just a few MDSs.

Every time that a MCS appears, the MDSs that cannot explain the conflict are replaced by one or more diagnoses that do explain those conflicts. These new MDSs must be supersets of the ones that were eliminated.

4.5. MDSs Identification Algorithm

The identification of MDSs is an incremental process in the sense that it uses the new MCSs and the older MDSs to obtain the new MDSs.

In our implementation, the candidate space is generated dynamically. This means that the diagnoses are built depending on the "complexity" (i.e., the number of components involved in the MDSs). This permits efficient memory management. In addition, in order to prevent a waste of memory at the beginning of the process, the candidate space is initially "restricted" to diagnoses that will not involve more than, say 3 simultaneous faulty components. Therefore, the number of candidates for a circuit with 10 components, which in theory would be $2^{10} = 1024$, is in the first step just the combination of 10 in 3 (120), plus the combination of 10 in 2 (60), plus 11, that is, 191 possible diagnoses, including the trivial case where there is no conflict in the circuit. This *a priori* level of expected "complexity" in the diagnosis can be adjusted depending on the user experience and on the application area. In the case that the diagnosis process requires a higher level of complexity, most of the lower-level candidates will have been already discarded from the candidate space.

Here is the formalization of the MDSs identification algorithm:

```
IF more than one new MCS
THEN
   INTERSECT all the new MCSs, I_c
   IF I_c ≠ empty set
   THEN eliminate all the subsets of I_c from the candidate
          space. If no other current MDS are left, also restrict the
          candidate space to proper supersets of I_c
LET C_s to be current level of "complexity" of the candidate space.
FOR each new MCS M_c
     OBTAIN the set O of components that do not explain M_c
     IF the "complexity" of O is ≥ than C_s
     THEN increase the "complexity" of the candidate space
            until C_s is greater than the "complexity" of O
     DISCARD from the candidate space all the proper subsets of O
                 that are not currently MDSs.
     FOR each subset O′ of O that currently is a MDS
          LET CO′ to be the "complexity" of O′
          SEARCH at the level of "complexity" CO′ + 1, in the candidate
                    space, all the possible diagnoses P_n that are supersets of O′
          FOR each P_n found, which is not a superset of any other current MDS,
               INCLUDE P_n as a new MDS
               DISCARD O' from the candidate space
```

In the first part of the algorithm, where the intersection of all the MCSs is empty, it means that those components resulting from the intersection are necessarily faulty, but there may be others. If we consider a system with a single fault, we can diagnose it by simply intersecting the MCSs.

In the second part of the algorithm, in order to obtain the set O of components that do not explain the conflict, the complement set of the MCS with the set F (which contains all the components or constraints that could be faulty) is computed. Also in this part, the checking of level of current "complexity" (C_s) is missing, so the algorithm is easy to read.

If during the process only one MDS is left, it becomes the final diagnosis.

4.6. Example

Suppose we have the following model:

$$\texttt{i} = ((\texttt{a} * \texttt{b}) * (\texttt{c} + \texttt{d})) + (\texttt{c} + \texttt{d}) \tag{1}$$

$$\texttt{h} = (\texttt{a} * \texttt{b}) - (\texttt{c} + \texttt{d}). \tag{2}$$

The system then proceeds to its transformations resulting in the constraint net show in Figure 2, where $\texttt{e} = \texttt{a} * \texttt{b}$, $\texttt{f} = \texttt{c} + \texttt{d}$, $\texttt{g} = \texttt{e} * \texttt{f}$, $\texttt{h} = \texttt{e} - \texttt{f}$, $\texttt{i} = \texttt{g} + \texttt{f}$, and each equation indicates the presence of at least one component (constraint). The candidate space for any fault in such a system is given by the lattice in Figure 3, where A13M2 means the component A1, A3 and M2.

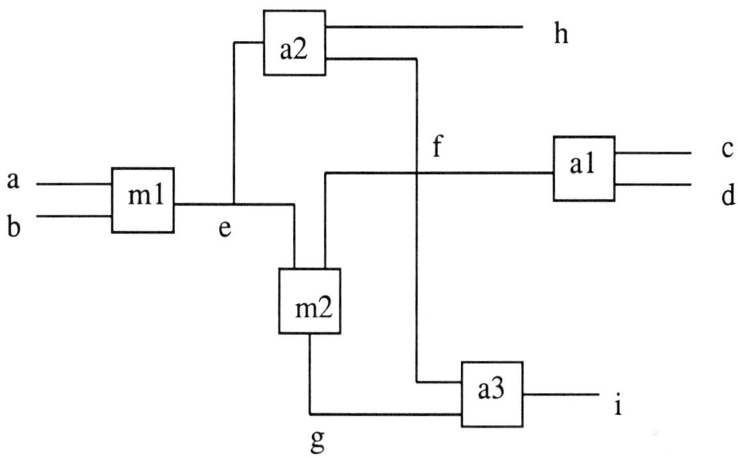

Figure 2. MFD example.

A123M12

A123M1 A123M2 A12M12 A13M12 A23M12

A123 A12M1 A12M2 A13M1 A13M2 A1M12 A23M1 A23M2 A2M12 A3M12

A12 A13 A1M1 A1M2 A23 A2M1 A2M2 A3M1 A3M2 M12

A1 A2 A3 M1 M2

[]

Figure 3. Candidate space lattice.

For example, consider the input to be

```
a = 1 [ ] 0
b = 2 [ ] 0.
c = 2 [ ] 0
d = 3 [ ] 0.
```

We need these initial values in order to "solve" the system (i.e., propagate through all the constraint net). These values do not depend on the correct functioning of any component, and we represent this with empty dependency sets. These initial values are propagated through the net, giving as a result the following values:

```
e = 2 [M1] 0
f = 5 [A1] 0
g = 10 [A1, M1, M2] 0
h = -3 [A1, A2, M1] 0
i = 15 [A1, A3, M1, M2] 0.
```

These propagated values show the predicted behavior of the system and are expected to agree with further measurements. Any discrepancy will indicate a conflict, and the dependencies of the values that are in conflict will give the set of components (MCS) in which at least one of them is faulty. In the beginning, the only MDS is [ ], which means that there is no faulty component in the system.

Suppose, that we measure `i = 11 [ ] 1`. This will produce a MCS of `[A1, A3, M1, M2]` (the dependencies list of `i = 15`). This MCS is then used to obtain the MDSs. The component `A2` is the only one that is not involved in the MCS (set O in the algorithm), i.e. the value of `i = 11` does not depend on the correct functioning of it. We can then eliminate

A123M12

A123M1 A123M2 A12M12 A13M12 A23M12

A123 A12M1 A12M2 A13M1 A13M2 A1M12 A23M1 A23M2 A2M12 A3M12

A12 A13 A1M1 A1M2 A23 A2M1 A2M2 A3M1 A3M2 M12

->A1 ->A3 ->M1 ->M2

Figure 4. Candidate space after i = 11.

[A2] and all its proper subsets from the candidate space. However, the only subset of [A2] is the current MDS [ ] (O' in the algorithm with $CO' = 0$), so it has to be substituted by its supersets in the "complexity" level 1. The supersets of [ ] in "complexity" level 1 (P_n in the algorithm) are [A1], [A3], [M1] and [M2] ([A2] has already been discarded). Now the MDSs tell us that the behavior of the system can be explained if any of the components A1, A3, M1 or M2 is faulty (see Figure 4, where → points to the current MDSs).

By propagating i = 11 [ ] 1, we obtain

```
g = 6 [A1, A3] 1                (from i = 11 and f = 5)
e = 6/5 [A1, A3, M2] 1          (from g = 6 and f = 5)
h = −19/5 [A1, A2, A3, M2] 1    (from e = 6/5 and f = 5).
```

Remember that the propagation will not take place in environments that are proper supersets of any MCS. If we take further measurements, say e = 2 [] 2, the system will identify that e is the last unknown (not measured) variable associated with component M1 and that the measured value is consistent with the predicted one. M1 is then assumed to be working correctly and is eliminated (as well as every superset of it) from the candidate space, reducing the candidate space to 50% of its original size. There is also a new MCS, [A1, A3, M2], from e = 6/5. The current candidate space is shown in Figure 5.

In this case, we do not need to propagate the value just measured, because it corresponds to the expected behavior of the system (we have already done the propagation) and all we have to do is eliminate the component M1 from the dependencies list of the values of the first propagation (i.e., the one with the first given values). The resulting values are

```
g = 10 [A1, M2] 2
h = −3 [A1, A2] 2.
```

Now, if we measure g = 7 [] 3, we have two new MCSs, [A1, A3] from g = 6 and [A1, A2] from g = 10. Here, the intersection of the new MCSs (I_c in the algorithm) is [A1], so we eliminate [A1] as a possible diagnosis. Due to the fact that [A1] is a current MDS, it has to be replaced by supersets of it in the next level of "complexity." However, we cannot restrict the candidate space to supersets of [A1] because there are two others MDSs.

The candidate space, after reducing it with the new information, is shown in Figure 6.

A123M2

A123 A12M2 A13M2 A23M2

A12 A13 A1M2 A23 A2M2 A3M2

->A1 ->A3 ->M2

Figure 5. Candidate space after e = 2.

A123M2

A123 A12M2 A13M2 A23M2

->A12 ->A13 ->A1M2 ->A3M2

Figure 6. Candidate space after g = 7.

4.7. Analysis

The above system was implemented on an IBM-PC-AT in PROLOG and tested with simple examples that involved a small number of components (Morales and Garcia-Mendoza, 1988a). As we modified our tests and tried the system with more complex examples, we soon realized that it was going to demand computational resources which were not available to us and that an alternative methodology was needed. In a real system, the complexity of the system may involve over 70 components. This implies that the candidate search space is $2^{70} \approx 1.2 \times 10^{21}$ diagnostics. Although we are not considering all, since simplifications are made during the process, this is just too big. Even if we restrict the number of possible faults, and eliminate components that are working correctly, we still have to deal with the constraint propagation process, which in the worst case would be storing half of the components on average for each new measurement and, depending on the connectivity of the system, the number of possible values could increase at an exponential rate. We believe that a model-based approach is the correct one, if the model is available. Our implementation was useful for systems where the number of components was around 10. For bigger systems, we need to reduce the complexity of the analysis in some other way.

5. Other Approaches

One primary concern when using model-based reasoning is the computational cost. There have been several suggestions on how to restrict the computations involved, and we mentioned very briefly some of them before introducing MFD2. Genesereth (1984) suggests possible ways to reduce the complexity for a single-fault diagnosis system, although his ideas are equally applicable to multiple faults. His system (DART) uses a logic-based representation language with an inference procedure based on resolution (Robinson, 1965). Due to the domain independence of the language, different design descriptions can be expressed to improve efficiency by reducing the number of components to consider. The most common example is a structural hierarchy where the system is described in terms of high-level components. A structural abstraction can keep small the number of components, and sometimes it is possible to diagnose faults without considering detail. This structural abstraction is usually accompanied by a behavioral abstraction where the behavior of the components at one level is often described in different terms from the behavior of components at other levels (e.g., for digital circuits, at one structural level we can express the behavior of the components in terms of voltage ranges, at an other level as binary numbers, and at yet another level as natural numbers). Davis (1984) proposes grouping components into larger modules, as in Genesereth (1984). The basic idea is to diagnose at the most abstract level in a hierarchy and only analyze the contents of a module when there is reasonable evidence to suspect it. Davis also proposes an enumeration and categorization of failures based on their likelihood. Troubleshooting should then proceed by considering first the more likely failures. In his system one has to define a language to describe the structure of the system, a language to describe the behavior, and a set of principles based on the above descriptions for troubleshooting. His system uses *constraint suspension* (similar to a backtracking algorithm) to propagate values: when an inconsistency is found, it backtracks to the point where there is no inconsistency.

Such abstractions are designed to reduce the computational overhead of the process. The diagnostic process could be made over the components at one level of the hierarchy and expanded into a next hierarchy level when a faulty component is recognized. Unfortunately, some components may not be so easily grouped into more generic ones and the behavior may not be easily abstracted without making strong assumptions. The generalization of a set of components into a high-level component, like every generalization, necessarily loses detailed information that may be required in some systems, such as digital circuits. Another limitation of the above systems is that they require all the design information which sometimes is unavailable. MFD, on the other hand, relies exclusively on the equations that describe the system. This has the advantage that if detailed information is available, this can be expressed in the model. If only a set of equations (a model) is available we can still approximate a diagnosis to enough detail. In MFD2, there is no need for describing the structure and hierarchy of the system. The model of a system expressed as a set of equations gives the information needed to perform the diagnosis and divide the system (see Section 3). Unlike the above systems, MFD can diagnose multiple faults using exclusively a set of equations.

6. Divide and Conquer (MFD2)

One of the main reasons for MFD limitations is the number of components that have to be considered in the constraint propagation process. In order to improve applicability, some reductions are needed. As mentioned in the previous section, one possible approach is to make a hierarchy of components where each component in the hierarchy can include several subcomponents. A more feasible approach is to divide the system into groups of components and diagnose the system by analyzing these regions. Note that we are not doing any structural abstraction since we are not removing any detailed information. What we are proposing is to divide the analysis by regions. This approach is more consistent with the way a diagnostician proceeds. Without simplifying the model of the system, a variant of constraint propagation, called *group propagation*, is introduced, which naturally focuses the diagnostic search and simplifies the computations involved. Before giving the details of group propagation, we suggest a possible method, based on a mathematical model, to divide the system into groups. The obvious approach is to divide the system across its "natural" physical boundaries. System components are usually grouped with other components that have a common purpose in the system. The specification of the divisions can be done easily with a user interface, allowing the user to decide where to divide. An alternative approach is to make the divisions based on the mathematical model of the system. With the model expressed as a set of equations, we can analyze the causality of the system (Simon, 1953; see also Iwasaki and Simon, 1986) to find out which sets of equations must be solved simultaneously and which ones depend for their solutions in solving others. This analysis could group sets of constraints that must be solved simultaneously based on their model. These groups will then be connected to other groups that depend on their solution. If the set of equations is still too big for our system, we can apply a *tearing* method (Westerberg *et al.*, 1979; Edgar and Himmelblau, 1988) to decide where to draw a border line (i.e., in the variables suggested by the tearing method).

Suppose that we have the following set of equations:

$$i = ((a * b) * (c + d)) + (c + d) \tag{1}$$

$$h = (a * b) - (c + d) \tag{2}$$

$$\mathtt{i} = (\mathtt{p} * \mathtt{n}) + (\mathtt{q} * \mathtt{o}) \tag{3}$$

$$\mathtt{h} = (\mathtt{m} * \mathtt{o}) + (\mathtt{p} * \mathtt{n}). \tag{4}$$

Equations (1) and (2) are the same as in the MFD example. We can see that equations (1) and (2) are "connected" (have common variables) with (3) and (4) by variables i and h. Suppose that we have initially the same input values as in the MFD example plus additional values for n, o and p.

$$\mathtt{a} = 1 \tag{5}$$

$$\mathtt{b} = 2 \tag{6}$$

$$\mathtt{c} = 2 \tag{7}$$

$$\mathtt{d} = 3 \tag{8}$$

$$\mathtt{n} = 4 \tag{9}$$

$$\mathtt{o} = 1 \tag{10}$$

$$\mathtt{p} = 2. \tag{11}$$

With these values we have 11 equations with 11 unknown values and we can "solve" the system. A system with N equations with N unknowns is called *self-contained.* Note that in general we do not need to have a self-contained system to start propagation, but without it we will not be able to propagate through all the net (which might not be necessary to reach a diagnosis). An initial given value is, by definition, a self-contained system. Looking more carefully at the set of equations, we can easily see that equations (1) and (2) with inputs (5), (6), (7) and (8) represent another self-contained set which can be solved independently of the rest of the equations, and which in fact needs to be solved before equations (3) and (4). Recognizing self-contained sets and the causality between them can be used to decide where to divide the system. Analysis of causality and grouping self-contained sets is not new, and there are several methods in the literature (e.g., Simon, 1953; Westerberg *et al.*, 1979; Edgar and Himmelblau, 1988). This causality analysis is graphically represented in Figure 7, where Region I (RI) is connected to Region II (RII) via h and i. In other words, with the above input values, in order to solve equations (3) and (4), we need to solve first (1) and (2).

6.1. A Theory of Group Propagation

Once a division has been decided and the border points between regions are localized, we can analyze the system by regions connected by common

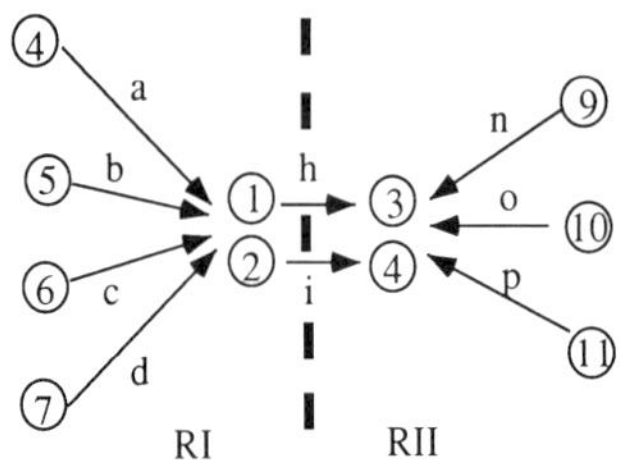

Figure 7. Causal relations.

variables. For each region, its neighbors are like black boxes with input/output terminals. Instead of propagating all the information about the constraints of the whole system, the propagation between groups involves only the values of the border variables and the region name of that value. A group of components "exports" the values used by other groups and "imports" inputs from other groups required for its analysis.

A region is called an *exporter* region when a variable value in the region is exported to a neighbor region. Similarly an *importer* region would be one that accepts a value from a connected region (its exporter). A value is called an *export* value if it is given by an exporter. Similarly, a value is called an *import* value if it is taken by an importer.

Each group looks at the other groups as input/output acceptors/donators of values. A group believes, at least at first instance, in the veracity of its imported inputs and uses them to propagate through its region. When a measurement is taken at the borderline, the involved groups can confirm or blame the predictions made by the other group, immediately splitting the analysis. If the predicted value from one region differs from the measured value, we can focus our diagnosis on that region. An imported value and all its inferences can be eliminated from the imported region if a measurement contradicts its value. The propagation process in each region is exactly the same as the propagation described in MFD; however, we add to the propagation the name of the region and the channel (variable name) by which the value was imported. In each region, the propagation only involves components of that particular region. In group propagation we do not always need to propagate across regions. The propagation process involves fewer components, as we export only names of the regions and the variables exported, and the propagation can sometimes be suppressed. We illustrate these ideas with the above example. Its constraint net is shown in Figure 8.

With the above values for a, b, c, d, n, o and p, we proceed with the propagation as before. We will call Region I (RI) the constraints in

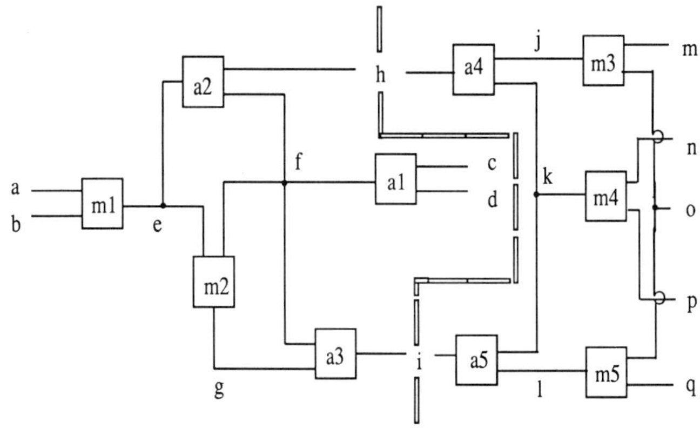

Figure 8. MFD2 example.

equations (1) and (2) and Region II (RII) the constraints in equations (3) and (4). In RI,

```
a = 1 [ ] 0
b = 2 [ ] 0
c = 2  [ ] 0
d = 3 [ ] 0
e = 2 [M1] 0
f = 5 [A1] 0
g = 10 [A1, M1, M2] 0
h = −3 [A1, A2, M1] 0
i = 15 [A1, A3, M1, M2] 0.
```

In region II we initially have

```
n = 4 [ ] 0
o = 1 [ ] 0
p = 2 [ ] 0
k = 8 [M4] 0.
```

In order to continue the propagation, we need to import values from RI to RII (in this case *h* and *i*). We express this by adding a parameter to the propagation consisting of the name of the region and the imported variable name.

```
h = −3 [ ], RI(h) 0
i = 15 [ ], RI(i) 0,
```

and consequently in RII,

```
l = 7 [A5, M4], RI(i) 0
j = −11 [A4, M4], RI(h), 0
m = −11 [A4, M3, M4], RI(h) 0
q = 7 [A5, M4, M5], RI (i) 0.
```

The immediate places to measure are m and q, which can be taken as the outputs of the system. If the measured values are the same as predicted we assume the system is working correctly and stop the analysis. Suppose that we measure `m`. If `m = −8 [ ] 1`, we have a conflict of `[A4, M3, M4], RI (h)` (due to `m = −11 [A4, M3, M4], RI (h) 0`). We proceed with the propagation with this new value in RII up to the border. The results are

```
j = −8 [M3] 1                          (from m = 8 and o = 1)
k = 5 [A4, M3], RI(h) 1                (from j = 8 and h = −3)
h = 0 [A4, M3, M4] 1                   (from j = 8 and k = 8)
l = 10 [A4, A5, M3], RI(i) 1           (from k = 5 and i = 15)
q = 10 [A4, A5, M3, M5], RI(i) 1       (from l = 10 and 0 = 1).
```

Note that the value for `h` depends exclusively on values and constraints from RII and is kept in the propagation. If we had done the propagation as in MFD, the conflict set would be `[A1, A3, M1, M2, A4, M3, M4]` and in principle we would continue the propagation to RI with `h = 0`. In group propagation, we stop the propagation at the border until a border value is measured. Note that the dependency list in each region will involve only components of that region. The other possible values for the variables are supersets of the conflict set and are not calculated (i.e., `h = −6 [A4, M3, M4], RI (h), k = 11 [A4, M3, M4], RI(h), j = −5 [A4, M3, M4], RI(h), i = 12 [A4, A5, M3, M4], RI(h, i)` and `i = 18 [A4, A5, M3, M4], RI(i)`). Note that although `h = 0` is kept in the propagation as it depends exclusively on values and constraints from RII, `h = −6` is eliminated since its dependency list is the same as the conflict set.

If we now measure `h` we have three cases:

1. If `h = −3` as predicted before by RI, we have a conflict in RII of `[A4, M3, M4]` from `h = 0` but now independently of `RI(h)`. Although this conflict is (almost) the same as before, we do not propagate anything with this value (we have already done it) and we can eliminate `RI(h)` from the analysis. The resulting values in RII will be the same except that we eliminate `RI(h)` from every dependency list and from the conflict.

    ```
    n = 4 [ ] 0
    o = 1 [ ] 0
    p = 2 [ ] 0
    i = 15 [ ] RI(i) 0
    k = 8 [M4] 0
    j = −11 [A4, M4] 0
    l = 7 [A5, M4], RI(i) 0
    q = 7 [A5, M4, M5], RI(i) 0
    m = −8 [ ] 1
    k = 5 [A4, M3] 1
    j = −8 [M3] 1
    l = 10 [A4, A5, M3] 1
    q = 10 [A4, A5, M3, M5], RI(i) 1
    h = −3 [ ] 2.
    ```

    We are "disconnecting" RI and RII via `h`. To proceed with the analysis, we will need to perform further measurements to find the faulty component(s). If we measure `i` and `i = 15`, we can forget about RI. If we first measure `q`, the propagation will produce a value for `i` depending only on components of RII. We will not be able to discard RI from the analysis until we measure `i`, nevertheless, we can assume that RI is working correctly without measuring `i` and still arrive at a diagnosis. We are not sure if there are faulty components in RI, but we are sure that there are in RII.
2. If `h = 0 [ ]`, we have a conflict in RI of `[A1, A2, M1]` from `h = −3 [A1, A2, M1] 0` and we need to propagate this value in RI (independently of RII). Since `h = 0` was predicted by RII relying

only on measures and components on RII, we can concentrate the analysis on RI. We can also eliminate the conflict [A4, M3, M4], RI(h) from RII since there is no reason to believe that there is a conflict in RII, as it came from h = −3. We eliminate as well all the values propagated with RII(h). This will leave RII as

```
h = 0 [ ] 2
j = −8 [M3] 1
k = 8 [M4] 0
i = 15 [ ] RI(i) 0
l = 7 [A5, M4] RI(i) 0
q = 7 [A5, M4, M5] RI(i) 0.
```

Recognizing wrong imported values is used to eliminate "spurious" propagation in importer regions and to concentrate the analysis in the exporter region. Note that there may still be a conflict in RII; all that we know is that there is a conflict in RI, and we concentrate our analysis on that region. Again we will need to perform further measurements. If i = 15, we can eliminate RI(i) from RII propagations and keep both regions independently. If i ≠ 15, we would eliminate all propagations made with i = 15 in RII and propagate this new value in both regions.

3. If h is different from −3 and 0, we have a conflict in both regions: [A4, M3, M4] in RII and [A1, A2, M1] in RI. We can again eliminate RI(h) and all the inferences made with h from RII (as in Case 2) but we would still have a conflict in RII (although it now does not depend on RI). We would then need to propagate this new h in both regions and continue with the analysis. In this case we are sure that there are faulty components in both regions.

With this approach we simplify considerably the computation of a system. We do not need to propagate through all the net all the time, we do not need to store all the dependencies, and we can focus the search easily towards the faulty components.

7. Conclusions

The introduction of artificial intelligence techniques to process engineering has automated diagnostic problems that were traditionally solved by domain experts. The shallow-knowledge approach has been the most common technique used so far. Unfortunately, it is difficult to test its reliability and the development time can be very slow. In reaction to this, a model-based approach has recently received considerable attention. In process engineering, where a mathematical model of a system is more likely to exist, a model-based approach should be preferred over an experiential or shallow-knowledge-based system. There have been several model-based techniques, some of which were mentioned here. One primary concern when using model-based systems has been the computational cost. The applicability of a model-based approach has been restricted by the number of components involved and the connectivity of the system. We have presented a multiple-faults, model-based diagnostic system in steady state using constraint propagation based on the system of de Kleer and Williams (1987). This kind of system is still restricted by the computational overhead involved in the process. Previous model-based systems have suggested structural and behavioral abstractions to reduce the complexity of the analysis by reducing the number of components involved. Unfortunately, every abstraction necessarily loses some detailed information that may be necessary to the analysis. In this chapter, we propose a variant of constraint propagation based on groups of constraints.

This group propagation technique reduces the computational cost involved in the process by dividing the system into regions. The propagation process in each region applies only to components of that particular region, simplifying the analysis. With this technique, we can concentrate on a particular region that looks faulty, forgetting temporarily the rest of the system. Unlike previous systems, MFD2 divides the system based on the mathematical model without doing any abstraction, keeping all the detailed information. We are currently working on a formalization of a group propagation theory and we are analyzing alternative methods to decide where to divide in cases where we do not have a self-contained system.

References

Banares R. and Morales, E. Herramientas Basadas en Tecnicas de Inteligencia Artificial usadas en la contruccion de Sistemas Expertos para la manipulacion de diagramas de flujo. *Curso International de Sistemas Expertos* CINVESTAV, Mexico (1987).

Bratko, I., Lavrac, N. and Mozetic, I. KARDIO-E An Expert System fopr electrocardiographic diagnosis of cardiac arrhytmias, *First Intl. Expert Systems Conference*, London, pp. 289–296 (1985).

Davis, E. Constraint Propagation with Interval Labels. *Artificial Intelligence*, 32, pp. 281–231 (1987).

Davis, R. "Shrobe, H., Hamscher, W., Wieckert, K., Shirley, M., and Polit, S." Diagnosis based on description of structures and function *Proc. of the Nat. Conf. Artificial Intelligence*, Pittsburgh, PA (1982).

Davis, R. Diagnostic Reasoning Based on Structure and Behavior. *Artificial Intelligence*, 24 (1984).

Edgar, T. F. and Himmelblau, D. M. *Optimization of Chemical Processes*. McGraw Hill, Chemical Engineering Series (1988).

Gallanti, M., Gilardoni, L., Guida, G., and Stefanini, A. Exploiting Physical and Design Knowledge in the Diagnosis of Complex Industrial Systems. *European Conference on Artificial Intelligence*, Brighton, U.K., pp. 335–349 (1986).

Genesereth, M. The use of design descriptions in automated diagnosis. *Artificial Intelligence*, 24, pp. 411–436 (1984).

Ginsberg, M. Counterfactuals. *Artificial Intelligence*, pp. 35–79 (1986).

Hamilton, T. P. HELIX: A helicopter diagnostic system based on qualitative physics. *Artificial Intelligence in Engineering*, 3 (3) pp. 141–150 (1988).

Iwasaki, Y., and Simon, H. Causality in Device Behaviour. *Artificial Intelligence*, 29, pp. 3–32 (1986).

Kramer, M. A. Integration of Heuristic and Model-based inference in Chemical process fault diagnosis. Cambridge, MA, MIT, Industrial Liaison Program, p. 5 (1986).

de Kleer, J. Local methods for localizing faults in electronic circuits. MIT AI memo 394, Cambridge, MA (1976).

de Kleer, J. An Assumption-based TMS. *Artificial Intelligence*, 28, pp. 127–162 (1986a).

de Kleer J. Problem Solving within the ATMS. *Artificial Intelligence*, 28, pp. 197–224 (1986b).

de Kleer, J., and Williams, B. Diagnosing Multiple Faults. *Artificial Intelligence*, 32, pp. 97–130 (1987).

Kuipers, B. Qualitative Simulation. Artificial Intelligence, 29, pp. 289–338 (1986).

Michalski, R. On the quasi-minimal solution of the general covering problem. *Proc. 5th. Intl. Symposium on Information Processing*, Vol.A3, Bled. Yugoslavia (1969).

Michalski, R., Mozetic, I., Hong, J., and Lavrac, N. The AQ15 learning system: An overview and experiments. *IMAL-1986*, Les Arcs, France (1986).

Milne, R. Fault Diagnosis through responsibility. *Proc. 9th. IJCAI*, Los Angeles, pp. 424–427 (1985).

Milne, R. Strategies for Diagnosis. *IEEE Transactions on Systems, Man, and Cybernetics*, 17, (3), pp. 333–339 (1987).

Morales, E., and Garcia-Mendoza, H. Diagnostico de Fallas Multiples basado en Modelos Fisicos. *V Reunion Nacional de Inteligencia Artificial*, Merida Yucatan, Mexico (1988a).

Morales, E., and García-Mendoza, H. Multiple Faults Diagnosis based on Physical Models. *Workshop of Artificial Intelligence Applications in Process Engineering, AAAI88*, St. Paul Mineapolis (1988b).

Mowforth, P. Some Applications with Inductive Expert Systems Shells. TIOP 86-002, Turing Institute, Glasgow, U.K. (1986).

Mozetic, I., Bratko, I., and Lavrac, N. Automatic synthesis and compression of cardiological knowledge. *Machine Intelligence 11* (J. Hayes, D. Michie and J. Richards eds.), Oxford University Press (1988).

Pazzani, M. J. Failure driven learning of faults diagnosis heuristics. *IEEE Transactions on Systems, Man, and Cybernetics*, 17 (3), pp. 380–394 (1987).

Pearce, D. The Induction of Fault Diagnosis System from Qualitative Models. Glasgow: The Turing Institute, TIRM-88-029 (1988).

Quinlan, J. The induction of decision trees. In *Machine Learning 1*, pp. 81–106 (1986).

Reggia, J. A., Nau, D. S., and Wang, Y. Diagnostic expert systems based on a set covering model. *Int. J. Man-Mach. Stud.*, 19, pp. 437–460 (1983).

Reiter, R. A Theory of diagnosis from first principles. *Artificial Intelligence*, 32, pp. 57–95 (1987).

Robinson, J. A machine-oriented logic based on the resolution principle. *J. ACM*, 12, pp. 23–41 (1965).

Rychener, M., Subramanian, E., Wiss, J., and Lasky, S. A hybrid system for helicopter engine troubleshooting. Engineering Design Research Center, CMU (1986)

Simon, H. A. Causal Ordering and Identifiability. In *Studies in Econometric Methods* W. Hood and T. Koopmans eds.), Wiley, N.Y. (1953).

Steele, G. The definition and implementation of a computer language based on constraints. AI Tech. Rept 595, MIT, Cambridge, MA (1979).

Sussman, G., and Steele, G. CONSTRAINTS: A language for expressing almost-hierarchical descriptions. *Artificial Intelligence*, 14, pp. 1–39 (1980).

Westerberg, A. W., Hutchinson, H. P., Montard, R. L., and Winter, P. *Process Flowsheeting*, Cambridge University Press, Cambridge (1979).

Xiang, Z., and Srihari, S. N. A strategy for diagnosis based on empirical and model knowledge. *6th International Workshop on Expert Systems and their Applications*, Avignon, France, Vol 1, pp. 835–848 (1986).

6 Modeling Real-World Processes: Deep and Shallow Knowledge Integrated with Approximate Reasoning in a Diagnostic Expert System

A. D'AMBROSIO

APEIRON
Glendale, California

1. Introduction

The representation of knowledge is a basic issue in the development of Expert Systems (ESs) for industrial applications. In fact, the performance of an ES heavily relies on the adequacy of its knowledge base.

There is a difference between theoretical and real-world knowledge. We acquire the former by studying the relevant theory, which is under the form of general laws and axioms. The latter is not completely acquired through theory: one also needs to gain insight into problems that may arise in actual situations and so acquire problem-solving capabilities in order to become an expert in a given real-world domain. Abstraction capability based on theoretical knowledge interacts with insight and problem-solving capabilities to form what we call real-world domain knowledge.

Real-world knowledge so defined calls for a distinction that is indeed well agreed upon; it is said to consist of two different components, *Deep* and *Shallow Knowledge*.

ISBN 0-12-480575-2

With Deep Knowledge, we mean the set of theoretical laws and axioms that form the basis for abstraction capability. It is the same kind of knowledge we learn at school.

Shallow Knowledge defines all that does not fall into the same category. Such knowledge will not be acquired from books, but will need mentors to provide insightful suggestions, experience to acquire problem-solving capabilities, and intuition to tackle unknown situations. Shallow knowledge is also called *expertise*, referring to the peculiar skills of an expert in a given domain. The outcome of shallow knowledge (or expertise) is also generally referred to as *heuristics* (or heuristic knowledge). This term refers to the so-called *rules of thumb*, meaning the judgment capabilities that an expert acquires through experience. Heuristic knowledge allows the expert to grasp the meaning of a given situation and to foresee the consequences even when lack of data or complexity of the domain do not permit a correct or complete theoretical analysis of the problem.

We will give an example to better clarify the above concept. In many domains we face the problem of dealing, at times, with large search trees. This is true in chess as in process diagnosis. There are several algorithms for optimizing the search technique, which can also be powerful. It may happen, though, that the situation does not allow exploiting such algorithms, either because of time constraints and/or hardware capability or because of lack or uncertainty of data.

This situation is common in the real world of industrial processes. Data are often unavailable or uncertain, and time constraints may be strict. On the other hand, the consequences of a missed identification of possible faults may be costly, dangerous, or both. Expert maintenance operators usually solve such situations by means of heuristic knowledge. Their insight and the rules of thumb they can immediately recall, can substitute for missing or uncertain data, thus solving a problem that would otherwise be too complex to be solved quickly.

Naturally, different tasks relevant to the same domain require different levels of knowledge, in a continuum from little theory and some hands-on experience to much abstraction capability and high-level expertise. We clearly distinguish between the manual operator performing routine procedure and experts who can perform diagnostics and rely on their insight to decide preventive maintenance or predict future business trends.

In our case (i.e., ES application to the real world of industry), we are involved with the level of expertise that allows a domain expert to identify quickly uncommon problems to find solutions, and to make do, if necessary, without explicit information and data. Hence the need for a domain-modeling methodology that can provide adequate means of including such complex knowledge in ESs. Theoretical (deep) knowledge can be represented by laws and axioms, but insight and problem-solving capabilities (expertise, shallow knowledge) cannot.

The aim of this chapter is to discuss issues relevant to the modeling of the real-world domain knowledge defined earlier as a combination of deep and shallow knowledge.

We will first give a general overview of the problems providing references to the literature available in the field. We will then depict the scenario in which such problems take place and illustrate a methodology for modeling a real-world domain knowledge. We will base the discussion on the development of an ES for diagnosing thermal power plant malfunctions, deepening the concepts already expressed in previous papers (D'Ambrosio *et al.*, 1988a, 1988b; D'Ambrosio, 1988).

2. Overview

Industrial real-world domains are complex to represent. The knowledge involved is of both the deep and shallow kind. Deep knowledge is present in engineering domains as physics of phenomena and mathematics to treat them; but the same emphasis must be given to the presence of shallow knowledge under the form of domain expertise. Truly qualitative models based on deep knowledge are not completely adequate to represent systems performing real-world tasks (e.g., diagnosing plant malfunctions) because

- Real-world industrial systems never provide all the information needed for exploiting models based on deep knowledge (incomplete data).
- Data from field may be incorrect (instrument failures).
- ESs dealing with real-world problems must provide reliable answers in a time "fast enough for use by the process being serviced" (definition of real-time in Marsh and Greenwook, 1986).

The knowledge representation (KR) issue is widely discussed in the literature. Earlier reports (Woods, 1975; Bobrow and Winograd, 1977) debate about how to better represent different kinds of knowledge (Winograd, 1975). From the recent literature we can refer to discussions on physical systems knowledge (Bobrow, 1985) and qualitative physics and modeling (deKleer and Brown, 1986; Iwasaki and Simon, 1986). The role of Knowledge Representation systems "is to manage a Knowledge Base for a Knowledge-Based system and present a picture of the world based on what it has represented in the Knowledge Base" (Levesque and Brachman, 1985). So the main effort in developing expert systems for industrial applications is usually aimed at finding a representation of the domain knowledge that can actually resemble the real world of an industrial plant where, indeed, both deep and shallow knowledge are exploited in all situations, such as process performance diagnosis, component trouble-shooting, design, operational training, maintenance, and planning.

Examples to overcome the problem of representing knowledge of engineering domains can be found in Kramer (1986), Dague *et al.* (1987), and Mavrovouniotis and Stephanopoulos (1987). They clearly discuss cases in which the purely qualitative approach is insufficient. They also show how to solve those cases by exploiting engineering knowledge (shallow, expertise).

An example of using such knowledge is given by the problem of validating field data for use by expert systems (no trivial issue): real-world processes do not always provide certain data, much less redundant instrumentation to validate them. This issue is discussed in Chandrasekaran and Punch (1987). The cost of instrumentation and data-aquisition systems must be kept as low as possible; asking for more (or redundant) instrumentation to validate data goes the wrong way. This is a situation where shallow knowledge is helpful. This kind of knowledge can often make do with uncertain data as well as with missing data. Data are instead necessary (although only qualitatively) in purely qualitative models of physical systems, not including shallow knowledge (expertise about the behavior of physical systems).

The ES development project on which the following discussion is based was carried out at ANSALDO—one of the world largest companies in power plant engineering and construction—by APEIRON, a consulting company in AI/ES R&D and application. The development project started with the prototype of PERF-EXS (D'Ambrosio *et al.*, 1988a), an expert system for diagnosing power plant process faults, and involved at the same time the search for an efficient Knowledge Representation and Acquisition methodology for industrial ESs. The project then resulted in the design and development of a general methodology for modeling industrial plant knowledge. The PERF-EXS prototype referred to one subsystem of a thermal power plant; the generic domain model defined was then used to develop the modules for all remaining subsystems of the thermal power plant and so complete the development of the entire system.

PERF-EXS and the basic model were implemented in KEE on a Symbolic 3640: the object-oriented knowledge representation (provided by KEE) was used to implement the system KB. KEE hooks to Common Lisp were used to implement procedural knowledge and computations where needed. Rules, ATMS (assumption-based truth maintenance systems), and worlds provided the means for implementing diagnostic reasoning (D'Ambrosio *et al.*, 1988a; Morris and Nado, 1986).

3. The Scenario

3.1. The Engineering Company: Its Deep Knowledge

The company deep knowledge background stresses accurate design providing lay-outs, detail drawings, manuals, and complex computational programs. Designing a power plant involves designing all single components in detail and having them on file together with all the information on design parameters and process variables.

3.2. The Engineering Company: Its Expertise

In the field the same emphasis is put on monitoring the actual situation accurately. As a matter of fact, the real world is different from drawing boards and the actual values of variables involved in real processes—although specified by design parameters—eventually depend on the actual operation. The operational expertise of plant experts allows them to understand how the actual situation—illustrated by variable behavior—affects design parameters, causing plant malfunctions and consequently poor overall performance. Feedback from field operations is also part of the company knowledge background. As a consequence, the generic domain model must be able to include and organize this sort of information and reasoning capabilities.

We will illustrate the process of designing the domain model based on the lessons learned during the search for optimal knowledge representation and organization. We will first briefly describe the domain of application (power plant) and the knowledge involved with diagnosing power plant malfunctions. Once the above-mentioned operations are clearly stated, it will be easy to understand the need for approximate reasoning of the diagnostic process as well as the requirements for the general knowledge modeling process.

4. A Real-World Domain: The Power Plant

A complex processing or manufacturing plant is usually thought of as consisting of different subsystems. Such a process is convenient, since it is possible to define subsystems according to the processes performed and the variables involved.

Power plant processes are described in detail in design documentation by means of process and instrumentation diagrams (P&IDs). Five separate subsystems are usually distinguished in a thermal power plant.

(1) Boiler
(2) Turbogenerator
(3) Condenser
(4) Water regeneration
(5) Electrical auxiliaries

Processes performed within each subsystem involve several process variables. The values of variables dependent on the processes performed within the current subsystem vary according to the actual behavior of the subsystem.

As mentioned earlier, only some process variables are measured; actual values of measured variables provide an indication of the actual behavior of the current subsystem. In fact each measured process variable has a proper reference value, given by the process design, which should consistently remain the same throughout the process performance. Therefore variations in such a value are significant, since they correspond to variations in process performance.

We call such variables (measured process variables) *index variables*. Their actual values are continuously tested and provide reference for the diagnostic process.

An example of an index variable within the condenser subsystem is given by the condenser pressure, P_c. Such pressure (vacuum) inside the condenser body has a value defined by design, which is to remain the same for proper performance of the condensing process. In other words, any variation in its value indicates that the process is not performing correctly and we need to diagnose the cause (or causes) of the malfunction.

The diagnostic system PERF-EXS is based on performance evaluation of the plant processes. It works on line with the process computer and the plant data acquisition system. The performance of each subsystem is described by the subsystem *index of performance*, which is *nominal* when the actual value of its index variables corresponds to their reference value.

Figure 1 shows PERF-EXS functional architecture.

- The system receives information from the field concerning measured process variables.
- The index of performance of each subsystem is continuously tested.
- When the index of performance of a subsystem deviates from its nominal value, the diagnostic process starts.
- The diagnostic process is performed by a general inference engine that operates on a KB (knowledge base) specific to each subsystem.
- The diagnostic process identifies the cause of malfunction and suggests a way of removing it.

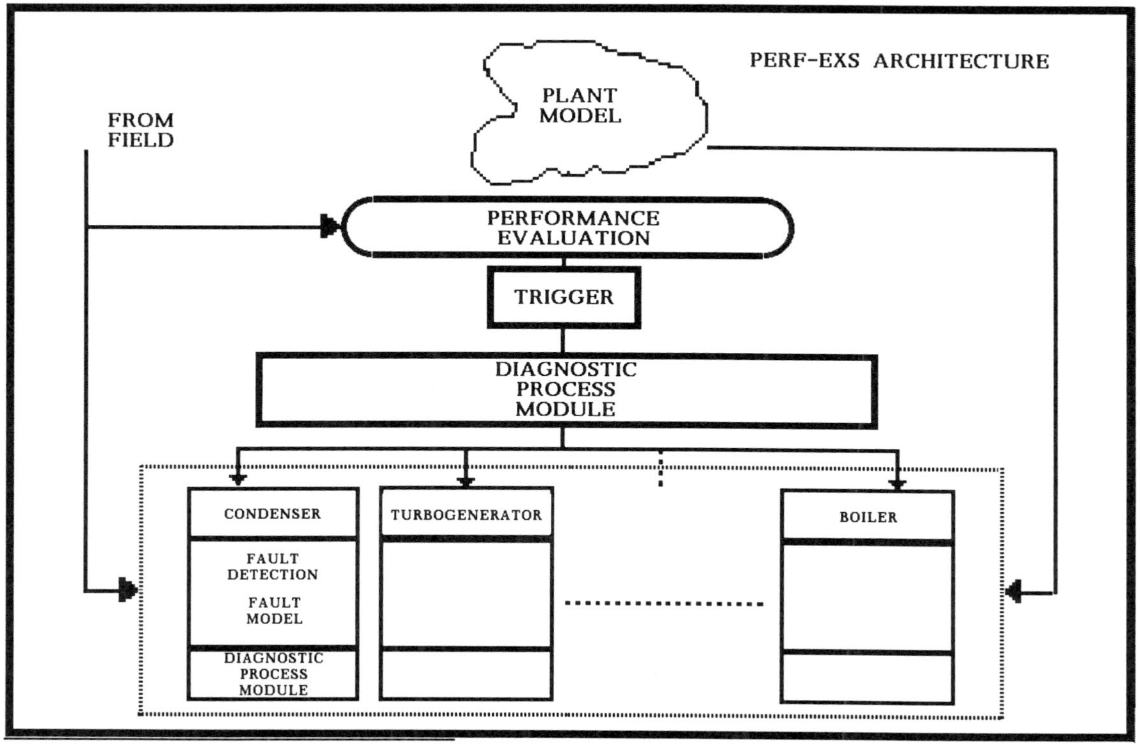

Figure 1. PERF-EXS functional architecture.

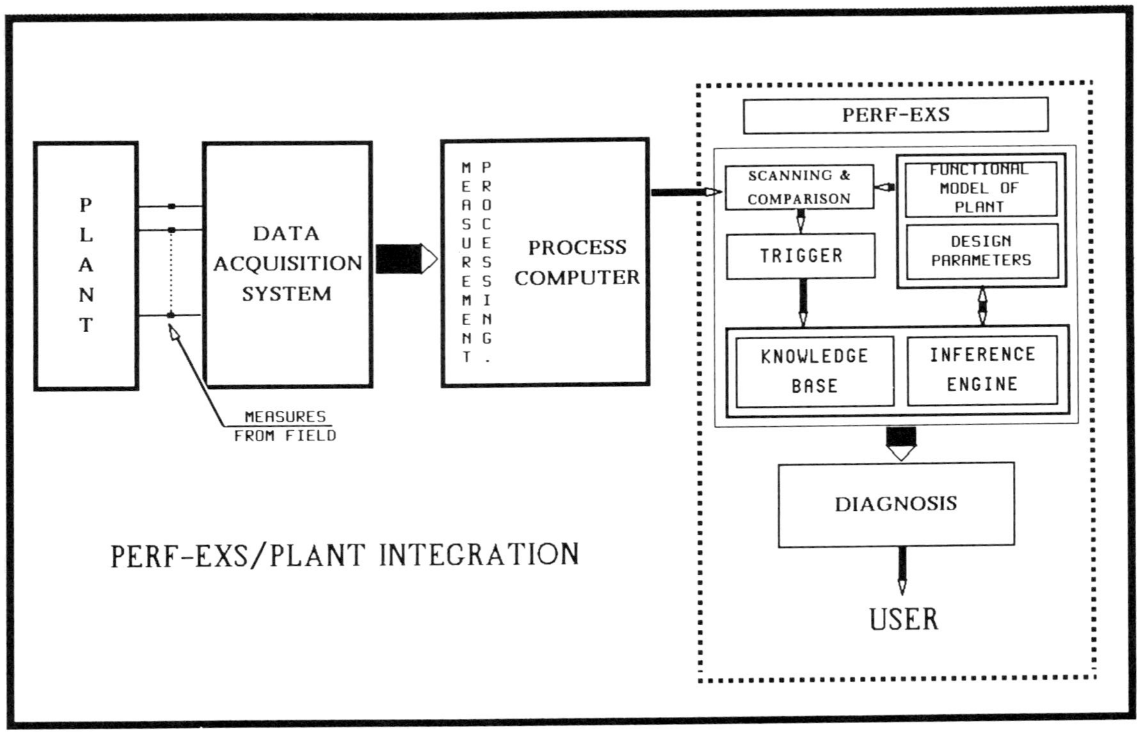

Figure 2. Integration with data acquisition system.

All subsystem KBs contain knowledge derived from the plant model according to the methodology described later in the chapter. We will describe how the final plant model was obtained so as to contain deep and shallow knowledge. As a consequence, the KBs of the resulting diagnostic ES contain both kinds of knowledge.

Figure 2 shows PERF-EXS integrated with the plant process computer and data acquisition system.

5. The Plant Model

To describe the plant model we will refer to a specific subsystem of the thermal power plant—the condenser. Here is the way the condenser works: Steam coming from the turbine discharge section is vacuum-drawn into the condenser body where cooling water flows through several tubes in cross-flow with the steam. Here the heat-exchange process takes place and the condensate is collected at the bottom hotwell. From the hotwell the condensate is pumped back to the boiler. The condenser can be thought of as consisting of several subsystems. For convenience, we will distinguish four of them.

(1) condenser body
(2) cooling water subsystem
(3) condensate extraction subsystem
(4) incondensable gas extraction subsystem

Each subsystem is described by different components: mechanics, hydraulics, and instrumentation.

Some examples of typical process data are

- pressure drop across the condenser
- heat transfer coefficients

Some examples of design data are

- characteristics of heat exchange surface
- number of tubes
- inner diameter of tubes
- cooling water nominal flow rate

As we discussed earlier, the model has to accomplish two tasks: (1) include all the information available from process and design data, as well as available expertise (deep and shallow knowledge); and (2) provide a general methodology for developing applications relevant to different industrial domains (and so provide a general frame on which to base the generic domain model). We need deep knowledge to describe the power plant processes and the behavior of process variables consequent to design parameters adopted. We also need expertise (shallow knowledge) to overcome problems related to unavailability of variable measurements or in situations where deep knowledge is too complex to be represented properly.

5.1. The Quantitative Deep Knowledge Model

The first step in defining the model was to build the equation-based model of a generic plant containing the equation-based model of each subsystem. In the case of the condenser, this model contains a set of equations involving process variables (typically, pressure, temperature, mass flow, and enthalpy), and design parameters (typically, number and diameter of condenser tubes, their material, and circuit geometry).

Process variables involved with the set of equations can be either exogenous or endogenous (Iwasaki and Simon, 1986), i.e., independent or dependent of the subsystem processes. Variables are classified as exogenous or endogenous according to the following definition:

> When the value of a process variable depends on processes performed within other subsystems and not the one we are currently considering, the variable is said to be exogenous to the current subsystem.
>
> When the value of a process variable does depend on processes performed within the current subsystem, the variable is said to be endogenous with respect to the current subsystem.

Exogenous variables define the boundaries of each subsystem.

Independent (exogenous) variables for the condenser are

- heat flow rate: imposed by the level of plant power output and by turbine characteristics (directly correlated to the steam flow rate)
- cooling water temperature: imposed by climate conditions (open circuit)
- heat flow rate from turbine seals
- motive steam flow rate (to air ejectors) and its characteristics—pressure and temperature.

Boundaries of the condenser subsystem are defined by the above exogenous variables.

Examples of dependent (endogenous) variables are

- condenser pressure (vacuum)
- pressure drop across the condenser
- cooling water flow rate

Design parameters have values imposed

- by design choices relative to other subsystems: such parameters *behave as exogenous (independent) variables*
- by system faults, which alter their design value: such parameters *behave as endogenous (dependent) variables.*

The latter subset of design parameters defines the so-called fault parameters. Faults refer to the deviation of one or more of these parameters from their nominal value, which in turn causes the process to deviate from the expected performance.

5.1.1. Characteristics of the Quantitative Model

This physical-mathematical model contains the information coming from process and design data (P&IDs). It could be exploited for diagnostic purposes, since performance diagnosis is done under steady-state conditions. In fact, independent variables are stationary (by their definition). Design parameters may modify their design value, but, once modified, their value remains stationary. When the cooling water comes from a natural body of water, variation in its temperature is slow enough to cause only negligible transients. Variations in the performance of plant subsystems are small with respect to the range of validity of model relationships.

The physical-mathematical model consists of algebraic equations describing mass and energy transfer phenomena that take place inside the condenser. Design parameters involved with these equations appear with their nominal (design) values.

The model so defined is directly derivable from design documentation.

5.2. *Model Structured by Functional Blocks*

Each plant or plant subsystem component influences the overall plant process according to the functions it performs. Therefore each component can be described by one or more *functional blocks* describing each function and exchanging process variables with each other.

The different functions performed by plant (or subsystem) components can be described, for instance, in terms of energy conversion processes, energy dissipation processes, energy transfer processes, and mass related processes.

Such processes are always simultaneously present; however, functional blocks can be defined according to the prevalence of one of the above classes of phenomena (for example, blocks representing turbines, piping, valves, condenser body, and pumps).

Functional blocks that are connected by process variables will therefore contain the mathematical relations describing different processes and involving all design parameters.

Condensing heat exchange block, shell side

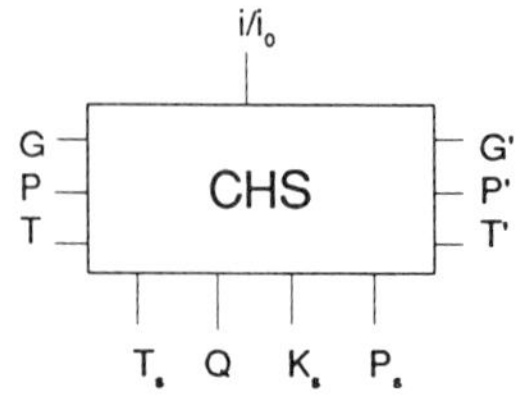

Relations

$G' = G$
$P' = P = P_s$
$T' = T = T_s$
$T_s = f_3\,(P_s)$
$K_s = K_{snom}\,f_4\,(i/i_0)$

Variables:
G = mass flow rate
P = pressure
T = temperature
Q = transferred heat flow rate
P_s, T_s = pressure and temperature of steam
K_s = condensation heat transfer coefficient
i/i_0 = incondensable gases vs. nominal concentration

Design and fault (*) parameters:
f_3 = state equation of the condensing fluid (saturation)
f_4 = function expressing the dependence of the heat exchange process on the concentration of incondensable gases, with respect to the nominal case
K_{snom} = heat transfer coef., nominal

Heat exchange surface

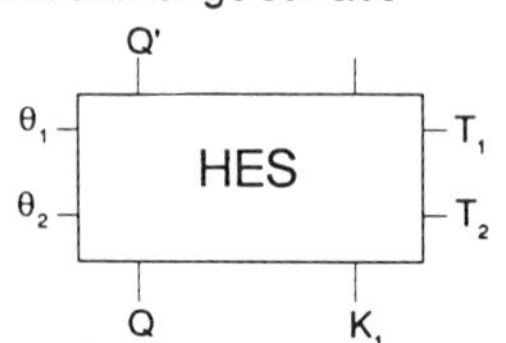

Relations

$$S = \pi \times D \times l \times n$$

$$\Delta T_l = \frac{(\theta_1 - T_1) - (\theta_2 - T_2)}{\ln \frac{(\theta_1 - T_1)}{(\theta_2 - T_2)}}$$

$$\frac{1}{K} = \frac{1}{K1} + \frac{1}{K2} + \frac{s}{\lambda}$$

$$Q = Q' = K\,S\,\Delta T_l$$

Variables:
T_1, T_2 = terminal temp. first fluid
θ_1, θ_2 = terminal temp. second fluid
Q = transferred heat flow rate
K_1, K_2 = Average heat transfer coef's on the two sides of the surface

Design and fault (*) parameters:
s(*) = thickness of tube
n(*) = number of tubes
D = average tube diameter
l = length of tubes
λ = conductivity of tube material

Monophase heat exchange block

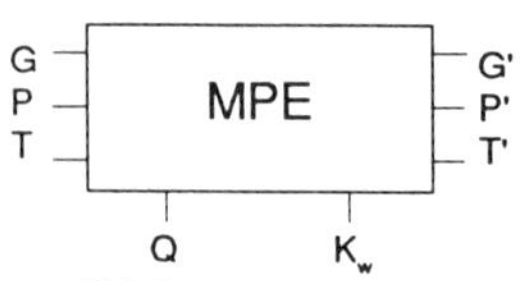

Relations

$G' = G$
$P' = P + \rho\,gh - rG$

$$T' = T + \frac{Q}{c\,G}$$

$$v = \frac{4\,G}{\pi\,D_i^2\,n}$$

$K_w = K_0 v^x$

Variables:
G = mass flow rate
P = pressure
T = temperature
Q = transferred heat flow rate
K_w = convective heat transfer coef.

Design and fault (*) parameters
D_i (*) = internal diameter of tubes
R (*) = global pressure drop factor
h = static head
c = fluid specific heat
ρ = fluid mass density
K_0, x = heat exch. coef.par.
n = number of tubes

Figure 3. Examples of functional blocks.

The structures so defined are independent of the particular system under study, so the blocks can be prototyped once and for all. The model of a given plant or plant subsystem is defined by connecting prototype functional blocks appropriately.

Examples of functional blocks are given in Figure 3, showing process variables, design and fault parameters, and internal mathematical structure.

The complete condenser model by functional blocks is shown in Figure 4. If we had measurements available for all process variables (dependent and independent), the model would be sufficient for any ES building requirements. And the model would also suffice for simulating the plant process. Real-world plants, though, do not have measuring points available for all process variables; some, such as condensing heat flow rate or cooling water flow rate, would require expensive pieces of instrumentation, and some, such as heat transfer coefficients, are simply nonmeasurable. For this reason we needed to produce a different model that could further provide means of detecting faults and performing the diagnostic task on the basis of usually available process data.

The new model contains experts' expertise (shallow knowledge). The

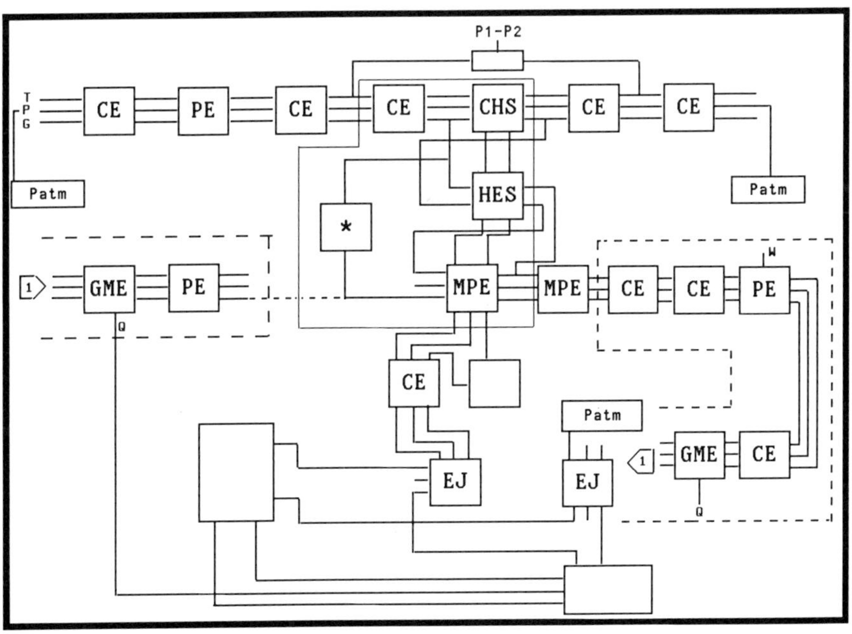

Figure 4. Model by functional blocks of the condenser subsystem.

diagnostic module of the expert system works on the basis of the knowledge contained in the model. The new model's expertise substitutes for missing measurements; thus, the diagnostic module is now able to draw diagnostic conclusions. We called this new model *the interpretation model.* It was obtained through the definition of a semiqualitative model.

5.3. The Semiqualitative Model

The semiqualitative model is derived from the model by functional blocks according to the following criteria:

- to supply the diagnostic system with useful information about the plant subsystem under consideration, including instrument performance
- to be derivable from the plant model independently from the subsystem under consideration. This is possible because of the way the model by functional blocks was defined
- to work with a minimum number of measurements from the plant. Measurements must be typically available through the plant process control system
- therefore to allow and facilitate the use of shallow knowledge necessary for filling in deep knowledge voids

The semiqualitative model is defined on the basis of the above, in the following way:

- Functional blocks involving nonmeasured process variables are grouped together.
- The physical process described by the set of their interacting mathematical equations is expressed by means of relations that we call *functional relations*.
- Functional relations are built from functional blocks by solving through approximate relations (shallow knowledge) critical points where unavailability of data does not allow quantitative solution but where purely qualitative treatment of phenomena does not suffice to resolve ambiguities.

Functional relations so obtained may contain dependent or independent measured variables; design parameters, including fault parameters; and nonmeasured variables.

An example of a functional relation (FR) is shown in Figure 5. The FR links the measured variable P_s (steam condensing pressure) to measured variables:

T_1—cooling water inlet temperature

G_c—condensate flowrate

nonmeasured variables:

G_w—cooling water flowrate

K_w—heat exchange coefficient, water side

K_s—heat exchange coefficient, steam side

design parameters, in this case geometrical ones, including fault parameters.

The known relationship exploited to group together several blocks is given by Mollier:

$$P_s = f(T_s),$$

which corresponds to the overall solution for the blocks grouped by the FR.

Mollier's relationship correlates thermodynamic properties of the water, such as temperature, pressure, and enthalpy. Such relationships have been calculated at discrete intervals for different values of steam pressure and temperature. The results are available under the form of steam tables. Therefore solutions of the above analytical function are known for any given value of P_s and T_s. In other words, given P_s, we know from the steam tables the value of T_s that satisfies the relationship and vice versa. It follows that, given the value of P_s (measured variable), we can obtain the correct value of the function $f(T_s)$ from the steam tables, even if we do not know the values of all the variables involved. Therefore we can exploit this further piece of knowledge to obtain information about the behavior of such unknown variables, if not their values. To do that we assign to design parameters and to nonmeasured (unknown) variables involved with the FR their Nominal (design), value. In this way we obtain what we call a *correlation* among measured variables involved with the FRs. We then repeat the same process with all the FRs that can be found to group together any given number of functional blocks.

The outcome of the above is a set of relationships among measured variables (correlations) that completely describe the process. These correlations use heuristic (shallow) knowledge to integrate deep knowledge where data are insufficient to describe completely the plant processes.

To clarify the discussion we will follow with examples of two correlations. The first refers to the FR consisting of Mollier's function and the related correlation. The second one describes a correlation that defines a fault model, that is, a correlation involving one or more fault parameters (see the definition in Section 5.1).

5.3.1. Correlation between the Process Variables: Condensing Pressure-Cooling Water Inlet Temperature

In Figure 5 the two variables T_s and P_s are linked together through a series of physical processes. These processes are described by mathematical equations, and each process corresponds to a functional block in the plant model. The blocks involved in this relationship are: 1) the condensing heat exchange block, shell side (CHS); 2) the heat exchange surface block (HES); and 3) the monophase heat exchange block (MPE). The group of these three blocks is shown by the solid box in Figure 4. Each block contains mathematical relationships among the process variables, as discussed earlier. The blocks describe energy transfer, mass transfer, and heat transfer phenomena.

We will consider the thermodynamic point of view and get the following relationship from block CHS (Figure 3):

$$T_s = f(P_s), \tag{1}$$

where

P_s = steam condensing pressure

T_s = steam condensing temperature

f = Mollier function.

The process variable P_s is measured and the values of the function f are known from steam tables.

From the same block we also know that a certain amount of heat Q flows from the steam through the shell side of the condenser with heat transfer coefficient K_s.

From block MPE (Figure 3) we know that the same amount of heat Q flows into the cooling water, arising its temperature according to the equation

$$T_o = T_i + \frac{Q}{c \times G_w} \tag{2}$$

where

T_i = cooling water inlet temperature

T_o = cooling water outlet temperature

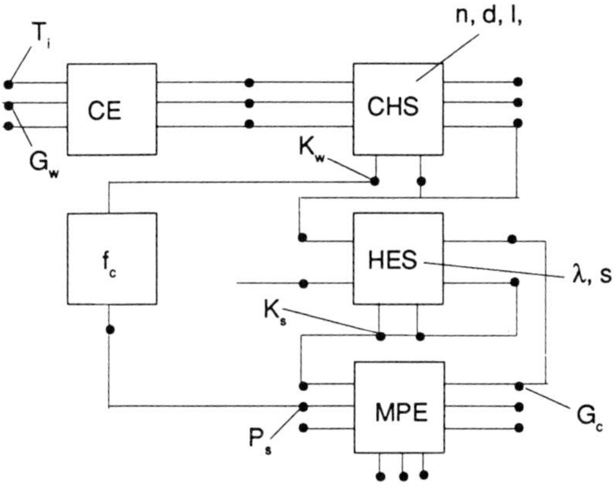

MEASURED VARIABLES: T_i, G_c, P_s

NON-MEASURED DEPENDENT VARIABLES: G_w, K_w, K_s

DESIGN PARAMETERS: n, d, l, λ, s

$$P_s = f^{-1}\left(T_i + \frac{\dfrac{f_c^{-1}(G_c)}{c\,G_w}}{1 - e^{-KS/cG_w}}\right)$$

$$S = \pi \times d \times l \times n$$

$$K = \frac{1}{\dfrac{1}{k_s} + \dfrac{s}{\lambda} + \dfrac{1}{k_w}}$$

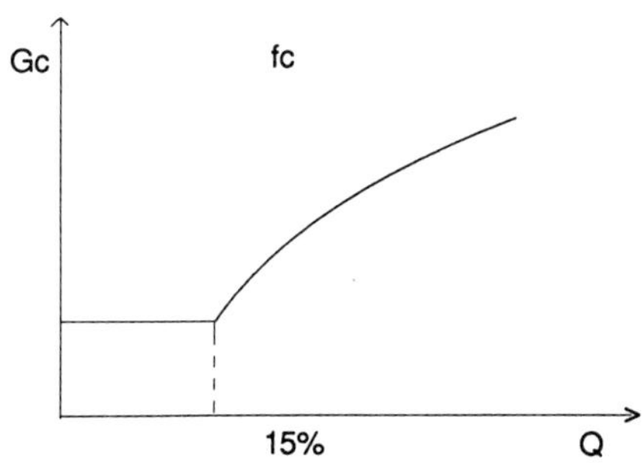

T_i = Cooling water inlet temp.

G_c = Condensate flow rate

P_s = Condensing pressure

G_w = Water flow rate

K_w, K_s = Heat transfer coefficients

n = Number of tubes

d = Diameter of tubes

l = Length of tubes

s = Tube wall thickness

λ = Conductivity of tube walls

f = Mollier function

f_c = analytical function

f

Ps	Ts
0.03	23.7
0.032	24.5
0.034	25.3
....	....
0.05	32.5

Figure 5. Example of functional relation.

c = water specific heat (ave.)

G_w = cooling water mass flow rate

T_i, and T_o are both measured variables.

From block HES (Figure 3) we derive that Q must equal the heat flow going into the water through the heat-exchange surface (pipe wall material with heat transfer coefficient s/λ) according to the equation

$$Q = K \times \pi \times d \times 1 \times n \times \Delta T_1, \tag{3}$$

(see block HES in Figure 3 for variable definitions, with ΔT_1 = log-mean temperature difference) or

$$Q = K \times \pi \times d \times l \times n \times \frac{(T_i - T_s) - (T_o - T_s)}{\ln \frac{(T_i - T_s)}{(T_o - T_s)}}, \tag{4}$$

where

$$K = \frac{1}{\frac{1}{K_s} + \frac{1}{K_w} + \frac{s}{\lambda}} \tag{5}$$

$$S = \pi \times d \times l \times n. \tag{6}$$

By recalling the distinction of variables as being exogenous or endogenous to a given subsystem, we can further say that Q is an independent variable with respect to the condenser subsystem (i.e., it is exogenous to the condenser). In fact its value depends upon the plant power level as well as the performance of plant subsystems other than the condenser, typically the turbine-generator subsystem. Q is therefore an endogenous variables with respect to the turbogenerator subsystem, while it is exogenous to the condenser. Furthermore, Q is a nonmeasured variable.

To avoid misdiagnosis of malfunctions, it is therefore necessary to express the variable Q with the inverse of the relationship

$$G_c = f_c\,(Q). \tag{7}$$

It must be noted that equation 7 holds for plant power output at least 15 percent of nominal full power. This relation is contained in block (*) of the plant model (Figure 4), where

G_c = condensate mass flow rate for CHS, measured variable.

From Equations 1, 2, 4, 6, and 7, we have

$$P_s = f^{-1}\left(T_i + \frac{f_c^{-1}(G_c)/(c \times G_w)}{1 - e^{-(K\times S)/(c\times G_w)}}\right) \tag{8}$$

When Equation 8 is satisfied (i.e., when the values of the variables T and P correspond to the values that can be found in the steam tables), the overall process performance is nominal. Non satisfaction of equation 8 identifies a malfunction in the condenser process.

Therefore, Equation 8, which is functional relation 1 (FR1) for the condenser model, can be used for fault detection; its satisfaction corresponds to nominal plant performance.

We now have to define the corresponding correlation, which can be obtained nothing that

- f is a known function (Mollier diagram)
- f_c is a mathematical function obtained from design data
- P_s and G_c are dependent measured variables
- T_i is an independent measured variable
- S is given by the product of design parameters
- K and G_w are dependent nonmeasured variables

Therefore we give S, K, and G_w their nominal values $S(nom)$, $K(nom)$, and $G_w(nom)$ and obtain a correlation among measured variables P_s, T_i, and G_w.

The correlation so obtained is correlation 1 (CO1) in the model of the condenser subsystem. A correlation is satisfied when the variables directly involved with it have, when measured, their expected value. A correlation is nonsatisfied when the variables involved do not actually have, when measured, their expected value. A nonsatisfied correlation is indicative of a fault, which could be in: 1) the subsystem process involving the faulty variables; 2) the measurement chain; or 3) both the process and the measurement chain. Therefore the correlations defines the *fault models* for the subsystem.

Furthermore, the correlation involving the subsystem index variables indicates that a fault is present in the subsystem's overall process, thus identifying an overall malfunction of the subsystem in its whole. Therefore the fault model established by such a correlation is relevant to the entire subsystem and can be used by the overall performance evaluation module (see Figure 1) to detect which subsystem (or subsystems) of the entire plant is (are) malfunctioning. The correlation involving index variable therefore establishes the *fault detection* module. All the remaining correlations establish the subsystem *fault models*. Diagnosing possible faults is then performed by the diagnostic expert system by means of the fault models.

CO1 above involves the index variables of the condenser subsystem and therefore defines the fault detection module of the subsystem.

5.3.2. Fault Models

We will now give an example of a correlation establishing a generic fault model. We chose an example to illustrate the need for heuristic (shallow) knowledge to integrate deep knowledge when necessary.

Consider the condensate flowing out of the condenser body (shell side) to the boundary of the condenser subsystem. This process involves several blocks shown in the plant model and presented in the dotted-line box in Figure 4. They are several circuit elements (CE), a pump element (condensate extraction pump) (PE), and a generic mono-phase heat exchanger (GME).

We will derive from the plant model and the relevant blocks the correlation between condensate ingoing and outgoing temperature.

The condensate temperature increases due to energy dissipation within the extraction pump, heat transfer (from turbine seals), and heat transfer in air ejectors (from motive steam).

The blocks involved in this thermodynamic relation are PE and GME; the mathematical equation is

$$T_{co} - T_{ci} = \frac{H(1-\eta_\rho)}{\rho \times c \times \eta_\rho} + \frac{Q_t}{c \times G_c} + \frac{G_{v1} \times \lambda_1 + G_{v2} \times \lambda_2}{c \times G_c}. \tag{9}$$

where

T_{ci}, T_{co} = ingoing and outgoing condensate temperature

H, η_p = pump head and mechanical efficiency

ρ, c = water density and specific heat

Q_t = heat flow rate from turbine seals

G_c = condensate mass flow rate

G_{v1}, G_{v2} = heat flow rate from ejectors motive steam

λ_1, λ_2 = heat of vaporization for G_{v1} and G_{v2}.

The coefficients λ depend on the motive steam temperature (i.e., its pressure) according to the steam tables.

Equation 9 involves independent nonmeasured variables. They are exogenous to the condenser; therefore, their values cannot be computed by means of the information available within the condenser subsystem since their variations pertain to the turbine subsystem. However, the diagnostic process needs to be able to correlate and use information on variables, although pertaining to other subsystems. Therefore we apply to equation 9 the same concepts used above to obtain a correlation among measured variables.

We substitute for all variables and parameters on the right-hand side their nominal values. The result is a delta temperature corresponding to correct performance.

$$T_{co} - T_{ci} = \Delta(\text{Nom}). \tag{10}$$

The new relationship represents a new correlation. If the actual values of T_{co} and T_{ci} do not satisfy equation 10, any of the following is possible:

(1) condensing system malfunction
(2) instrument failure: T_{co}, or T_{ci} or both
(3) cause of malfunction exogenous to the subsystem (Q_t)
(4) any combination of the above

In such a case the diagnostic process cannot distinguish which cause is actually present, and the correlation is of no use.

However, if the correlation is satisfied we can add a piece of shallow knowledge; diagnostic expertise tells us that it is unlikely for both instruments to be faulty and compensating each other's fault. We can therefore deduce that both measurements T_{co} and T_{ci} are correct with a high certainty factor and use this deduction during the diagnostic process. The correlation is so used as a no-fault detector for instrument malfunctions. There is no risk for the diagnostic system to overlook possible simultaneous faults, since the remaining fault models would still be able to detect them during completion of the entire diagnostic process.

Moreover, the correlation obtained from equation 10 is important for further discriminating fault models within the condenser subsystem. From the plant model we can see that T_s (temperature of condenser shell side) is linked to T_{ci} by the relation

$$T_{ci} = T_s - \Delta, \tag{11}$$

where T_{ci} is measured, T_s is known from equation 1, and Δ is known from experience to be a few tenths of a degree Centigrade. Combining satisfied or nonsatisfied values of equation 10 (just its result, without considering the expertise as above) and equation 11 (shallow knowledge) allows us to discriminate between condenser malfunctions and instrument failures.

5.4. The Resulting Interpretation Model

Finally, we obtain the ***interpretation model***, which forms the basis for the diagnostic system. In fact, this model provides an adequate interpretation of the knowledge (deep and shallow) relevant to the plant and the plant processes. By "adequate interpretation" we mean that the real-world knowledge relevant to the plant has been processed from the standpoint of the diagnostic paradigm and the knowledge was so made available for direct processing by a diagnostic inference engine. The interpretation model consists of two modules:

Fault Detection

Fault Models

When the knowledge contained in the above two modules (see again Figure 1) is processed by the diagnostic inference engine, the functions proper to a diagnostic system are performed, i.e.,

The system detects the presence of a fault.

The system diagnoses the fault cause(s) through the processing of the related fault models.

The two modules contain the correlations which are required by the relevant functions performed by the process. As we saw in the previous sections, values of correlations cannot be calculated exactly; therefore the diagnostic process needs to exploit approximate reasoning in order to evaluate the status of each correlation.

The interpretation model (semiqualitative) forms the basis for a general diagnostic process: correlations containing fault parameters identify the process fault models; they are specifically instantiated in configuring the model for a specific process, when a number of design parameters are marked as endogenous.

At this point approximate reasoning is applied which is somewhat similar to the O[M] formalism in Mavrovouniotis and Stephanopoulos (1987):

Nominal variable values that satisfy each correlation are known (from design data).

All values are nominal within a fuzzy interval of validity.

From available measured variables it can be inferred when a correlation is satisfied, or the difference (sign and approximate quantity) between its actual value and the nominal one.

By reasoning on these values the system can infer approximate information on the remaining unknown entities and relevant fuzzy intervals involved with the FR.

The value of a Correlation can be *satisfied* under nominal circumstances and *nonsatisfied* otherwise, both within a fuzzy interval.

As a final result the interpretation model contains a set of relations among measured variables (correlations) that completely describe the process. Such relations use heuristic (shallow) knowledge to integrate deep knowledge where data are insufficient to describe the plant processes completely.

Although the value of each correlation cannot be calculated exactly, the reasoning mechanisms can perform approximate evaluation. Thus correlations may assume values of absolutely smaller than, moderately smaller than, equal to, moderately greater than, or absolutely greater than, with respect to their nominal value of satisfaction.

The diagnostic inference mechanisms exploit this information to infer the approximate behavior of nonmeasured variables and fault parameters involved with the correlation being evaluated. This mechanism is general as the functional blocks were. The fuzzy intervals of satisfaction are specifically configured for the correlations pertaining to the specific process. The same is true for the fuzzy intervals within which each variable or parameter is nominal. In other words, the interpretation model may easily be configured for the specific plant or process to be described.

6. The Diagnostic Expert System

We described above the process of structuring a semiqualitative interpretation model, obtained by interpreting the plant model from the standpoint of the diagnostic paradigm. The general structure of the model is suitable for modeling generic plant or plant process knowledge. The knowledge so obtained can be directly exploited in diagnostic expert systems. The diagnostic system PERF-EXS was, in fact, completed by filling in the semiqualitative model structure with the knowledge pertaining to the remaining subsystems of the thermal power plant.

A methodology was then devised to represent the knowledge contained in the interpretation model. Such methodology was described in a previous paper (D'Ambrosio *et al.*, 1988a). Since this is not the subject of this chapter, we refer the reader to this paper for details on the knowledge representation methodology that allows one to exploit directly the knowledge contained in the model. The diagnostic process itself was described in the same paper and in the following (D'Ambrosio *et al.*, 1988b). The diagnostic process, also generic, defines the inference engine of the expert system. Thanks to the general definition of the domain model structure, diagnostic expert systems can be developed for different domains (processes) just by filling in the model with the knowledge pertaining to the specific domain and even automating the knowledge acquisition process (D'Ambrosio, 1988).

In the following section we will instead present an overview of the implemented diagnostic expert system PERF-EXS and its operation.

6.1. PERF-EXS Operations

PERF-EXS runs either on-line (on the plant floor) or in simulation mode (off-line). Here it is how it works: The system has four fundamental work-areas to which one or more interactive screens correspond. The work areas are

(1) Configuration-simulation
(2) Plant schematics
(3) Subsystem diagnostics
(4) Working area of inference engine

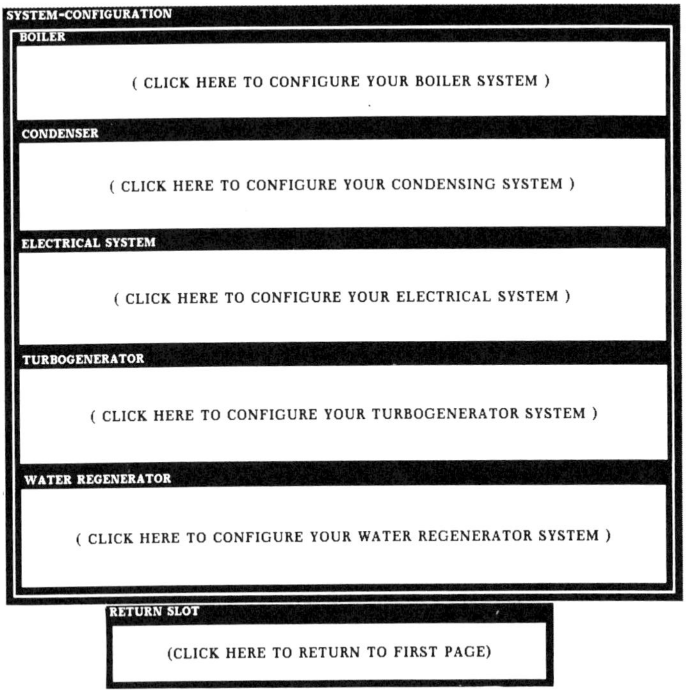

Figure 6. Access screen.

(1) Starting a system run, the user has the option to reconfigure the plant or single plant subsystem, inputting the values for design parameters. Figure 6 shows the access screen for plant subsystems, and Figure 7 shows the configuration windows for the condenser subsystem (control panel). Otherwise the user can skip this option and go directly to the simulation, which is accessed through the plant schematics screen.

(2) The plant schematics screen is shown in Figure 8, where we can distinguish two different parts. The upper part shows the overall plant schematics, while at the bottom we find a set of interactive windows from which to select the plant subsystem to be analyzed:

boiler
turbogenerator
electrical auxiliaries
water regeneration.
condenser

By mousing on the condenser window we access the screen of Figure 9 for the condenser subsystem diagnostics.

(3) The condenser diagnostic screen is again divided into two different parts:
   (a) On the left we have the condenser schematics with indication of dependent (white boxes) and independent (gray boxes) measured variables.
   (b) On the right we have
      (*i*) a series of windows (upper part) to perform manually several commands:
         go to condenser control panel (Figure 7)
         start diagnosis
         go to main menu
         return to plant schematics (Figure 8)
      (*ii*) a series of windows describing the variable currently being checked:
         name of the variable
         actual value of the variable
         history of the variable values
      (*iii*) a bar graph representing the current variable value with respect to its nominal range.

The condenser schematics screen becomes interactive for process simulation, when the system runs off-line. Thus in simulation mode the values of measured process variables can be inputted from this

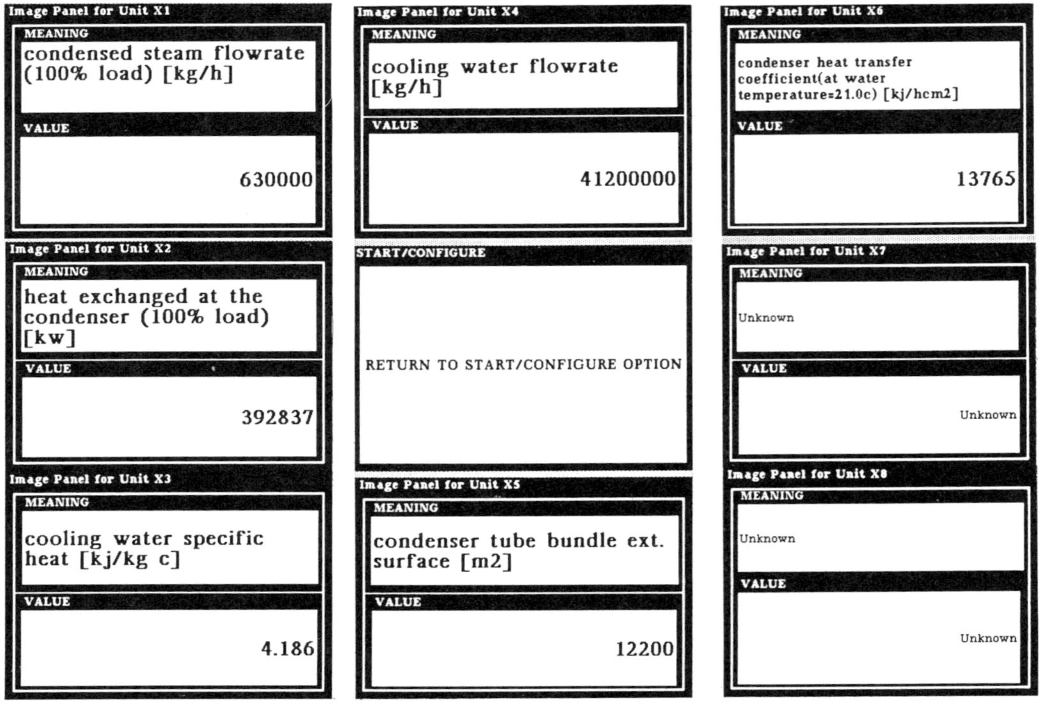

Figure 7. Condenser configuration screen.

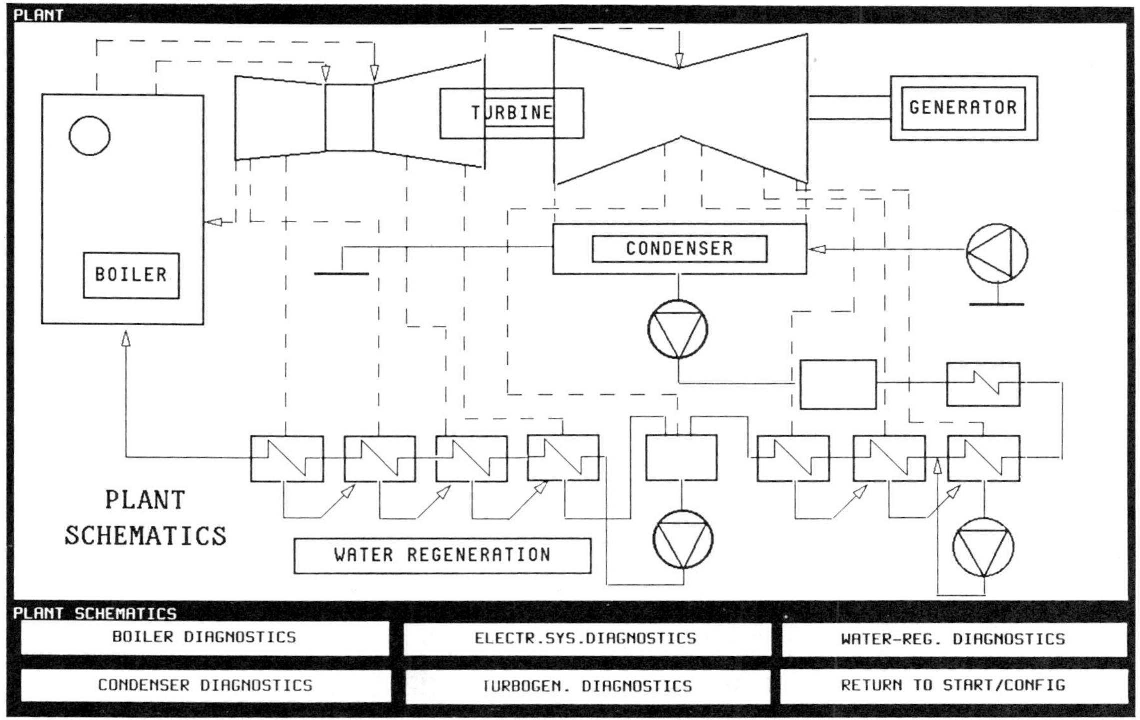

Figure 8. Plant schematics.

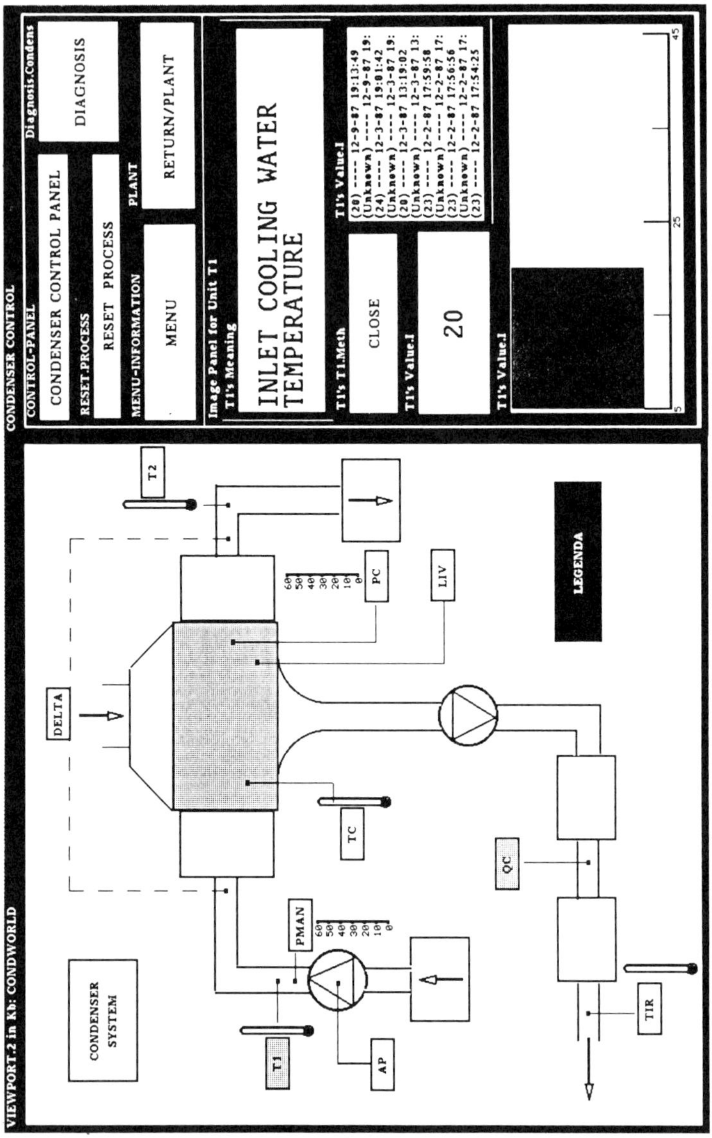

Figure 9. Condenser subsystem diagnostics.

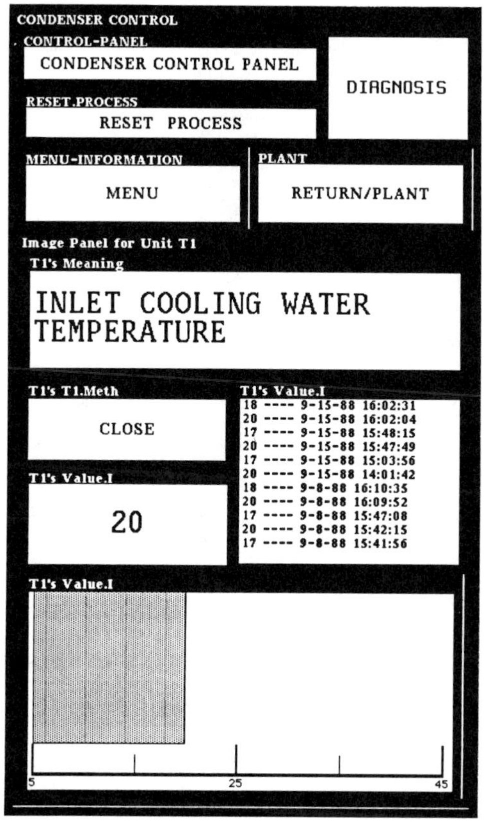

Figure 10. Example of variable input.

screen by mousing on the corresponding boxes. This operation opens a new interactive window (Figure 10 shows the input window for cooling water outlet temperature) through which the values can be inputted. When the values of all measurable variables have been inputted, the diagnostic procees is manually started by mousing on the diagnosis window on the upper right. In other words, during a simulation run the interactive windows are activated, the variable values are manually inputted and also the diagnosis is manually started (system off-line).

Vice versa, during an on-line run, there is no manual interaction. The system receives the values of measured variables directly from the field and the simulation of correct plant performance is automatically executed. In either case the simulation compares the actual

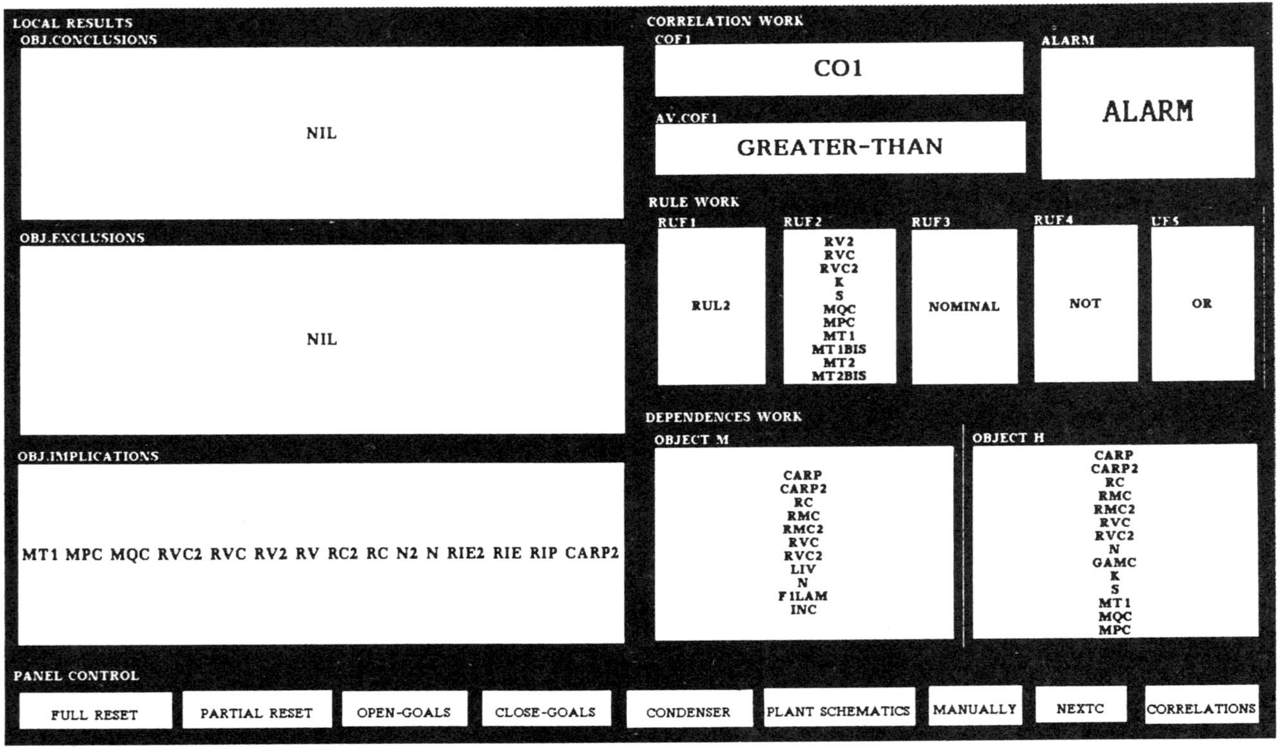

Figure 11. Inference engine working area.

value of the variables to their expected ones. Results of this comparison allow the system to identify the status (satisfaction or nonsatisfaction) of the correlations, thus identifying the fault models which are to be activated. The diagnosis is automatically started by the fault detection module.

The condenser diagnostic screen described above corresponds exactly to the semiqualitative plant model and the measurement points (indicated by variable boxes) correspond to the testing points of correlations, which determine the value of the corresponding fault model. The diagnostic process is visualized on the inference engine screen.

(4) The working area of inference engine screen is shown in Figure 11. This screen reports correlation by correlation the results of the comparison of variable values, i.e. the satisfied or nonsatisfied value of the correlation. To this screen the user can return 1) after the run is completed, if the system is on-line; or 2) at any moment, if running in manual and the system is off-line. This feature allows the user to check the diagnostic results and the system reasoning process.

Figure 12 shows the final screen with diagnosis, comments and actions suggested in a given case.

7. Conclusions and Future Work

The diagnostic expert system PERF-EXS was extensively tested and its results (diagnoses) successfully compared to actual diagnostic operations performed by experts in the field. PERF-EXS is now used in the field and provides excellent results. It has proved capable of detecting deviations in plant process performance as low as 0.1% of the nominal value, suggesting actions to correct them. Detecting a deviation of 0.5% in the overall plant performance provides for a saving of approximately 10 Cal/Kwh, with consequent saving in fuel consumption. If this figure is multiplied for the power output (Kwhs) and the lifetime of the plant (about 30 years), the reader can readily appreciate the economical outcome of the fielded system.

The general model structure and knowledge representation, together with the generic diagnostic inference engine, were then further exploited to form the basis for a knowledge acquisition tool (D'Ambrosio, 1988) aimed to the development of diagnostic expert systems for industrial applications.

We plan in the future to investigate the possibility of broadening the knowledge representation and acquisition methodology for building knowledge bases for expert systems running under consultation paradigms other

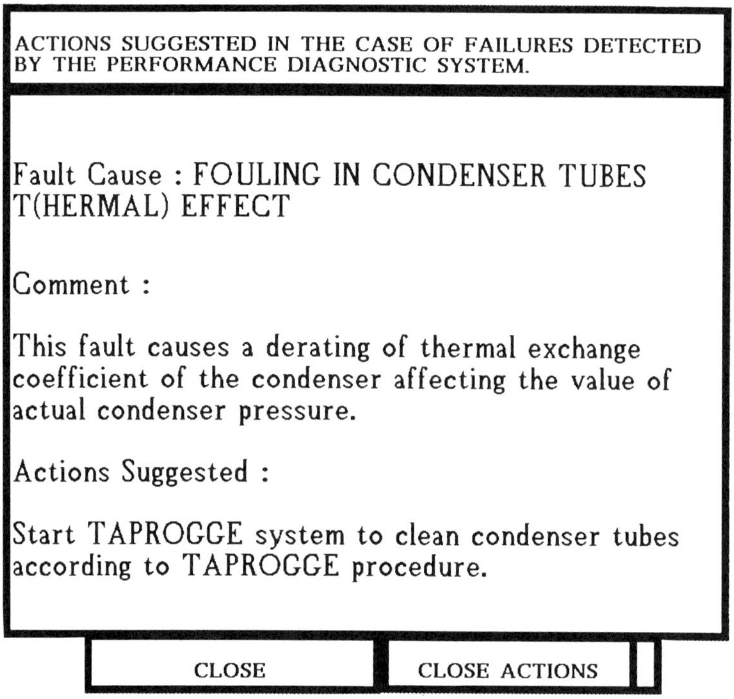

Figure 12. Output screen.

than diagnostics. This operation involves deriving from the same semi-qualitative plant model different interpretation models on which to base the reasoning for such different paradigms. However, this part of the project is still to be investigated.

Acknowledgments

The author gratefully acknowledges the significant contribution to the project of ANSALDO fellows Alfonso Cavanna, Mirco Oriati and Antonio Serventi, from ANSALDO Power Plant Engineering Company.

The development of PERF-EXS was partially carried out within the frame of ESPRIT Project 820 of the EEC.

References

Bobrow, D. G. "Qualitative Reasoning about Physical Systems: An Introduction. *AI Journal*, 24 (1984).

Bobrow, D. G., ed. *Qualitative Reasoning about Physical Systems.* The MIT Press, Cambridge, MA (1985).

Bobrow, D. G., and Winograd, T. An overview of KRL, a Knowledge Representation Language. *Cognitive Science*, 1 (1), pp. 3–46 (1977).

Boose, J. "Personal construct theory and the transfer of human expertise. In *Proceedings of the 4th AAAI* (1984).

Chandrasekaran, B., and Punch III, W. F. Data Validation During Diagnosis, a Step Beyond Traditional Sensor Validation. *Proceedings of AAAI-87*, Seattle, WA (1987).

Dague, P., Raiman, O., and Devès, P. Troubleshooting: when modeling is the trouble. *Proceedings of the Sixth AAAI*, Seattle, WA (1987).

D'Ambrosio, A. Deep and Shallow Knowledge Representation in Industrial Domains: Building Plant Models with PERF-KAT, a Knowledge Acquisition Tool. *Proceedings of the AI-88 Australian Joint Artificial Intelligence Conference* (1988).

D'Ambrosio, A., Serventi, A., and Oriati, M. AI in the Power Plants: PERF-EXS, a PERFormance diagnostics EXpert System. *Proceedings of the 4th IEEE Conference on Artificial Intelligence Applications*, San Diego, CA (1988a).

D'Ambrosio, A., Oriati, M, and Serventi, A. Approximate Reasoning and Semi-qualitative Model: PERF-EXS, an ES for Thermodynamic Process Fault Diagnosis. *AAAI88—Workshop on AI in Process Engineering*, St. Paul, MN (1988b).

deKleer, J., and Brown, J. S. Theories of Causal Ordering. *AI Journal*, 29 (1), (1986).

Eshelman, L., and McDermott, J. MOLE: a knowledge acquisition tool that uses its head. In *Proceedings of the 5th AAAI* (1986).

Iwasaki, Y., and Simon, H. A. Causality in Device Behaviour. *AI Journal*, 29 (1), (1986).

Kahn, G., Nowlan, S., and McDermott, J. MORE: an intelligent knowledge acquisition tool. In *Proceedings of Ninth International Conference on Artificial Intelligence* (1985).

Klinker, G., Boyd, C., Genetet, S., and McDermott, J. A KNACK for Knowledge Acquisition. *Proceedings of AAAI-87*, Seattle, WA (1987).

Kramer, M. A. Integration of Heuristic and Model-Based Inference in Chemical Plant Fault Diagnosis. *IFAC Workshop on Fault Detection and Safety in Chemical Plants*, Kyoto, Japan (1986).

Levesque, H. J., and Brachman, R. J. A Fundamental Tradeoff in Knowledge Representation and Reasoning (revised version). In *Readings in Knowledge Representation* (Brachman and Levesque, eds.), Morgan Kaufmann Publishers, Inc., Los Altos, CA (1985).

Marcus, S., McDermott, J., and Wang, T. Knowledge acquisition for constructive systems. In *Proceedings of Ninth IJCAI* (1985).

Mavrovouniotis, M. L., and Stephanopoulos, G. Reasoning with Order of Magnitude and Approximate Relations. *Proceedings of the Sixth AAAI*, Seattle, WA (1987).

Minsky, M. A Framework for Representing Knowledge. In *Mind Design* J. Haugeland, ed. pp. 95–128, MIT Press, MA (1981).

Morris, P. H., and Nado, R. Representing Actions with an Assumption-Based Truth Maintenance System. *Proceedings of the 5th AAAI* (1986).

Raiman, O. Order of Magnitude Reasoning. *Proceedings of the 5th AAAI* (1986).

Winograd, T. Frame Representations and the Declarative/Procedural Controversy. In *Representation and Understanding*: Studies in Cognitive Science. (Bobrow and Collins, eds.), pp. 185–210, Academic Press, NY (1975).

Woods, W. What's in a Link: Foundations for Semantic Networks. In Representation and Understanding: Studies in Cognitive Science (Bobrow and Collins, eds.) pp. 35–82, Academic Press, NY (1975).

7 XIMKON—An Expert Simulation and Control Program

A. FERIT KONAR
BHAVANI M. THURAISINGHAM AND
PAUL E. FELIX

Honeywell, Inc.
Sensor and System Development Center
Golden Valley, Minnesota

Abstract

Artificial intelligence (AI) and object-oriented techniques are applied to the development of XIMKON, an expert software system for simulation and control engineering of dynamic processes. As an enhancement to a broad FORTRAN base of control design tools, XIMKON incorporates the principles of AI planners and advanced user interface technologies in an effort to automate the procedures involved in modeling, analysis and synthesis of control systems. A methodology for control system design is discussed, followed by a description of XIMKON's current implementation for process control applications.

1. Introduction

Current trends in controls point to more complex systems because of higher-level system integrations and stringent quality control requirements for economical operation. Unified control design methodologies and automated control design tools are being developed to enable the engineer to design more complex control systems in a shorter time frame, with reduced risk. In light of these trends, we present an overview of an evolving control design methodology and a control design tool developed at Honeywell for industrial control system applications.

Our goal is to apply artificial intelligence and object-oriented techniques to provide a more effective tool that aids the control engineer throughout

A short version of this work has appeared in "1989 ACC Proceedings."

ISBN 0-12-480575-2

the design process and performs many of the iterative and difficult tasks involved in the generation of a solution.

In Section 2, we describe our approach to the control system design process. Section 3 provides a general description of XIMKON and its use of AI planner technology. Section 4 presents elements of XIMKON's Expert Process Modeler and advanced user interface implemented to facilitate the modeling process. Section 5 describes the current state of our development of an Expert Controller Designer. Finally, we provide concluding remarks in Section 6.

2. Control System Design Process

In general, a complex process control system design involves the definition of system requirements, process modeling and optimization, controller design and performance analysis. More specifically, six steps may be defined as illustrated in Figure 1. These six steps form the basis for the organization of XIMKON.

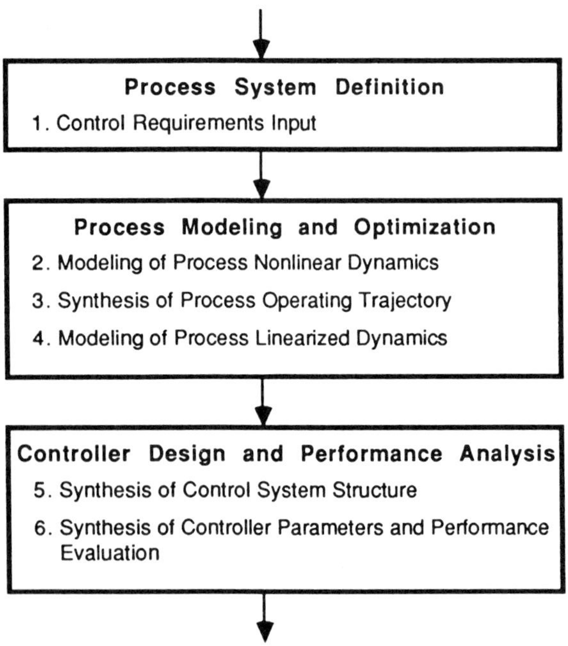

Figure 1. Control system design process.

2.1. Control Requirements Input

This step involves the mission statement and specification data for the controlled system. Most often this data includes the operating envelope and limits of the controlled system, specific command and disturbance inputs, their types, ranges, and expected responses, parameter uncertainty ranges in the process model, steady-state performance levels, and actuator activity and limits. Depending upon the nature of the specific control system, additional requirements may be imposed (i.e., maximum temperature, pressure, structural stress, and so on). These requirements are met by an appropriate system architecture and control strategy.

Most often this architecture is in the form of multilevel and distributed control and may contain robust functional modules such as controllers, estimators, predictors, identifiers, optimizers, etc. The key requirement here is that these modules perform reliably over a wide range of operating conditions and application scenarios.

Figure 2 shows a block diagram of a typical system architecture for an advanced process control system. It consists of nine levels. The first level includes the process dynamics. Most often the elements of process subsystems dynamically interact with each other due to mass and energy transfer between them. The second level represents the equipment dynamics involved: compressors, pumps, etc. The third level includes the final control elements, such as valves, actuators, and sensors. A system composed of levels 1, 2, and 3 is called a *process plant.*

The fourth level is assigned to the local controllers, controlling the process variables such as temperature, flow, concentration, etc. The set points of the local controllers are driven by the supervisory controller at the fifth level. The supervisory controller develops the optimal command profiles for batch processes or optimal operating points for continuous processes.

Allocation of local costs among the subsystems is performed at the sixth level by the local coordinator to optimize the unit operating cost.

The filtering of the measured plant signals and the estimation of the inferred plant and ambient system variables and their prediction into the future based on the data available up to the present time are carried out at the seventh level. This level consists of observers and linear and nonlinear Kalman filters of appropriate size. They contain the plant and ambient dynamic models in their structure. In addition, models of deterministic as well as stochastic disturbances into the plant are included. The model structure may be based on first principles or on black-box autoregressive moving-average methods. These models are computed in the eighth level.

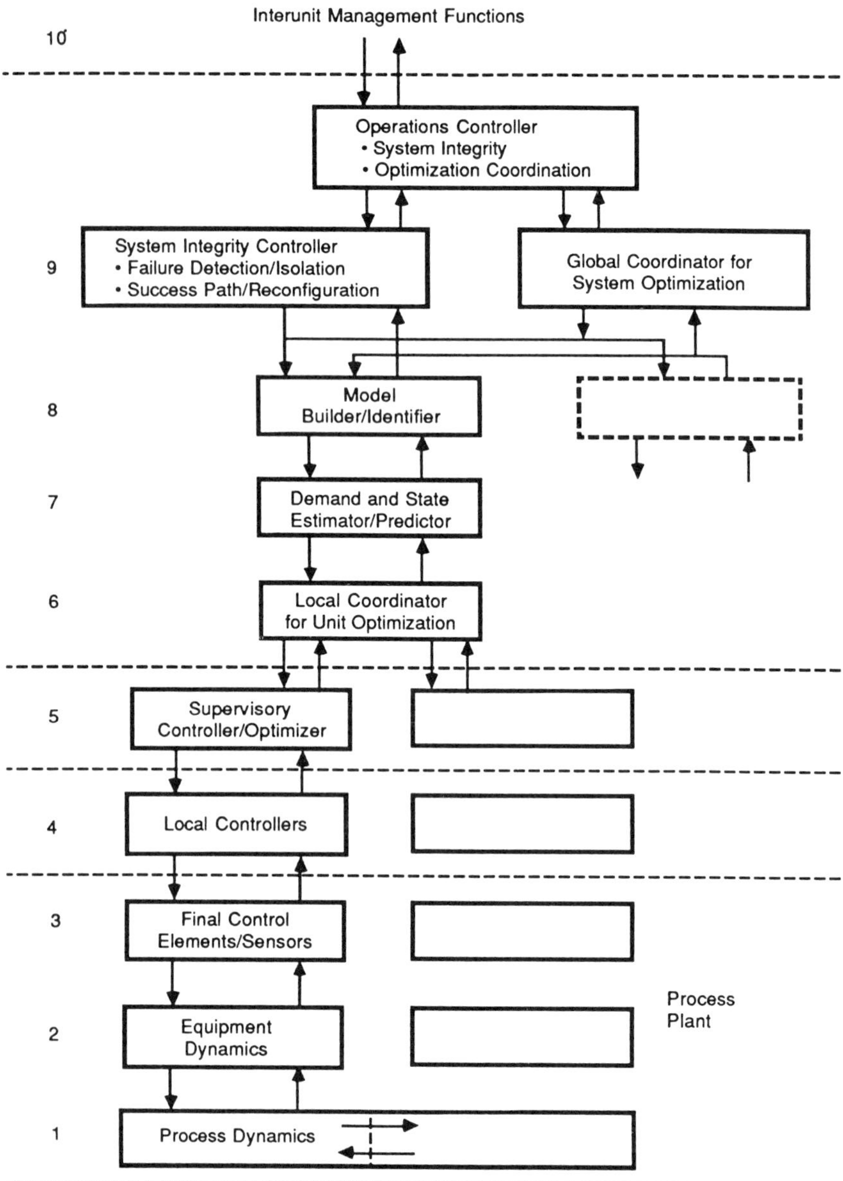

Figure 2. Multilevel process control system architecture.

Because of the availability of the on-line model parameters of the load and the plant, plant status can be monitored. This is done in the ninth level. The current parameter estimates are compared with the normal operation parameters and any significant deviation is detected and isolated. The operator is informed, success paths to recover from the fault are suggested, and the control system is reconfigured for gracefully degraded operation until the failure is repaired.

The ninth level also contains a global coordinator for the optimization of the total operating cost of the factory. Because of the interaction of the unit processes at the plant level, this coordinator adjusts the goals of the local optimizers on the lower level to achieve the best overall operating efficiency.

2.2. Modeling of Nonlinear Process Dynamics

Accurate dynamic modeling of chemical process systems is by no means a trivial task. The processes are generally nonlinear, time varying, and have variable time delays and other non-minimum-phase effects. In addition they are often interactive and are perturbed by changes in feedstock and operating conditions. All of these factors make dynamic modeling and control somewhat different from that of typical electrical and mechanical systems where linear, time-invariant stochastic models often give sufficient flexibility and accuracy.

The dynamics of a physical process are assumed to be characterized by the following set of differential/algebraic equations:

$$\frac{d}{dt}[\phi(x, \theta)] = \Psi(\dot{x}, y, x, u, \theta) \tag{1}$$

$$y = g(\dot{x}, y, x, u, \theta) \tag{2}$$

$$r = h(\dot{x}, y, x, u, \theta), \tag{3}$$

where

x = state vector of the process, n_x-vector

y = algebraic/summing point variables, n_y-vector

r = response/output variables, n_r-vector

u = control/disturbance input variables, n_u-vector

θ = parameter vector of the process, n_p-vector

Application of the first principles to dynamic processes yields this form of equations (Konar, 1985; Luyben, 1973). For the dynamic simulations of the process, the equivalent form of Equation 1 may be used:

$$\dot{x} = f(\dot{x}, y, x, u, \theta). \tag{4}$$

The dynamic system parameters, such as the reaction rates and heat transfer coefficients in chemical process dynamics, are assumed to be represented either in the form of analytical expressions (i.e., y-variables) or in the form of experimental tables with multiple arguments.

The measured data history is used to extract model parameters, which may be done in batch or recursively in time. The parameter estimation function is accomplished by a repertoire of algorithms such as recursive and generalized least squares, extended, modified Kalman filters, maximum likelihood estimators, etc. Figure 3 shows a block diagram of a parameter estimator. The process model parameters and Kalman filter gains are changed on the basis of the likelihood function obtained from the residuals.

In process control systems there often may be significant time delays due to the distributed nature of the process dynamics, which normally are characterized by the partial differential equations. To accommodate these subsystems within the framework of ordinary differential equations, a mixture of lumped-parameter models with appropriate time delays is utilized. In this case, the process equations given by Equations 1, 2, and 3 are augmented with additional variables having delayed arguments, i.e.

$$u_d = u(t - \tau_d), \qquad y_d = y(t - \tau_d), \qquad r_d = r(t - \tau_d). \tag{5}$$

The software modules characterizing each algebraic and/or dynamic subsystem are then interconnected to form the total system simulation

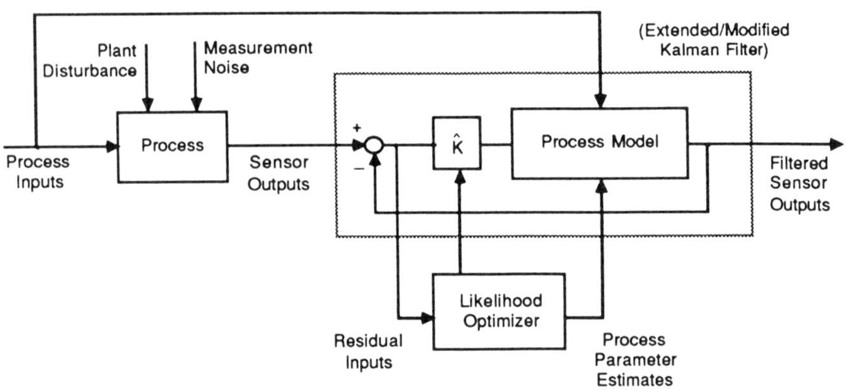

Figure 3. Block diagram of a parameter estimator.

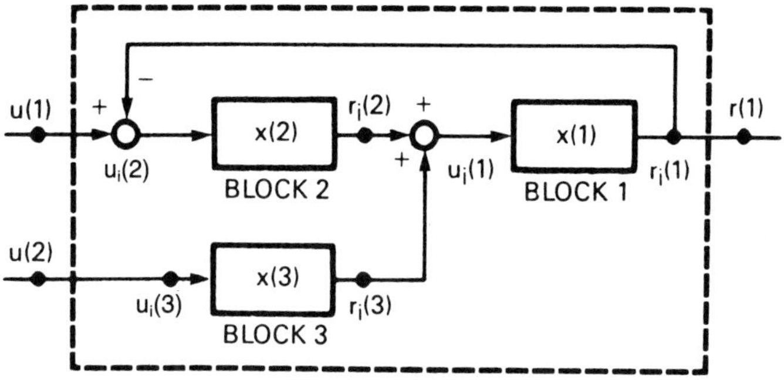

Figure 4. Interconnection of subsystem blocks.

model, as shown in Figure 4. This approach provides great flexibility in forming very large dynamic systems.

The interconnection equations are assumed to be in the following form:

$$u_{\mathrm{i}} = Pr_{\mathrm{i}} + Qu \tag{6}$$

$$r = Rr_{\mathrm{i}} + Su, \tag{7}$$

where

u_{i} = internal input vector

r_{i} = internal output vector

u = external input vector

r = external output vector,

and (P, Q, R, S) is called the *interconnection data* of the system. Figure 5 shows a typical overall system interconnection diagram.

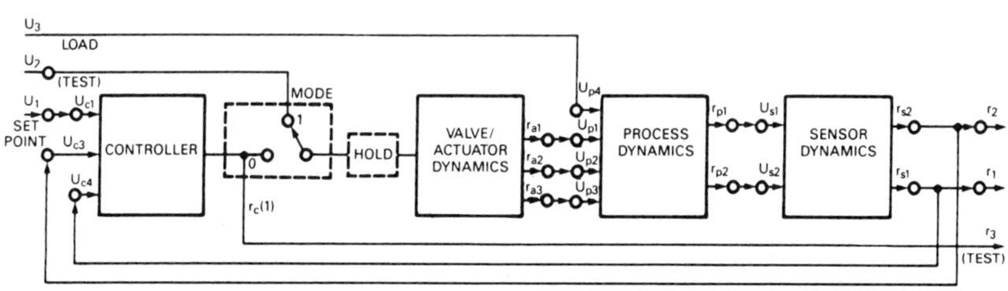

Figure 5. Overall system block diagram.

2.3. Synthesis of Process Operating Trajectory

This step is most often concerned with the economics of process operations and involves a large-scale, nonlinear constrained-optimization task. Examples are the definition of temperature trajectory of an exothermic batch reactor, operating point of a distillation column, and so on. Two types of optimization may be carried out for an operating trajectory like the one shown in Figure 6. In the dynamic optimization, the optimum profiles of the process variables are determined over a prescribed period:

$$[\dot{\bar{x}}(t), \bar{y}(t), \bar{x}(t), \bar{u}(t)], \qquad T_0 \leq t \leq T_f. \tag{8}$$

In static optimization, the optimal steady-state operating point is determined by setting $\dot{x} = 0$ in Equation 4 and computing

$$[0, \bar{y}, \bar{x}, \bar{u}] \tag{9}$$

In practice, the nominal trajectory definition step is normally separated from the controller design task. In the case of process control, it is known as the set-point optimization or supervisory control task. Optimal control theory and optimization theory are used to generate the nominal trajectory data and the nominal set-point commands to the control system (Ray and Szekely, 1973; Schuldt and Smith, 1971).

The integration of these functions for on-line variable trajectory generation and control has been pursued in aerospace scenarios (Comegys and Hanson, 1981; Jones *et al.*, 1981) and process control (Horn, 1978).

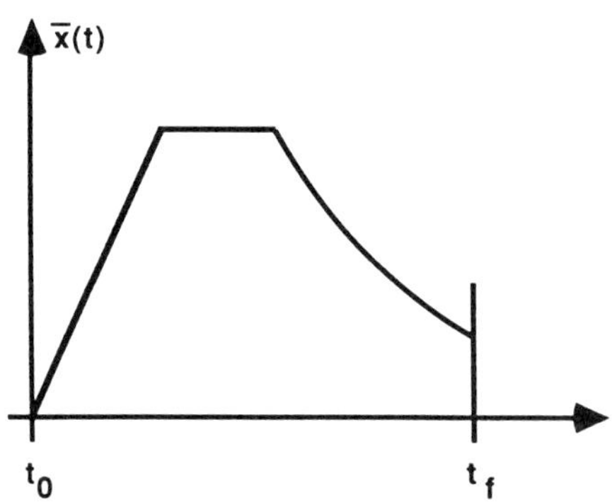

Figure 6. Nominal operating trajectory.

2.4. *Modeling of Process Linearized Dynamics*

Designing a perturbation control system that will keep the dynamic process variables about the prescribed trajectory or an operating point under the influence of external disturbances, set-point changes, measurement noises, and system parameter uncertainties requires a perturbation model of the process.

The process dynamics given by Equations 1, 2, and 3 are linearized about the nominal trajectory defined by

$$\dot{\bar{x}} = f(\dot{\bar{x}}, \bar{y}, \bar{x}, \bar{u}, \theta) \tag{10}$$

$$\bar{y} = g(\dot{\bar{x}}, \bar{y}, \bar{x}, \bar{u}, \theta) \tag{11}$$

$$\bar{r} = h(\dot{\bar{x}}, \bar{y}, \bar{x}, \bar{u}, \theta), \tag{12}$$

yielding

$$\delta\dot{x} = F_{\dot{x}}\,\delta\dot{x} + F_y\delta y + F_x\delta x + F_u\delta u \tag{13}$$

$$\delta y = G_{\dot{x}}\,\delta\dot{x} + G_y\delta y + G_x\delta x + G_u\delta u \tag{14}$$

$$\delta r = H_{\dot{x}}\,\delta\dot{x} + H_y\delta y + H_x\delta x + H_u\delta u. \tag{15}$$

The coefficient matrices in Equations 13, 14, and 15 are gradient matrices evaluated at the nominal point. These matrices are obtained either by analytical or numerical means. When a portion of the process variables is characterized by experimental tables, the numerical linearization becomes mandatory.

These equations subsequently are manipulated into the standard state-space form:

$$\dot{x} = Ax + Bu \tag{16}$$

$$r = Cx + Du, \tag{17}$$

where the set of four matrices (A, B, C, D) is called the *quadruple data* of the dynamic process (Konar *et al.*, 1976; Konar *et al.*, 1974). In general, the process quadruple is time varying. For control design, a subset of selected time points and the associated data are used. These finite points, referred to as *operating conditions*, represent the frozen-point linear process model within the operating envelope.

2.5. *Synthesis of Control System Structure*

The synthesis of a control system structure depends upon the nature of the process to be controlled, the effector and measurement points and qualities, the implementation considerations, and the controlled system performance requirements.

The selection of the actual control law is often iterative and based on intuition and insight to select a feasible solution from the vast number of alternatives. Some algorithmic procedures to generate these alternatives have been developed (Morari *et al.*, 1980; Govind and Powers, 1982).

The selection of the effectors (manipulated variables) and measurement points is also based on process knowledge. Algorithmic approaches for optimal sensor locations are given in Romagnoli and Stephanopoulos (1980).

Virtually all chemical process are characterized by multiple inputs and outputs. In practice, the multivariable problem is decomposed into a series of single-input/single-output problems. To determine the best input/output pairings, several procedures have been developed (Tung and Edgar, 1981).

For preliminary design, the optimal systems control in the form of a predictor and Kalman filter with state-estimated feedback as shown in Figure 7 provides a unified framework for design trade-off studies (Marshall, 1979; Kleinman, 1969).

Smith predictors, complementary control, self-tuning regulators and predictors, PID, feedforward, extended horizon, model algorithmic, dynamic matrix, internal model, two-port controllers and multivariable controllers are some of the frequently used concepts in the process control industry today (Stephanopoulos, 1984; Ogunnaike and Ray, 1979; Stephanopoulos

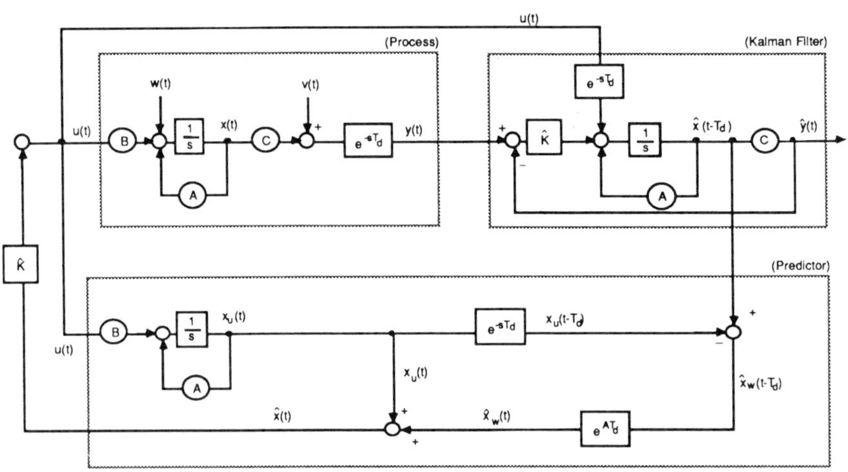

Figure 7. Optimal feedback system for a process with transport delay.

and Huang, 1986; Garcia and Morari, 1982; Joseph and Brosilow, 1978). Most often a control system must be modified to meet a set of steady-state requirements (i.e., temperature regulation for constant disturbance) by adding integrators, cross-feeds, and roll-off filters. If controllers are implemented using a digital processor, they are most often designed to operate in a multirate format due to different bandwidths of the process subsystems and finite word-length arithmetic operation (Konar *et al.*, 1981). At the end, the overall controller is characterized by transfer functions or differential (difference) equations and interconnections between them, as shown in Figure 4.

2.6. Synthesis of Controller Parameters and Performance Evaluation

In the case of optimal controller configuration, automated tools have been developed for determining controller and estimator parameters over a period of time for a nominal operating point (Konar and Ward, 1972; Konar and Mahesh, 1977). For classical controller configurations, controller parameter tuning procedures are available (Luyben, 1973). Parameter tuning can also be accomplished by parameter optimization (Borow *et al.*, 1972).

Most often, design of complex processes requires a multiperformance evaluation capability to facilitate a quantitative analysis of all parameters that effect the design. Performance measures include transient response, poles and zeros, singular values, open- and closed-loop frequency responses, stability margins, multivariable root locus, covariance response and power spectral density response. These responses are computed as functions of common design parameters such as gains, control filters/laws, noise, stochastic disturbances and operator inputs, multisample rates, word lengths, and process and computational delays. Final design is selected using the multiperformance data base generated within the operating envelope of the system (Konar *et al.*, 1974; Konar, 1973).

3. XIMKON

XIMKON (Expert Simulation and Control), whose components are illustrated in Figure 8, is a computer-aided design methodology for large dynamic process control systems (Konar, 1987; Konar *et al.*, 1987). XIMKON contains a set of procedural algorithms (totaling approximately 50,000 FORTRAN statements) based on state-of-the-art simulation and control technologies. The algorithms are the result of many years of

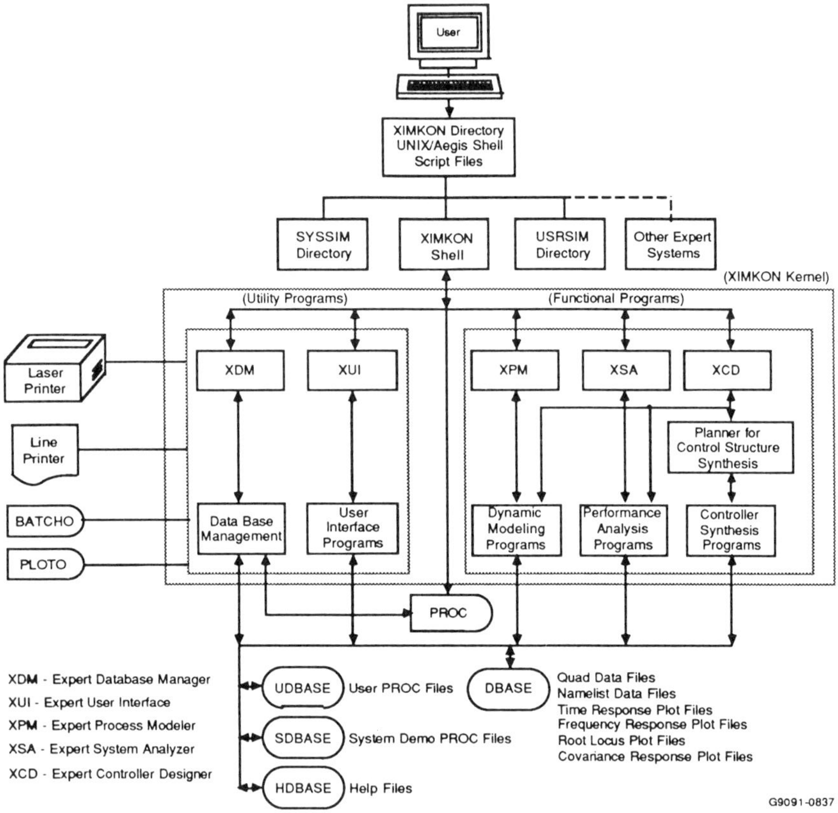

Figure 8. Organization of XIMKON software system.

program development at Honeywell, including SIMKON V, DIGIKON IV, ADAP and KONPACT. XIMKON also contains object-oriented modules using AI and advanced user-interface techniques, written in about 10,000 lines of Objective-C.

3.1. XIMKON's Foundation

The FORTRAN modules, which serve as the foundation for XIMKON, provide a wide range of capabilities, as summarized below.

- Coupled, nonlinear and linear differential equation input
- Multiple transfer function input
- Generalized interconnection input
- Multisample rate control system structure

- Process transport and computational delays
- Poles, zeros, pseudo-zeros, singular values
- Frequency response in *s*, *z*, *w* planes
- Transient response and nonlinear simulation
- RMS response, power spectral density and power, nonlinear covariance analysis
- Multivariable root locus, sensitivity analysis
- Interactive parameter optimization
- Optimal control and estimation synthesis
- Batch/recursive parameter estimation

The following example illustrates a typical application of XIMKON. Figure 9 shows a control system for a continuous stirred tank reactor (CSTR). The proportional-plus-integral (PI) controller maintains the reactor temperature at a prescribed level by manipulating the flow of the jacket water.

We may analyze the response of the system to a step disturbance at the inlet concentration, Ca0, by performing a nonlinear dynamic simulation. The CSTR is modeled with a FORTRAN subroutine containing nonlinear differential equations based on first principles. The PI controller is modeled with a transfer function and discretized using the Z transform. (The modeling process will be discussed further in Section 4). Figure 10 shows the results of the simulation with a plot of the reactor temperature response.

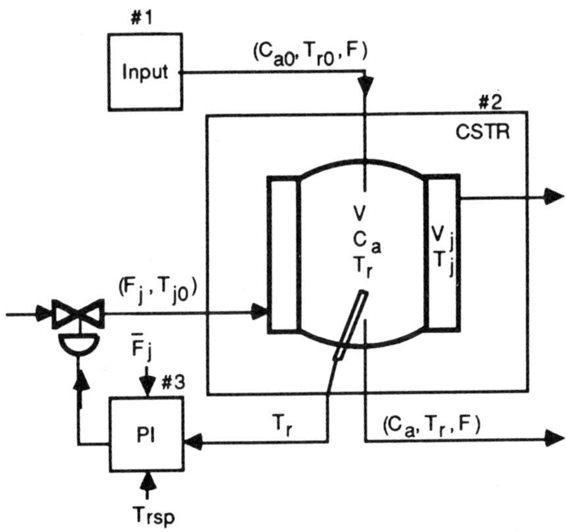

Figure 9. CSTR control system.

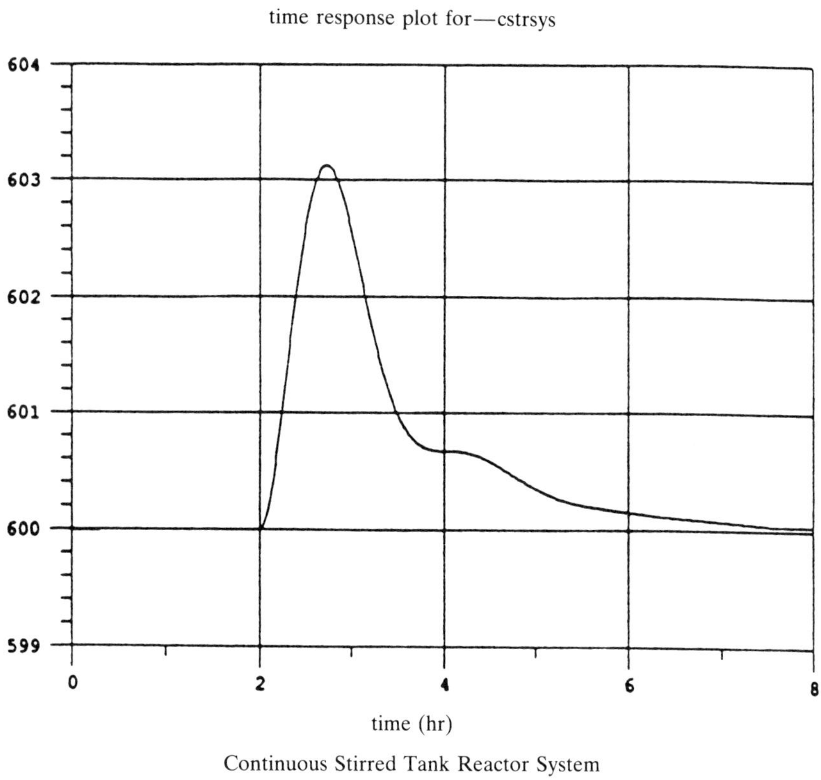

Figure 10. Reactor temperature response.

Figure 11 is the nonlinear simulation PROC (command) file for this example. XIMKON's FORTRAN modules will read the PROC file and execute its instructions. Statement 3 of Figure 11 is the nonlinear simulation command, with subsystem names and time parameters. Statements 4 through 22 define the internal and external interconnections. Statements 23 through 39 define a namelist for external inputs and outputs. Statements 40 through 43 specify initial conditions. Statements 44 asks for a linearization of the CSTR process at two time points. When the simulation is finished, Statements 47 and 48 request a listing of the quadruple data for the two resulting linear CSTR systems. The rest of the statements are for time response plots of the variables.

This example shows the power of XIMKON as a computer-aided engineering (CAE) tool for automating some of the steps outlined in our control system design process, however, it becomes evident that the user of XIMKON at this level must be well versed in PROC file generation and

```
print,cstrsys
 rgpt
 Continuous Stirred Tank Reactor System
 cnsm(cstrsys,cstr,dpic,dt=.001,pt=.01,tf=8,tu=hr)
 ui1(1)=u(1)
 ui1(2)=u(2)
 ui1(3)=u(3)
 ui1(4)=u(4)
 ui1(5)=ri2(1)
 ui2(1)=u(5)
 ui2(2)=u(6)
 ui2(3)=ri1(2)
 r(1)=ri1(1)
 r(2)=ri1(2)
 r(3)=ri1(3)
 r(4)=ri1(4)
 r(5)=ri1(5)
 r(6)=ri1(6)
 r(7)=ri1(7)
 r(8)=ri1(8)
 r(9)=ri1(9)
 r(10)=ri2(1)
 end
 u1,ca0,moles/ft3
 u2,tr0,deg R
 u3,f,ft3/hr
 u4,tj0,deg R
 u5,fj,ft3/hr
 u6,tsp,deg R
 r1,ca,moles/ft3
 r2,tr,deg R
 r3,tj,deg R
 r4,k,/hr
 r5,qj,btu/hr
 r6,qg,btu/hr
 r7,cadot,moles/ft3 hr
 r8,trdot,r/hr
 r9,tjdot,r/hr
 r10,fj,ft3/hr
 end
 ic(u1=step(.50,0,.1,2))
 ic(u2=530,u3=40,u4=530,u5=49.9,u6=600)
 ic(x1=.245,x2=600,x3=594.6)
 end
 lin(cstr,0.,.5)
 end
 end
 quad,nlcs1
 quad,nlcs2
 ptr(p,ca,ca0)
 ptr(p,tr,tr0)
 ptr(p,tj,tj0)
 ptr(p,fj,f)
 ptr(p,qg,qj)
 ptr(p,k)
 ptr(cp,tr,k)
 ptr(p,cadot,trdot,tjdot)
```

Figure 11. Nonlinear simulation PROC file.

FORTRAN programming. Indeed, the user must initially call upon his or her own expertise and knowledge in choosing the type of control structure to be implemented in the design.

In order to eliminate the manual chores and automate the rest of the steps in the design process, more expertise and knowledge must be inserted into the program, thus allowing the non-expert engineer to successfully complete the design process.

3.2. XIMKON's Expert System Enhancements

The enhancements to XIMKON required for further automation of the control system design process are implemented using the object-oriented language, Objective-C (Stepstone, 1988). The result is a hybrid software system containing low-level, computationally intensive FORTRAN modules, and a top layer of object-oriented modules in Objective-C.

The XIMKON Planner is the main driving force behind our insertion of additional expert knowledge. Based on classical AI planners (Wilkins, 1983; Cohen and Feigenbaum, 1982; Chapman, 1987), it is designed to supervise the execution of the modeling, analysis and synthesis tasks inherent in the control system design process. Use of AI planners for process control applications has been considered by other investigators. For example, the design of a knowledge-based expert system using a planner to supervise adaptive control systems in real time is reported in Trankle and Markosian (1985).

A planner is used in problem-solving tasks for developing a sequence of actions to achieve a goal. In most cases, the order of the actions is important, because one action may rely on the successful completion of a previous action. For the control system design process, the six steps, shown in Figure 1, may be thought of as a plan of six actions executed in sequence to achieve the goal of designing a control system. These six actions define the basis for XIMKON's supervisory planner. The completion of each action will involve the generation of subplans.

XIMKON's planning methodology uses two types of planners: 1) a forward chaining or "script based" planner, and 2) a backward chaining planner.

A forward chaining planner can be used when the selection and sequence of actions is initially defined and fixed (a script). XIMKON's supervisory planner is forward chaining, since it uses the six steps of the control system design process.

A backward chaining planner is used when there are decisions to be made about the selection and ordering of actions. For this type of planning, a tree data structure is used with the root node of the tree defining the goal

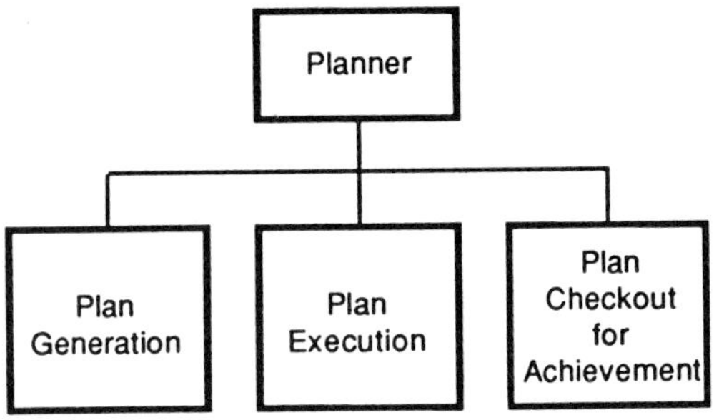

Figure 12. XIMKON planner.

of the plan. The planner begins by selecting an action that will satisfy the goal. However, the action may require other goals to be satisfied before it can be executed. These goals become subgoals and, consequently, child nodes of the tree structure. Actions are applied, and the tree is expanded until a selected action requires no subgoals. Plan execution will start at the last action selected and end at the first action selected. The backward chaining planner is used for subplanning in step 5 of our six steps for control system design, namely, "Synthesis of Control System Structure." This planner will be described further in Section 5.

As shown in Figure 12, the XIMKON planner consists of plan generation, plan execution and plan checkout. In most cases, plan execution will translate into the automatic creation and execution of a XIMKON PROC file, such as the one in Figure 11. During plan checkout, the results of plan execution will be examined and, perhaps with input from the user, will be either accepted or rejected in favor of another plan.

As shown in Figure 8, XIMKON's expert system enhancements are divided into five major components:

- Expert Process Modeler
- Expert System Analyzer
- Expert Controller Designer
- Expert Database Manager
- Expert User Interface

The Expert Process Modeler uses first principles and test data from the specific plant to model the structures of the equations and the parameters for a particular process unit. This expert system has access to XIMKON's dynamic models database as well as modeling commands including batch recursive nonlinear regression and maximum likelihood estimation procedures.

The Expert System Analyzer makes use of the models developed by the Expert Process Modeler. The expert system analyzer supervises the analysis algorithms for selecting appropriate analysis procedures and has access to all XIMKON commands and databases. It can also be activated by the Expert Controller Designer to evaluate particular intermediate designs developed.

The Expert Controller Designer makes use of the models developed by the Expert Process Modeler and also the performance evaluation capabilities in the Expert System Analyzer. The Expert Controller Designer supervises XIMKON for selecting a candidate control law, manipulating and controlling variables, synthesizing the parameters of the law, and evaluating and analyzing resulting designs.

The Expert Database Manager component of XIMKON is envisioned to perform the functions of many expert database systems that are currently being designed and developed (Brodie and Mylopoulos, 1986). These include intelligent query processing and consistency checking. All of the data/knowledge that are requested by the other components of XIMKON are stored in the database and processed by the object-oriented model (Banerjee *et al.*, 1987).

The Expert User Interface provides pull-down menus, dialog boxes and direct manipulation capability, making the program much easier for the non-expert engineer to learn and use.

Note that other expert system tools such as Design-Kit developed at Massachusetts Institute of Technology (Stephanopoulos, 1987; Leon and Henning, 1988) can be interfaced with XIMKON (as illustrated in Figure 8. For example, such a tool could be used to represent the information pertaining to a process system flowsheet. This representation could be used by the expert controller designer and expert system analyzer components of XIMKON for controls.

4. Expert Process Modeling

This section focuses on our approach toward automating the steps in the control system design process (Figure 1) outlined for process modeling and optimization.

4.1. Nonlinear Process Modeling

Step 2 involves modeling the nonlinear process dynamics. In the example described in Section 2, the CSTR was modeled by creating a FORTRAN subroutine containing differential equations based on first principles. The equations used are listed in Figure 13.

Reactor Total Continuity:

$$\frac{dV}{dt} = F_0 - F$$

Reactor Component Continuity:

$$\frac{d}{dt}(V C_a) = (F_0 C_{a0} - F C_a) - V k C_a^n \therefore k = k_0 e^{-\frac{E}{RT_r}}$$

Reactor Energy Equation:

$$\rho C_p \frac{d}{dt}(V T_r) = \rho C_p (F_0 T_{r0} - F T_r) + \lambda V k C_a^n - UA(T_r - T_j)$$

Jacket Energy Equation:

$$\rho_j C_j \frac{d}{dt}(V_j T_j) = \rho_j F_j C_j (T_{j0} - T_j) + UA(T_r - T_j)$$

Figure 13. Nonlinear dynamic modeling equations for CSTR.

For XIMKON's nonlinear simulation, the structure of a FORTRAN subroutine is illustrated in Figure 14. Formulating the modeling equations and coding them into a FORTRAN subroutine of this form requires a large amount of time and expertise. As a result, many engineers will decline and choose another route for solving their problem. Therefore, it is desirable to extract this expertise from the user and computerize it so that a non-expert engineer can use the system.

A hierarchical approach designed to lift the user from the low level modeling tasks is illustrated in the plan of actions shown in Figure 15. The plan represents three stages of nonlinear process modeling, conceptual, block diagram and FORTRAN subroutine.

With this approach, the user has the option of defining the model at any stage, and then calling upon XIMKON to translate the model to the lower stage(s). For example, a model defined in the conceptual stage would automatically be translated to a block diagram model and a FORTRAN subroutine.

This methodology forms the basis for development of XIMKON's Expert Process Modeler. Currently, the block diagram and FORTRAN subroutine modeling stages have been implemented in XIMKON. The conceptual stage represents work in progress. Our approach to the conceptual and block diagram stages are described below.

4.2. Conceptual Modeling

The conceptual level provides the best level of communication for the non-expert engineer. In the case of the CSTR example, it would be much

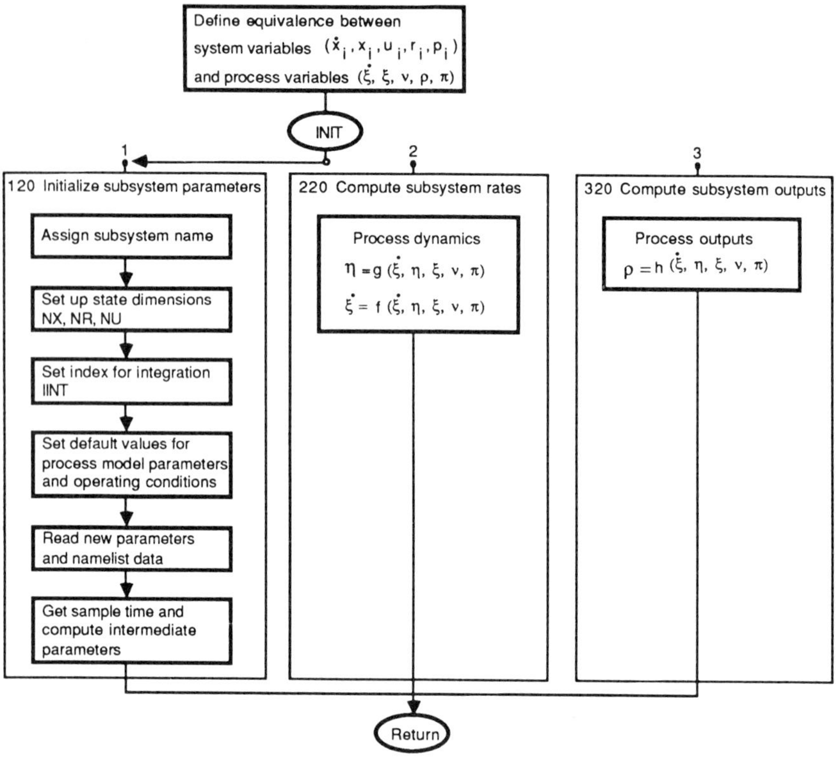

Figure 14. Nonlinear process model FORTRAN subroutine structure.

more productive to allow the user to define the CSTR model in terms of a reactor tank and a cooling jacket (shown as graphical icons), rather than in terms of differential equations coded into a FORTRAN subroutine.

Providing process modeling on a conceptual level requires a system representation of the entities used in the model, such as reactor tank. An object-oriented representation is most appropriate for this application. The advantages of object-oriented programming have been well documented (Pascoe, 1986; Cox, 1986).

As stated in Banerjee *et al.*, (1987), all conceptual entities are modeled as objects in an object-oriented system. A group of objects with similar properties form a class that is also an object. A class could be a system-defined class such as a class of integers, reals or Boolean values, or it could be a user defined class such as a class of Blocks or CSTR. Associated with each class is a set of instance variables that describe properties of the

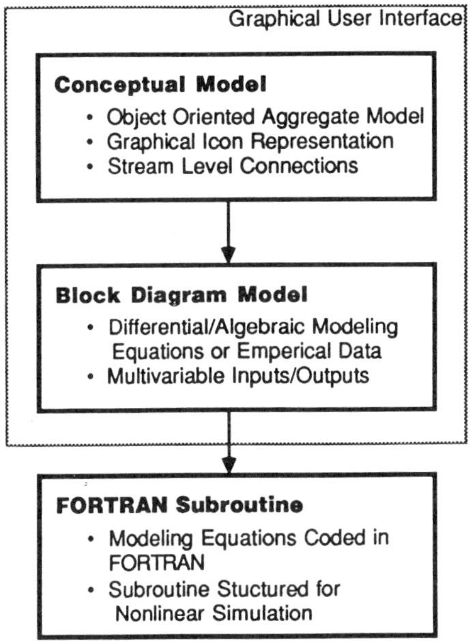

Figure 15. Plan for modeling of nonlinear process dynamics.

objects belonging to that class. For example, a class of Blocks could have the block's name, inputs and outputs as its instance variables. An instance variable is also an object whose domain may be another class.

A class has methods associated with it. These methods encapsulate the behavior of the object. Methods consist of code that can be executed when appropriate messages are sent to the class. For example, a message may be sent to the class of blocks to retrieve the inputs of a particular block.

Two types of object hierarchies may be defined. One is the IS-A hierarchy where a class has subclasses associated with it. The subclass inherits all of the instance variables contained in its superclass. For example, Figure 16 shows a class structure in which the highest level is the class *ProcessModel*. The class *CSTR* is a subclass of *ProcessModel.* That is, any instance of the class *CSTR* is also an instance of the class *ProcessModel*. If a class has more than one superclass, and two or more of the superclasses have the same instance variable names, then the value of the instance variable inherited by that class depends on some enforced rule.

The second class hierarchy is the IS-PART-OF hierarchy. Here, an object in a class is considered to be an aggregation of a set of objects, each of which belongs to some class. For example, as shown in Figure 17, an

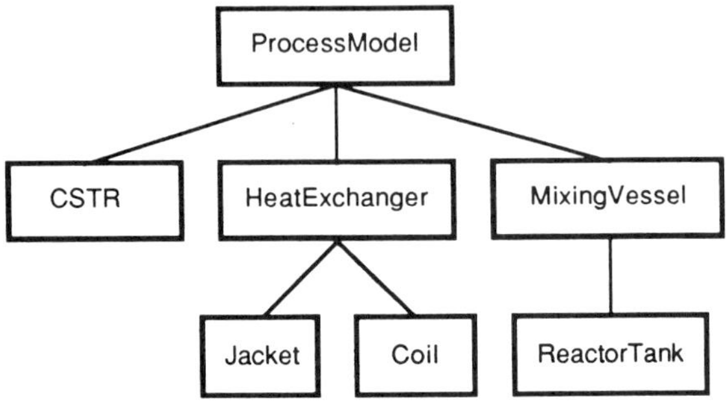

Figure 16. IS-A object representation (class structure).

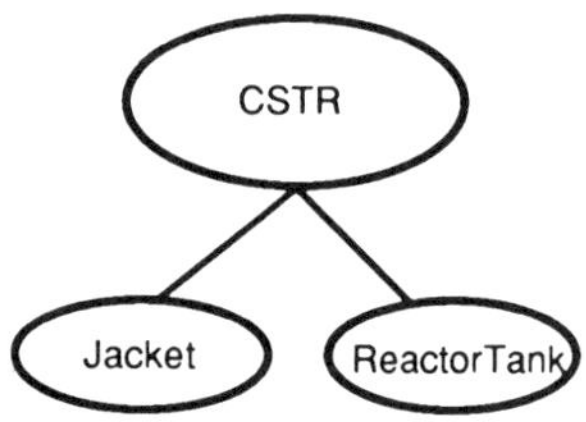

Figure 17. IS-PART-OF object representation (aggregate object structure).

object of class *CSTR* may be an aggregate containing an instance of an object from class *Jacket* and an instance of an object from class *ReactorTank*.

In addition to the features above, a process model is subject to many constraints, which are special types of rules, such as heuristic rules and processing rules. Each rule may be modeled as an object, and classes of rules may be defined, such as class *HeuristicRule* and class *ProcessingRule*. Associating rules with classes is uncommon in most object-oriented models. However in Beech (1987), suggestions are given on how to incorporate them.

Inherent in the object representation of a process model are rules that specify the differential equations for the model. The CSTR aggregate object combines the equations for the total and component continuity, as well as the reactor and jacket energy equations, as shown in Figure 13. Associated with these equations are rules that specify the operation of the reactor. For example, for perfect level control, the value of dV/dt in the first equation is zero. Therefore, if the user requests perfect level control, the rules will be examined and, subsequently, dV/dt will be set to zero.

Again, communicating in terms such as "perfect level control" is much more desirable than "$dV/dt = 0$."

When the conceptual modeling is completed, XIMKON will continue in the modeling procedure by generating a multivariable block diagram model to be used for controller design, and a FORTRAN subroutine model for simulation.

A conceptual, object-oriented approach to process modeling has been implemented by the LISPE team at M.I.T. (Stephanopoulos, 1987; Leone and Henning, 1988). The software called Design-Kit uses symbolic manipulation to generate the final form of the modeling equations from the component objects.

4.3. Block Diagram Modeling

The block diagram modeling stage is designed to generate a model in one of two ways; 1) by acquiring the modeling equations from a conceptual model, or 2) by allowing the user to enter the modeling equations directly. In both cases, the result is a multivariable, nonlinear block to be used for simulation, linearization and controller design. The first case represents a situation in which the user has generated a model at the conceptual stage. For the second case, XIMKON's Expert User Interface has been developed to facilitate direct entry of the modeling equations.

The Expert User Interface, with pull-down menus and dialog boxes, has been developed on an Apollo workstation using Objective-C (Stepstone, 1988), Apollo's user interface toolkit DOMAIN/Dialog (Apollo, 1987), and Apollo's GPR graphics (Apollo, 1985). The Expert User Interface is based on a subset of the Model-View-Controller (MVC) paradigm used in the Smalltalk-80 (Krashner and Pope, 1988) object-oriented environment. Specifically, an object class View is used as the base class from which many other objects are derived. A view is an object on the screen that graphically represents a data model associated with it. A view has a bounding box in which subviews can be displayed, and each subview may have its own subviews.

For block diagram modeling, the user interface provides a menu of five iconic primitives: block, block variable, branch point, summer and gain. Figure 18 is a screen shot of the user interface.

Each primitive is a unique data model, and therefore each has its own class. Figure 19 shows the IS-A class structure implemented for the primitives. Each box represents a view class that is directly or indirectly derived from class *View*. Inside each box is the view's class name and a list of the pertinent instance variables for the class. For example, class *Block* is

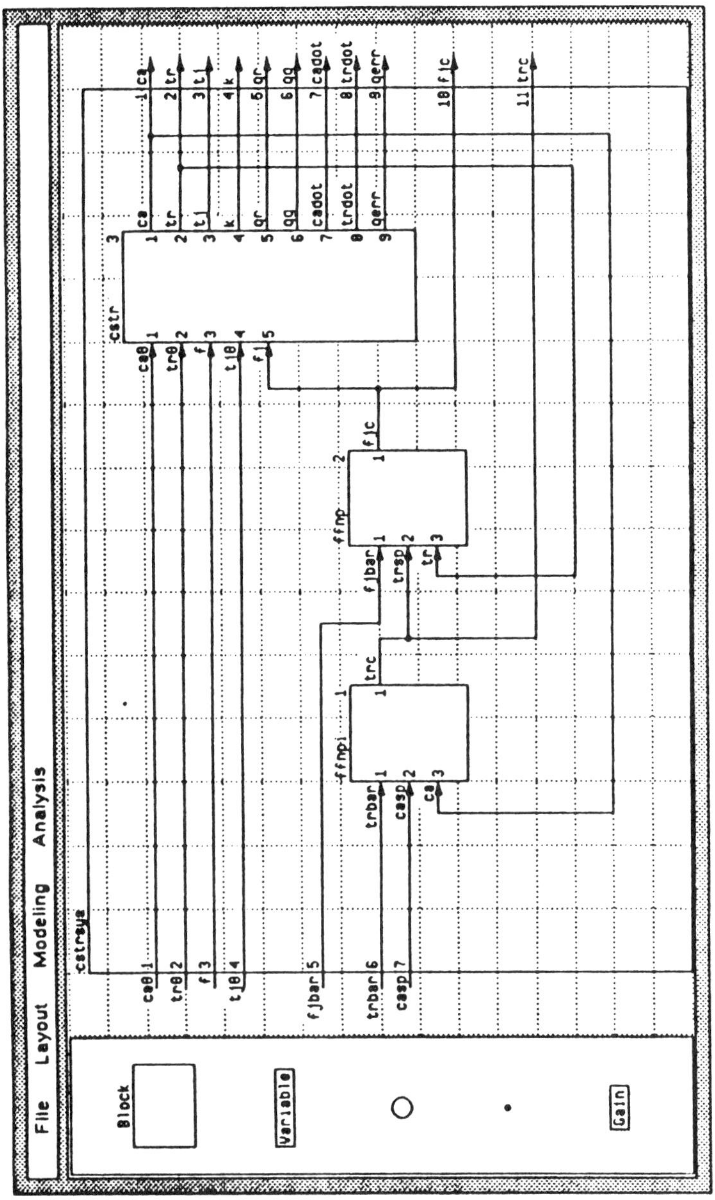

Figure 18. Screen shot of XIMKON user interface.

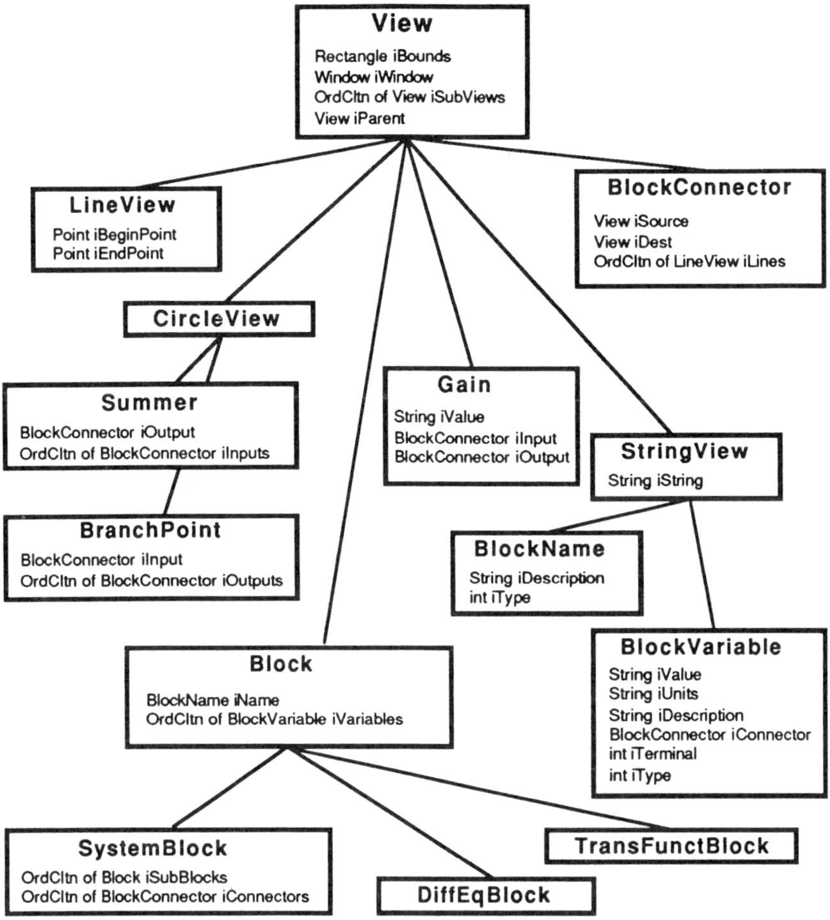

Figure 19. IS-A class structure for block diagram primitives.

directly derived from class *View*, and contains a name from class *BlockName*, and a list of variables of class *BlockVariable*. Class *SystemBlock* is derived from class *Block*, and contains a list of subblocks of class *Block* and a list of connectors of class *BlockConnector*, which connect input/output variables between blocks.

A nonlinear model is created by selecting a block from the menu, placing it in the work area, setting its type to be a differential equation block, and entering the modeling equations, as illustrated in Figure 20, for the CSTR example. At present, the modeling equations are entered as FORTRAN statements. The equations used for the CSTR are shown in Figure 21.

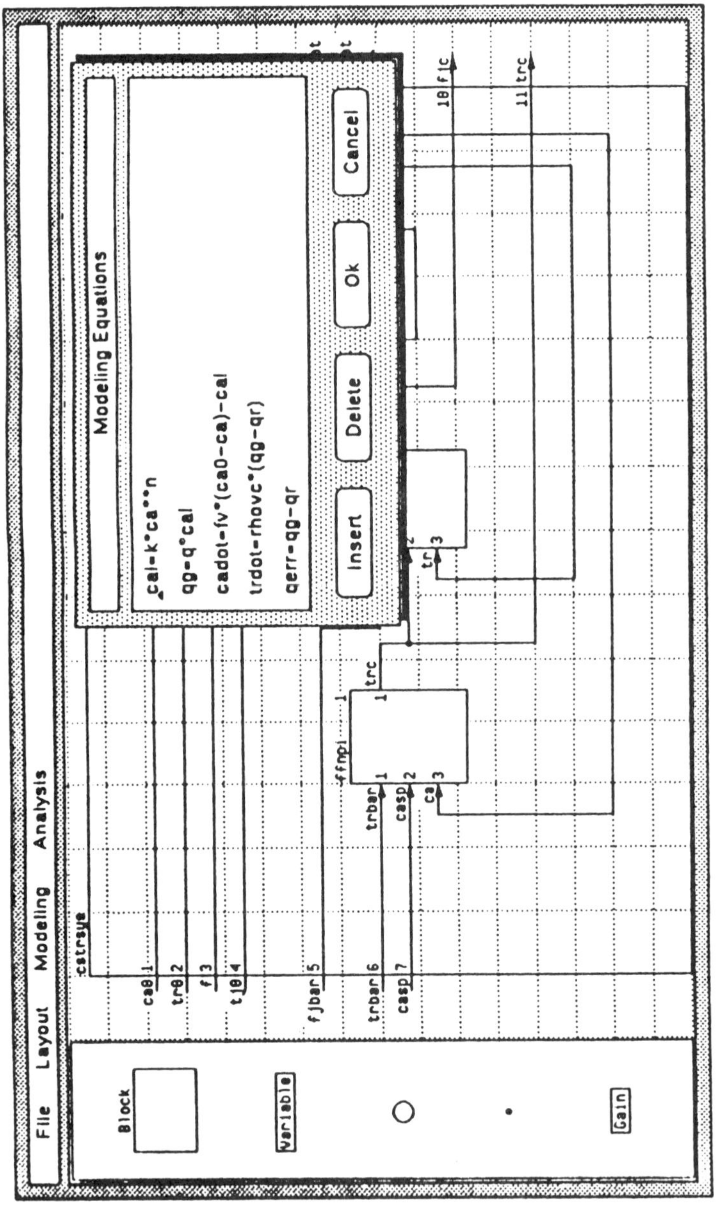

Figure 20. Entry of nonlinear modeling equations.

Parameter Equations

```
er=e/rr
rhovc=1.0/(rho*v*cp)
rhovcj=1.0/(rhoj*vj*cj)
ua=uj*a
q=lamda*v
rhocp=rho*cp
rj=(rhoj*cj)/ua
```

Modeling Equations

```
fv=f/v
fvj=fj/vj
k=k0*exp(-er/tr)
pj=1/(1+rj*fj)
tj=pj*tr+(1-pj)*tj0
qj=ua*(tr-tj)
qf=rhocp*f*(tr-tr0)
qr=qj+qf
cal=k*ca**n
qg=q*cal
cadot=fv*(ca0-ca)-cal
trdot=rhovc*(qg-qr)
qerr=qg-qr
```

Figure 21. Modeling equations for CSTR.

XIMKON will parse the variables introduced in the equations and load them into the block. The input and output variables of the block can then be graphically interconnected to the inputs and outputs of the controller block. Internally, XIMKON will automatically create a FORTRAN subroutine representing the CSTR block, and a PROC file of the overall system to be used for the nonlinear simulation and linearization.

In this example, the CSTR was modeled as one block, which includes the reactor dynamics, jacket dynamics and reactant concentration dynamics. However, the individual process components may also be modeled by separate block, with appropriate interconnections to form the overall model of the process.

4.4. Process Operating Trajectory and Linearized Dynamics

Steps 3 and 4 of the control system design process involve synthesis of the process operating trajectory and modeling of the process linearized dynamics. At present, development of the XIMKON Planner for process operating trajectory is in its preliminary stage. A candidate optimization package, MINOS, has already been identified for Honeywell internal use and is scheduled to be integrated into the XIMKON system (Murtagh and Sanders, 1987). Minos was developed by Stanford University with U.S. government funds. It is a large-scale optimization system for the solution of linear and nonlinear programs with nonlinear objective functions and constraints. The source code is written in FORTRAN and requires 500K bytes of storage.

As shown in the CSTR example in Section 2, the nonlinear process model can be automatically linearized at desired time points during non-linear simulation, resulting in (A,B,C,D) quadruple data for each time point. The quadruple data is stored with the process block in the block diagram, and made available to the user for the further performance analysis, using linear tools in XIMKON, such as frequency response and root locus analysis.

5. Expert Controller Design

This section focuses on our approach toward automating steps 5 and 6 in the control system design process (Figure 1) outlined for controller design and performance analysis.

5.1. Control Structure Synthesis

Control structure synthesis is a problem-solving task. As an example, consider a synthesis problem of controlling the reactor concentration, Ca, of a CSTR. Basically, a set of answers to the following questions define the type of plan we can use to arrive at a solution.

(1) What is wanted?
(2) What are the ways to achieve it?
(3) What are the requirements for achieving it in the chosen way?

Figure 22 (a) shows an application of these questions to the main problem at hand. Symbols used for the CSTR system variables are defined below.

- Ca-reactor concentration
- Tr-reactor temperature
- Fr-reactor flow rate
- Tj-jacket temperature
- Fj-jacket flow rate

Four options are available. However, an option can be used only if its requirements are met. For instance, option 1 requires that the reactor flow rate, Fr, be a controllable input to the CSTR system. If option 2 is selected, a subproblem is created, namely, controlling the reactor temperature of CSTR. Figure 22(b) shows a similar development for this subproblem.

Along with the selection of the manipulated and measured process variables, a selection of the type of controller (control law) must be made. Figure 23 shows a subset of possible controllers.

(a)

	What is wanted? : Control Ca of CSTR	
Options	What are the ways to achieve it?	What are the requirements for achieving it in the chosen way?
Option 1	Manipulate control variable Fr of CSTR	Fr is controllable
Option 2	Manipulate control variable Tr of CSTR	Tr is controllable
Option 3	Manipulate control variable Tj of CSTR	Tj is controllable
Option 4	Manipulate control variable Fj of CSTR	Fj is controllable

(b)

	What is wanted? : Control Tr of CSTR	
Options	What are the ways to achieve it?	What are the requirements for achieving it in the chosen way?
Option 5	Manipulate control variable Tjof CSTR	Tj is controllable
Option 6	Manipulate control variable Fj of CSTR	Fj is controllable

Figure 22. Planning table for (a) main goal, and (b) subgoal.

Figure 24 shows resulting control structure block diagrams for two solutions to our problem. The following symbols are used in the block diagrams.

- PV-process variable: C_{am}, T_{rm}
- SP-set point: C_{asp}, T_{rsp}
- CV-control variable: T_{rc}, F_{rc}, F_{jc}
- FF-feed forward nominal: T_r(bar), F_j(bar)

Figure 24(a) represents a plan utilizing option 1 of the planning table. In this case, the reactor concentration, Ca, is controlled by controlling (manipulating) the reactor flow, Fr, using a simple proportional-plus-integral controller. Figure 24(b) represents a plan utilizing options 2 and 6. In this case, two controllers are implemented in a cascade structure. For the master controller, Ca is controlled by controlling the reactor temperature, Tr, using proportional-plus-integral control. For the slave controller, Tr is controlled by controlling the jacket flow, Fj, using proportional control. For both controllers, a steady-state operating value of the control variable is used as a feed-forward nominal input.

The above procedure conceptually describes our approach to the development of the XIMKON Planner for control structure synthesis.

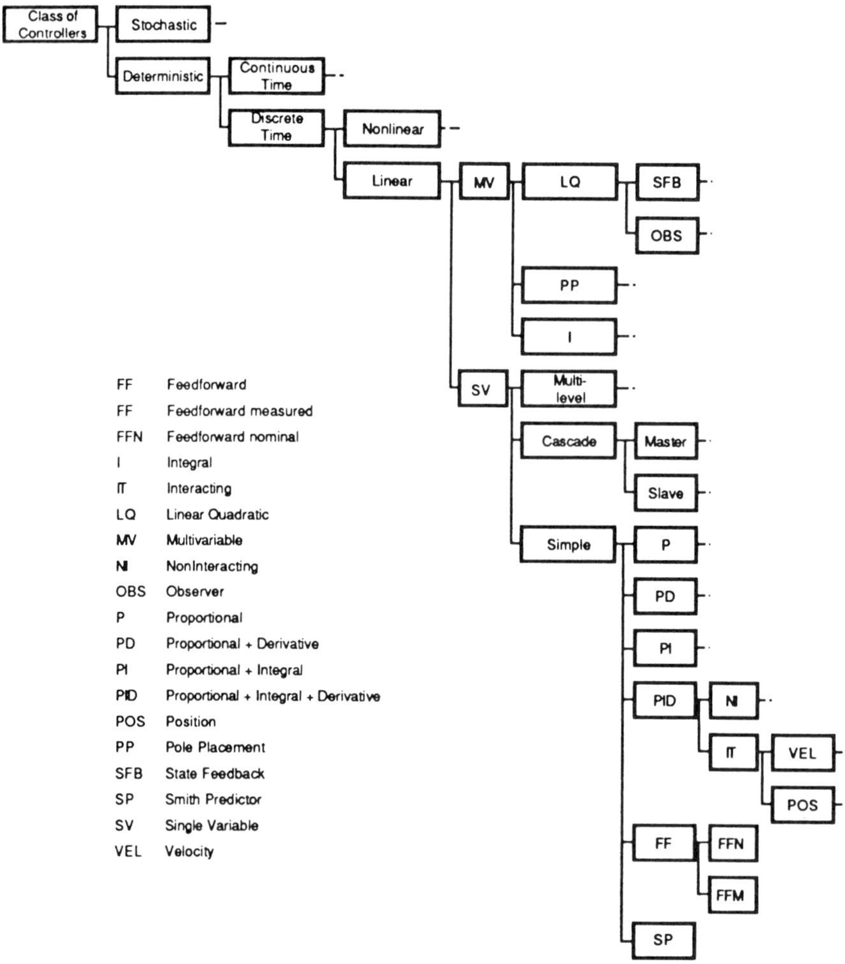

Figure 23. Subset of possible controllers.

5.2. *XIMKON Planner Framework*

Like XIMKON's user interface, an object-oriented approach with Objective-C is used to develop the XIMKON Planner for backward chaining planning. Also, as described earlier, the XIMKON Planner is based on some of the fundamental techniques common to many of the traditional AI planners (Wilkins, 1983; Cohen and Feigenbaum, 1982; Chapman, 1987).

Within the framework of expert system technology, a solution to a problem of the type described above calls for the development of a planner that generates a plan to accomplish the desired goal. The planner has a set

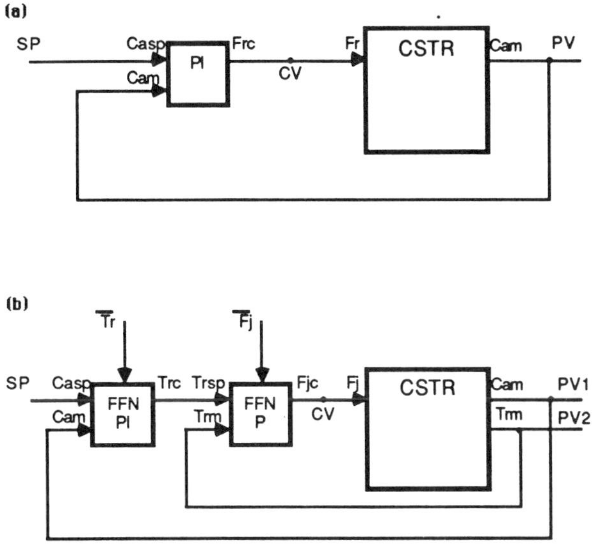

Figure 24. Control structure block diagrams of the reactor concentration control system.

"operators" (actions) that are employed to achieve goals. Generally speaking, an operator can be utilized for a particular goal if execution of the operator changes the "state of the world" by making the goal true. These changes to the state of the world are called the operator's *effects*. When an operator has requirements, or *preconditions*, that must be met before it can be executed, the preconditions become subgoals for which other operators may be employed.

These basic elements that make up the XIMKON Planner can be equated to our original questions that were posed in the above problem.

(1) What is wanted? . . . . goal/effect
(2) What are the ways to achieve it? . . . . operators
(3) What are the requirements for achieving it in the chosen way? . . . . preconditions

The planner is responsible for generating a planning tree by applying appropriate operators to the desired goal and newly created subgoals until a set of initial goals is reached, i.e., goals that already exist in the state of the world at the start of plan execution. The planning-tree generation can be perceived as a backward chaining or inverse problem solving task, because the planner starts at the desired goal and works its way down to the initial goals.

When the planning tree has been generated, the planner will attempt to search the tree to find one or more feasible plans that will solve the of problem. Multiple plans may be found if a goal has more than one operator that will make it true. A successful plan consists of an ordered sequence of operators which, when executed, will achieve the desired goal. In other words, the success of a plan depends on its ability to go from an initial state of the world to a state where the desired goal is true.

To determine a plan's success or failure, the planner will simulate the execution of the plan by starting with a list of initial states and recursively stepping through the plan, adding each operator's effects to the states list (if the operator is successful), until it gets to the last operator, which will add the desired goal to the states list. If any operator is not successful, the plan fails. The planner can also be given a list of rules that must be adhered to when a plan is evaluated. A plan fails if a rule is broken during the plan's simulated execution.

Actual execution of a plan can be performed by executing each operator's action with associated arguments. For control structure synthesis, our goal is to have operators' actions produce a block diagram of the control system by creating block(s) representing the control structure, and connecting the inputs and outputs to the process model block. This will allow the user to view and evaluate automatically generated block diagrams of the candidate control systems that the planner has generated.

5.3. XIMKON Planner's Object-Oriented Structure

The XIMKON Planner consists of two main object classes: 1) *Planner* and 2) *Operator*. The *Planner* is the executive for the planning system. Given a desired goal and a list of operator objects of class *Operator*, the *Planner* is responsible for generating the planning tree as described above. Using these two object classes, the basic planner was designed to be domain-independent.

To make the planner specific to a particular domain or application, the object-oriented technique of subclassing is utilized. A subclass of class *Planner* is created for the domain, and subclasses of class *Operator* are created for each type of operator used in the domain. The *Planner* subclass will create instances of all the *Operator* subclasses and add them to its list of operators used to generate the planning tree. The *Planner* subclass will also acquire the domain specific initial states and rules.

The class *Operator* is provided as a template from which all domain-specific operators are subclassed. Each *Operator* subclass will uniquely specify an operator name, action, effects and preconditions. A subclass of class *Operator* can in turn be subclassed, to allow the creation of an operator that will inherit the data defined in its parent operator, and introduce the new data that differentiates itself from its parent.

Figure 25 shows the IS-A class structure generated for the domain of

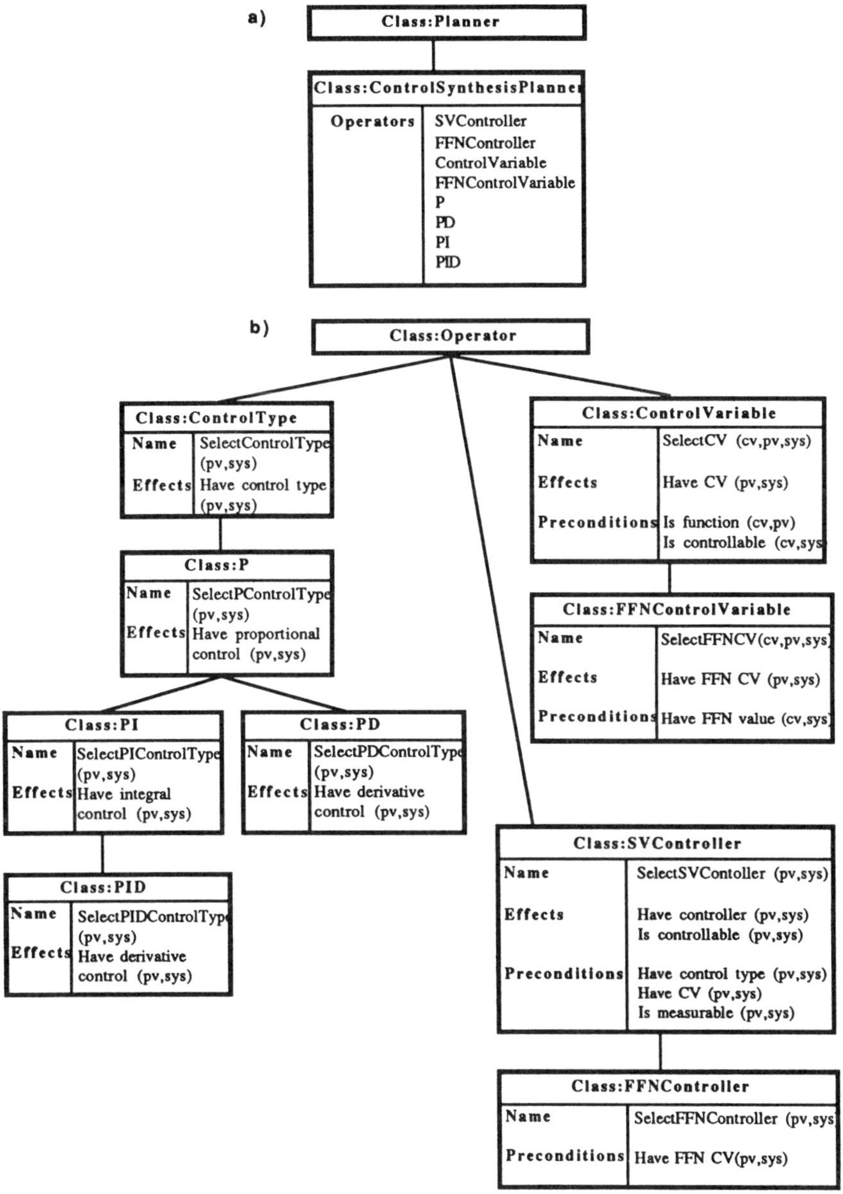

Figure 25. Planner class and subclass structure. (a) Planner. (b) Operator.

control structure synthesis. Each box represents a particular class and contains the pertinent instance variable data needed to generate the planning tree. The lines connecting the boxes define the superclass-subclass relationships. Figure 25(a) shows the subclassing of class *Planner* to create class *ControlSynthesisPlanner*. The new class will create instances of all the classes of operators used in our demonstration of control structure synthesis planning. Figure 25(b) shows the different classes of operators used. As an aid in understanding Figure 25, further description of the Operator *SVController* is as follows:

Class: *SVController*

Superclass: *Operator*

Name: SelectSVController (pv,sys)

Description: This class defines an operator for selecting a single variable controller for a process variable (pv) of a system (sys).

Effects:
(1) Have controller (pv,sys) - We have a controller for pv of sys.
(2) Is controllable (pv,sys) - pv of sys is now controllable.

Preconditions:
(1) Have control type (pv,sys) - A control type must be selected (P,PI, PID,etc.)
(2) Have CV (pv,sys) - A control variable to be manipulated must be selected.
(3) Is measurable (pv,sys) - pv must be a measurable output.

5.4. *XIMKON Planner Operation/Results*

The XIMKON Planner has been developed and tested using the operators described above and the goal of our original problem: Control Ca of CSTR. The other information required to run the planner are the initial states which define the state of the world at the start of plan execution. For the CSTR example, we have defined the following set of initial states.

• Is measurable (Ca,CSTR)	(Ca is a measurable output of CSTR)
• Is measurable (Tr,CSTR)	(Tr is a measurable output of CSTR)
• Is measurable (Tj,CSTR)	(Tj is a measurable output of CSTR)
• Is controllable (Fj,CSTR)	(Fj is a controllable input to CSTR)
• Is controllable (Fr,CSTR)	(Fr is a controllable input to CSTR)
• Is function (Tr,Ca,CSTR)	(Ca is a function of Tr)
• Is function (Fr,Ca,CSTR)	(Ca is a function of Fr)
• Is function (Fj,Tr,CSTR)	(Tr is a function of Fj)

- Have FFN value (Tr,CSTR) (Have feed-forward nominal value for Tr)
- Have FFN value (Fj,CSTR) (Have feed-forward nominal value for Fj)

Some of the initial states, such as "Is function ()" can be automatically extracted from the modeling equations. The initial states "Is measurable ()" and "Is controllable ()" pertain to those process variables that are feasible candidates for measuring or manipulating. Identifying those variables would involve a query session with the user.

We have also implemented a set of rules that restrict the choices of control types in certain situations. For a simple controller and for the master controller in a cascade structure, we want the control type choices to be either PI or PID. Also, for the slave controller(s) in a cascade structure, we do not want integral control at all, and we do not want derivative control if the master controller has derivative control. This results in the following rules syntax:

```
IF Have integral control
   NOT SelectPIControlType
   NOT SelectPIDControlType
ELSE
   NOT SelectPControlType
   NOT SelectPDControlType
ENDIF
IF Have derivative control
   NOT SelectPDControlType
ENDIF
```

A set of rules, as shown above, represents expert knowledge or expert preferences, which are defined and inserted into the planner to further enhance or customize the planner's selection process.

A subset of the generated planning tree for this problem is shown in Figure 26. The elliptical nodes of the tree represent goals and subgoals. The rectangular nodes represent operators used. This representation of the planning tree can be thought of as an AND/OR tree. The goal nodes are OR nodes which means we have the choice of going down one branch or the other, since either branch leads to an operator that will achieve the goal. The operator nodes are AND nodes, because all branches leading from an operator node represent preconditions, so all the branches must be traveled and proven successful before the operator can be applied.

The simple controller solution in Figure 24(a) is defined by the following plan of operators:

(1) SelectPIControlType (Ca,CSTR)

(2) SelectCV (Fr,Ca,CSTR)
(3) SelectSVController (Ca,CSTR)

The cascade controller solution in Figure 24(b) is defined by the following plan of operators:

(1) SelectPIControlType (Ca,CSTR)
(2) SelectPControlType (Tr,CSTR)
(3) SelectFFNCV (Fj,Tr,CSTR)
(4) SelectFFNController (Tr,CSTR)
(5) SelectFFNCV (Tr,Ca,CSTR)
(6) SelectFFNController (Ca,CSTR)

The cascade solution is also represented in Figure 26 as the highlighted path.

5.5. *Controller Parameter Synthesis and Performance Analysis*

Controller parameter synthesis involves the iterative task of tweaking or tuning the controller parameters such that the control system responds within desired specifications, during a particular performance analysis. For example, the planner generated cascade control system shown in Figure 24(b) contains three controller parameters that must be tuned. The master controller has a proportional and an integral parameter, and the slave controller has a proportional parameter. The control system maintains the reactor concentration at a prescribed operating level by manipulating the jacket flow. The problem is to tune the controller parameters using a step disturbance at the inlet concentration.

Figure 27 shows the block diagram of the control system generated in XIMKON. The CSTR block was modeled as described in Section 4, and the cascade control structure was synthesized by the XIMKON Planner, as described above. There are a number of performance analysis tools that can be used for parameter synthesis, including root locus analysis, frequency response and nonlinear least squares optimization of the time response. Tools such as root locus analysis, however, require a linearized model of the process. The linearized quadruple data generated from the CSTR model, via nonlinear simulation, is illustrated in the interactively drawn transfer function block diagram shown in Figure 28.

Given the control system described above, our goal is to implement a forward chaining planner in XIMKON that will utilize a performance analysis tool to automatically tune the controller parameters. The choice of

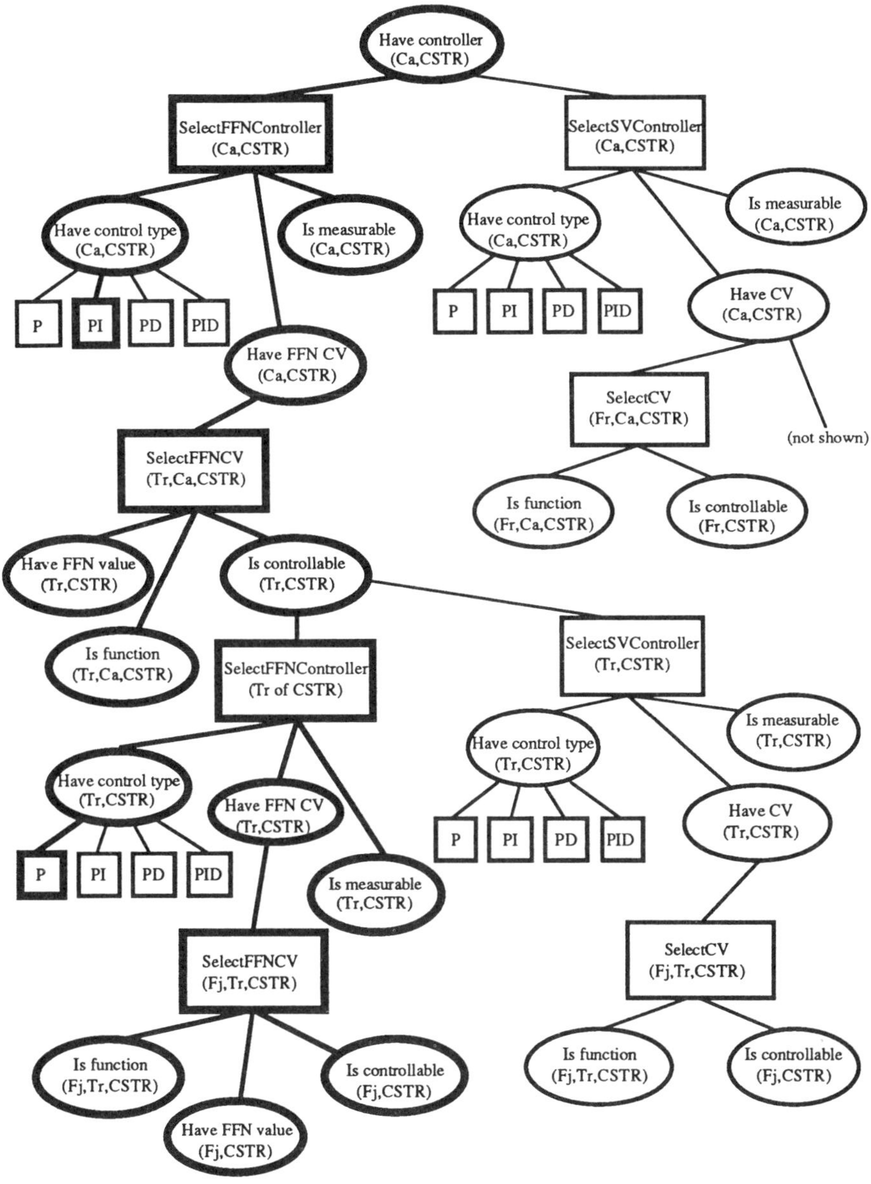

Figure 26. Subset of planning tree for control structure synthesis CSTR example.

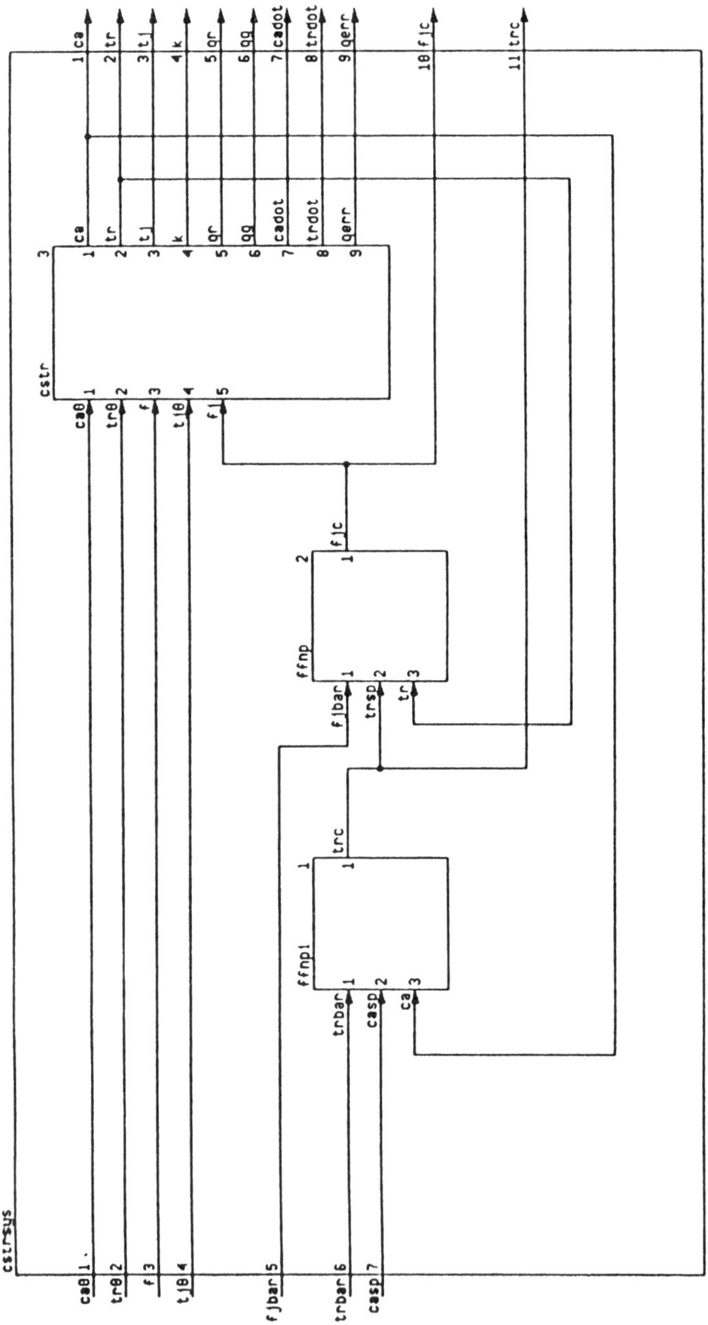

Figure 27. Block diagram of CSTR control system.

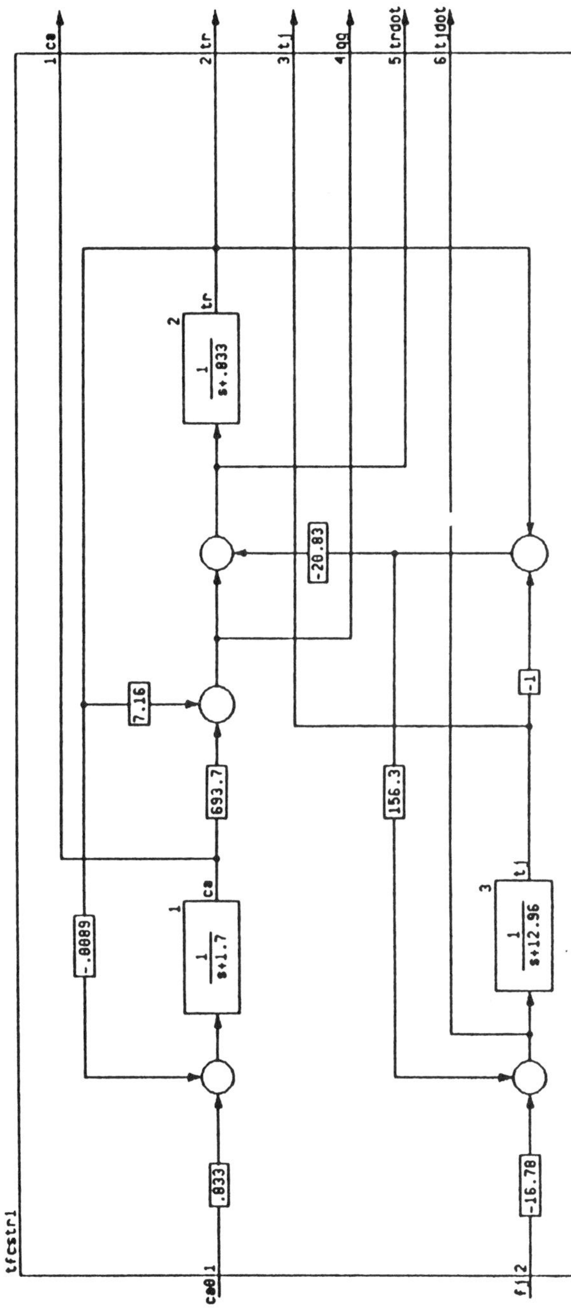

Figure 28. Transfer function block diagram of linearized CSTR model.

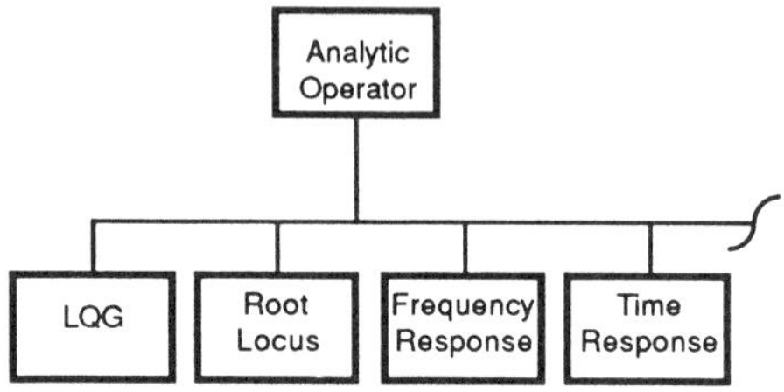

Name	LQG
Arguments	(A, B, C, D) Plant Quadruple Cm, Cd Measurement and Design Output Matrices QR Weighting Matrices
Preconditions	(A, B) Stabilizable Q≥0, R> 0, Symmetric
Effects	Stable Closed-Loop System State/Estimator Gains Computed
Action	Use XIMKON Commands CSFG ( ) CKEG ( )

Name	Time Response
Arguments	System Model Time Parameters Initial Conditions
Preconditions	System Interconnections Complete External Inputs/Outputs Defined
Effects	Simulation Time Response Data Computed
Action	Use XIMKON Commands CNSM ( ) - Nonlinear CTR ( ) - Linear

Figure 29. XIMKON planner analytic operators.

the analysis tool may be dictated by the user or by some defined rules resident in the planner. The structure of a planner operator provides an ideal template for inserting and automating a performance analysis tool. Figure 29 shows a subset of the analytic operators available to the planner and the structure for two of the operators: 1) an operator to compute the state feedback and estimator gains for optimal control, and 2) an operator to compute the time response of the system.

For the CSTR control system, the XIMKON Planner may use a root locus operator to synthesize the controller parameters for critically damped dominant time response. Figure 30 shows the time response of the reactor concentration to a step input in inlet concentration. The solid line shows the response of the system with initial controller parameters. The dashed line shows the response of the system with tuned controller parameters.

6. Conclusion

We have described XIMKON, an evolving expert simulation and control program. Object-oriented programming and AI planner technologies have been implemented in an effort to automate the modeling, analysis and

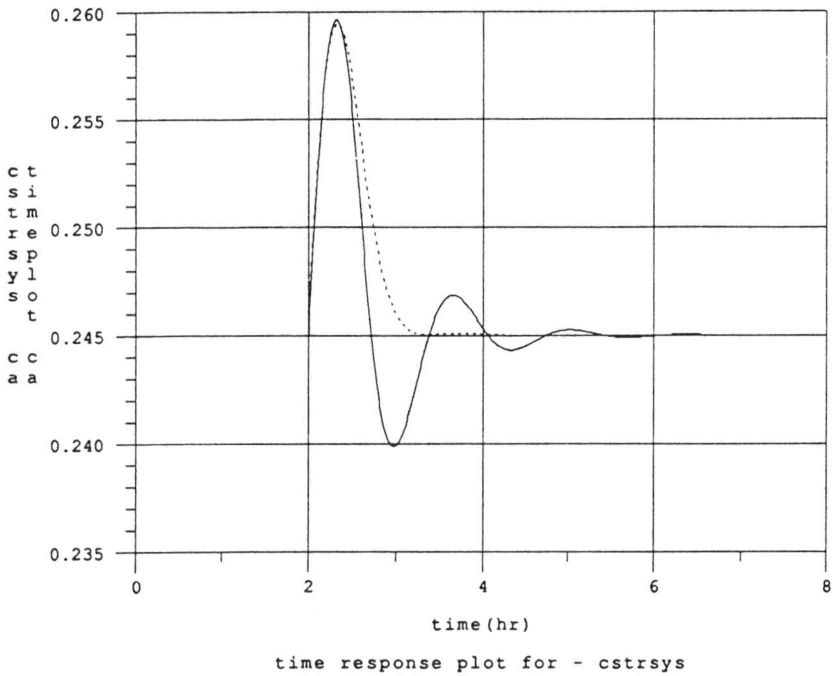

Figure 30. CSTR reactor concentration response

synthesis tasks. The early results of XIMKON's development serve to lay the groundwork for our goal of complete automation of the control system design process.

Our future work will focus on further development of the XIMKON Planner for the individual steps in the design process. For example, development of the planner for control structure synthesis will involve the addition of operators that will allow the planner to search over a wider range of control structures, including multivariable and optimal controllers. We also have plans for the development of a conceptual modeling stage for XIMKON's Expert Process Modeler and/or the development of an interface to other expert systems that provide conceptual modeling capabilities, such as Design-Kit.

Acknowledgments

Several investigators have contributed to the results presented here. In particular, the authors would like to thank J.K. Mahesh and M.D. Ward for their continued efforts in supporting the development of XIMKON.

References

Banerjee, J., *et al.* Data model issues for object-oriented applications. *ACM Transactions on office Information Systems*, 5 (1), p 66 (1987).

Beech, D. *Groundwork for an Object Database Model, in Research Directions in Object-Oriented Programming*, editors: B. Shriver and P. Wegner, MIT Press, Cambridge, MA (1987).

Borow, M. S., *et al.* Navy digital flight control system development. Honeywell report 21857-FR GAP (1972).

Brodie, M., and Mylopoulos, J. *On Knowledge Base Management Systems*, Springer Verlag (1986).

Chapman, D., Planning for Conjunctive Goals. *Artificial Intelligence*, 32 (3) (1987).

Cohen, P., and Feigenbaum, E. Strips and abstrips. Chapter 15B of the *Handbook of Artificial Intelligence*, Vol 3, Los Altos, CA: W. Kaufman Publishing Co., pp. 551–556 (1982).

Comegys, G. L., and Hanson, C. R. Integrated flight trajectory control development program. AFWAL-TR-81-3077 (1981).

Cox, B. *Object-Oriented Programming—An Evolutionary Approach*, Addison-Wesley, Reading, MA (1986).

Garcia, C. E., and Morari, M. Internal model control. *Ind. Eng. Chem. Process Des. Dev.*, 21 (1982).

Govind, R., and Powers, G. J. Control system synthesis strategies. *AIChE Journal*, 28 (1) (1982).

Horn, B. C. On-line optimization of plant utilities. *Chemical Engineering Progress*, pp. 76–79 (1978).

Jones, A. L., *et al.* Integrated control design techniques. AFWAL-TR-81-3074 (1981).

Joseph, B., and Brosilow, C. B. Inferential control of processes. *AIChE Journal*, 24 (3) (1978).

Kleinman, D. L. Optimal control of linear systems with time-delay and observation noise. *IEEE Trans., Auto, Cont.* (1969).

Konar, A. F. Precision weapon delivery. *Trans. First NWC Symposium on the Application of Control Theory to Modern Weapon System* (1973).

Konar, A. F. Simulation and control workshop notes. Honeywell Technology Strategy Center (1985).

Konar, A. F. Methodology for large-scale process control system design. *IFAC Proceedings Series*, No. 7 (1987).

Konar, A. F. and Mahesh, J. K. KONPACT program manual. Honeywell reports 77SRC42 and 77SRC43 (1977).

Konar, A. F., and Mahesh, J. K. Computer-aided analysis and design of large-scale control systems. *IFAC Workshop Proceedings*, Ankara, Turkey (1982).

Konar, A. F., and Ward, M. D. Development of weapon delivery models and analysis programs. AFFDL-TR-71-123, Vol. I, II, III (1972).

Konar, A. F., Mahesh, J. K. and Jacobson, R. D. SIMKON command description list. Honeywell Corporate Technology Center (1983).

Konar, A. F., Mahesh, J. K. and Kizilos, B. Digital flight control systems for tactical fighters. AFFDL-TR-73-119.

Konar, A. F., Mahesh, J. K., Stone, C. R., and Hank, M. Active control synthesis for flexible vehicles. AFFDL-TR-75-146, Vol. I, II, III (1976).

Konar, A. F., et al. Dynamic process simulation and control. Technical Report No. G87375, Honeywell Corporate Systems Development Division (1987).

Krashner, G., and Pope, S., *A Cookbook for Using the Model-View-Controller User Interface Paradigm in Smalltalk-80*. ParcPlace Systems (1988).

Leone, H. and Henning, G. *Design-Kit, A User's Guide*, LISPE, MIT, Dept. of Chemical Engineering (1988).

Luyben, W. L. *Process Modeling, Simulation, and Control for Chemical Engineers*. McGraw-Hill Co., New York (1973).

Mahesh, J. K. and Konar, A. F. Interactive flight control system analysis program. NASA contract report NAS 1-16438, Vol. I, II, III, IV (1983).

Marshall, J. E. Control of time-delay systems. *IEEE Control Engineering*, Series 10 (1979).

Morari, M., Arkun, and Stephanopoulos, G. Studies in the synthesis of control structures for chemical processes. *AIChE Journal*, 26 (2) (1980).

Murtagh, B. A., and Saunders, M. A. *MINOS 5.1 User's Guide*, Technical report SOL 83-20R, Stanford University, Stanford, CA (1987).

Ogunnaike, B. A., and Ray, W. H. Multivariable controller design for linear systems having multiple time delays. *AIChE Journal*, 25 (6) (1979).

Pascoe, G. Elements of Object-Oriented Programming, *BYTE Magazine*, August, 11 (8) (1986).

Ray, H. W., and Szekely, J. *Process Optimization*. John Wiley & Sons, New York (1973).

Romagnoli, J. A., and Stephanopoulos, G. On the rectification of measurement errors for complex chemical plants. *Chem. Eng. Sci.*, 35 (5) (1980).

Schuldt, S. V., and Smith, F. B. An application of quadratic performance synthesis techniques to a fluid cat cracker. *J.A.C.C.*, Paper No. 3-E4 (1971).

Stephanopoulos, G. *Chemical Process Control*. Prentice Hall, New Jersey (1984).

Stephanopoulos, G. Future of expert systems in chemical engineering. *Chemical Engineering Progress*, pp. 44–51 (1987).

Stephanopoulos, G., and Huang, H. P. The 2-port control system. *Chemical Eng. Sci.* (1986).

Stepstone. *Objective-C 3.3 Reference Manual*. Productivity Products International, Inc. (1988).

The DOMAIN/Dialog User's Guide. Apollo Computer, Inc. (1987).

The DOMAIN System Call Reference. Apollo Computer, Inc., Vol. 1 (1985).

Trankle, T. L., and Markosian, L. Z. An expert system for control system design. *Proceedings of IEEE International Conference on Control* (1985).

Tung, L. S., and Edgar, T. F. Analysis of control-output interactions in dynamic systems. *AIChE*, 27 (4) (1981).

Wilkins, D. Representation in a domain independent planner. *Proceedings of the Eighth International Joint Conference on Artificial Intelligence*, Karlshrue, West Germany, Vol. 2, pp. 733–737 (1983).

8 Exothermic Batch Chemical Reactor Automation via Expert System

DANIEL R. LEWIN AND RAM LAVIE

Department of Chemical Engineering
Technion, I.I.T.
Haifa, Israel

Abstract

This chapter describes a proposed automation system that has been designed to supervise and control an exothermic batch chemical reactor. The proposed control system has a learning mode that emulates operator reasoning, gradually abstracting essential features of the process which are then used to upgrade control, while satisfying a safe "optimal" trajectory. Analysis is utilized in rule formulation, through generalized heat and mass balances. Implementation on a simulated plant indicates excellent adaptability.

1. Introduction

Batch processing is probably the oldest form of technological activity. The characteristics of a batch process, such as activity planning, object preparation, start-up, path following and ending, appear to fit naturally into the way man thinks. This has changed remarkably little from the times when prehistoric man lit a fire to cook his meal, to modern batch processing of chemicals. Unlike the relatively recent concept of continuous processing (which needs much abstraction and whose development proceeded with the help of science in general, and system theory in particular), batch processing enjoys the privilege of decomposition in time. Decomposition into manageable tasks allows the human operator to participate in the process decisions, to guide and control it, and also to improve it from one batch to the next by empirical learning.

ISBN 0-12-480575-2

The advent of modern continuous processing gradually eroded the status of traditional batch processes. Those were mostly replaced, wherever possible, by their continuous equivalent, thus minimizing the expenditure of two vital resources: manpower and time. However, this evolution had to face its eventual limit. This was brought about by economics (relative costs and profits) or by performance considerations; some processes are too complex or poorly understood to permit automatic continuous generation of a product at strict quality standards. This is not to say that automation does not apply to batch processing. But batch process control evolved along lines that are distinctly different from those prevailing in continuous, steady-state, process control. The issues concerning the maintenance of a steady state, distributed in space, in a continuous process (such as feedback dynamics, disturbance rejection and interation decoupling) play a relatively minor role in batch process control where scheduling, interlocking and guidance along a path in time prevail. Oddly enough, the digital computer, which appeared on the scene in the second part of this century, while ideally suited to deal with sequential logic, was first developed as a computation tool, and then applied at great effort to the control of continuous processes, before being harnessed to the control of batch processes (in the form of Programable Logic Controllers). This did provide substantial automation of those elements in batch processing having to do with (possibly complex combinations of) simple tasks such as pump or valve motor activation and sequencing, etc. Automation of more complicated decisions had to wait for the development of Artificial Intelligence tools which are now nearing some form of maturity and are increasingly applied in the form of Expert Systems.

In this chapter, a typical batch chemical reactor and the essential problems associated with its control are delineated. One particular aspect, that concerning the thermal reactivity of the chemical reaction and its exponential effect on reaction rate, poses a control problem of considerable difficulty and importance. Controlling the temperature in such a batch reactor implies a capability to provide a wide range of heat supply or removal capacity. Conventional feedback control systems are not usually fit to handle such a wide range of operating conditions. Worse still, economical design of the reactor will not bear such extreme capacity, so that the actual cooling/heating capacity will be constrained. Manual guidance of the operating conditions is often possible. This appears to lead naturally to an expert control system approach. Analysis of the problem indicates the desirability of its decomposition in time with flexible bounds. Analysis of the method leads to the definition of process states, conditions and parameters used in a generalized information feedback scheme, handling immediate control actions within the time frame of a particular batch as well as learning from one batch to the next, such as to improve performance with time and to overcome longer time scale changes in the process and its environment.

The chapter is organized on the following lines. Firstly, in Section 2 we present a review of alternative approaches that have in the past been applied to the control of batch reactors. In Section 3 a generalized presentation of the problem is offered, in which we identify the elements of the processing sequence that characterize all batch reactor control problems. The expert system approach that we have taken to the solution of the control problem is presented in Section 4. Integrated into the rule-based control system are conventional control algorithms which can be invoked as required. These are briefly discussed in Section 5. In Section 6, we present a typical example, demonstrating by simulation how the expert system strategy relates to practice. Finally, we conclude with a statement of our future goals and research directions.

2. Review

The previous section has summarized the problem of controlling exothermic batch reactors in which the removal of the heat released by the reaction poses a significant task. The safety aspects invovled have led to significant academic interest (Lake, 1977; Bonvin and Saner, 1988; Hugo *et al.*, 1988; Gygax, 1988; Lewin and Lavie, 1989). As in all control problems, the more information we have about the system to be controlled, the better the results that can be expected. However, as in all batch processes, the process parameters may be time varying, and because of the transient nature of the reactor trajectory, accurate process knowledge cannot always be assumed. The highly nonlinear dynamics brought by the rate of reaction compounds the issue further. This explains why batch reactor control has been widely studied as a potential beneficiary of alternative control strategies rather than of simple conventional feedback control.

An excellent review on the current status of batch process control was presented by Juba and Hamer (1986). Deterministic applications have covered a broad spectrum, from combinations of fixed parameter feedforward and feedback schemes to the application of adaptive controllers. We should note at this point that the variation of system parameters, even if significant, does not in itself necessarily mean that constant parameter controllers cannot be tuned to do the job. Indeed, Åström (1983) criticizes the blanket use of adaptive controllers for applications where well tuned PID feedback could have performed as well. Robust process control (Morari and Zafiriou, 1989) has emerged as a viable alternative, even for open-loop unstable processes in which there is significant parametric uncertainty, a classification that applies to our particular problem. Jutan and Uppal (1984) report the successful theoretical testing of a controller where feed-forward action compensates for reaction heat release, which is considered as an external disturbance. However, most of the recently reported controllers proposed for batch reactor systems have involved adaptive schemes, and we give here only a flavor of the published material. Theoretical studies on the applications of adaptive controllers to batch reactors have been made by Shah and his co-workers (Kiparissides and Shah, 1983; Tzouanas and Shah, 1985) and by Takamatsu *et al.*, (1985). The latter based their design method for a self-tuning PID controller on Internal Model Control methods. There are also reports of industrial applications: Åkesson (1985) reports the use of a commercial ASEA NOVATUNE adaptive regulator to an operating batch reactor, while Niederlinski *et al.* (1985) describe the use of adaptive methods to a PVC batch convertor.

The implementation of an adaptive scheme is included to ensure robust behavior in normal operating conditions without the need for continuously retuning the feedback algorithm. In any case, the start-up ramp must be handled by some alternative method, since a conventional scheme will still only track a desired setpoint. Philips *et al.* (1988a, 1988b) considered the control of the start-up of a fluidized bed reactor heated by a sand bath. Since no cooling was available, the reactor was controlled in the initial stages of the operation by the application of a pulse of a duration predicted to project the temperature trajectory exactly onto the target level. This was done assuming a second-order model to describe the dynamics of the sand bath. The fine tuning of the temperature trajectory was then handled by an adaptive control scheme. Since the sand bath dynamics are indeed near-linear and of second order, it is not surprising that the prediction scheme to determine the time at which the heating pulse is extinguished apparently works well. In our case, the start-up trajectory largely depends on the nonlinear reaction kinetics whose parameters are assumed to be unknown. This uncertainty suggests a heuristic approach to controlling the start-up path.

The application of expert system technology to process control in general (Bailey, 1986; Umeda and Niida, 1986; Kane, 1986) and to the control of batch reactors in particular (Nisenfeld and Turk, 1986; Blickley, 1987; Cram and Clarke, 1988) has been with us for some time. By definition, expert systems are meant to handle poorly defined problems which do not lend themselves to conventional control methods because of their complexity and for which some expertise in human operation is available. The control of batch reactors is a problem particularly suited to these conditions. Various approaches to the problem have been presented in the literature, including at one extreme the application of fuzzy logic (King and Mamdani, 1977; Tong, 1977; Zadeh, 1984), where uncertainty is expressed in a statistical framework. At the other extreme, analytical reasoning has been applied in "Robust Optimal Control" (Sripada *et al.*, 1986), where time-optimal (bang-bang) control is used. This, however, necessitates a substantially accurate model.

The approach suggested here tries to steer a middle course whereby general characteristics of the process are clearly defined, but no detailed model is relied upon. A rule-based expert system is applied to diagnose the current status of the process and decide which control action is most appropriate. It also gradually learns from previous experience and delineates the most appropriate future trajectory. The choice of controller action is selected from an armory of possible alternatives, including the implementation of either a robustly tuned or an adaptive feedback scheme. Thus, this approach matches the control strategy to the prevailing conditions. The alternative conventional strategies that are to be invoked under normal operating conditions are outlined in Section 5.

3. The Generalized Batch Reactor Control Problem

Generalization of batch chemical reactors as a single category is not easy since their crucial element, the chemical reaction, may differ widely from one case to another, with implications for objectives and performance criteria. Thus the production of polymers or other reactive materials involving highly exothermic and sometime autocatalytic reactions poses problems that are different from those related to more benign reactions aiming at high selectivity or product purity. Nevertheless, an attempt at the formulation of a conceptual model, including sufficient elements such as to allow the representation of a wide variety of typical situations of interest as special cases, is well justified. In so doing, it is worth separating the relatively simple, preparatory and post-processing stages (such as recipe formulation, reactant loading and product discharging). These steps are characteristically process-specific and largely independent of the central issue concerning the carrying out of the chemical reaction itself, which will be the subject of the following discussion.

The choice of an expert system approach to our problem implies that there are some aspects of the problem that are not fully known or understood or that are not easily formulated in a deterministic mathematical form. This is not to say that all parts of the problem need belong to this category. Indeed, one expects a well balanced approach to utilize to the fullest a deterministic mathematical model for those parts of the problem where such is feasible, and let the expert system do what it is best fit to do. The limits between the parts of the problem that are best handled by either approach are often arbitrary and are determined by the judgement of the modeler. This is particularly evident in the formulation of assumptions that underly every deterministic model.

The model will be based on the following assumptions:

(a) The reactor content is well mixed at all times. While this is not necessarily always the case in practice (in particular in large reactors and in cases where the reaction mass is viscous or otherwise difficult to mix), this is at least an objective.
(b) Jacket heating/cooling, with heating/cooling medium being well mixed in the jacket.
(c) The operating procedure along the reaction path follows a predefined recipe concerning amounts and timing for the addition of reactants, solvents, catalysts, moderators etc. The amounts (n_{ik}) and timing (t_k) may be defined as a preprogrammed sequence, or as a control action in response to conditions developing along the reaction path (feed-back). This also applies to the amount and timing for the heating/cooling rate in the jacket.

This leads to the following simplified material and energy balances:

$$\frac{dn_i}{dt} = \delta(t - t_k)n_{ik} + \sum_{j=1}^{J} r_j(n_i, T, P), \qquad n_i(0) = 0 \tag{1}$$

$i = 1, \ldots, I$ (species)

$j = 1, \ldots, J$ (reactions)

$k = 1, \ldots, K$ (shots)

$$\frac{dT}{dt} = \delta(t - t_k)n_{ik}\frac{n_{ik} C_{pi} T_k}{\Sigma_j C_{pj} \Sigma_i n_{ik}} + UA(T_i - T) + \sum_{j=1}^{J} \Delta H_j r_j(n_j, T, P), \qquad T(0) = T^0 \tag{2}$$

$$\frac{dT_j}{dt} = \gamma_c U_c(T_c - T_j) + \gamma_h U_h(T_h - T_j) - \beta UA(T_j - T), \qquad T_j(0) = T_j^0 \tag{3}$$

$$P = \Psi(n_i, T) \tag{4}$$

$$\xi_l = \phi_l(n_i, T, P), \tag{5}$$

where

$\delta(t - t_k)$ = unit impulse function,

n_i = the amount of species i at time t,

n_{ik} = the amount of species i added to the reactor at time t_k,

r_j = the rate of reaction j,

T = the temperature of the reacting mixture,

T_k = the temperature of shot added at time t_k,

ΔH_j = the heat of reaction j,

U_c = the desired cooling level at time t, $0 \leq U_c \leq U_c^{max}$,

U_h = the desired heating level at time t, $0 \leq U_h \leq U_h^{max}$,

P = the pressure,

ξ_l = one of l reaction extent indicators.

The model just defined presents a generalized framework that can be used to analyze the behavior of a typical batch chemical reactor without specifying the particular reaction system. The expert system could use the model to *abstract* qualitative measures for such unknown or uncertain quantities as heat of reaction or reaction intensities. This approach will be apparent in the example given in Section 6, where a simplified mathematical model is used to construct a measure of the heat of reaction based on observations of the process state and control variables.

Major considerations in the formulation of a desired reaction path include (see Figure 1)

(a) Starting up the reaction. The transition from cold start to the level of "optimal" reacting conditions.
(b) Bringing the reaction to substantial completion at "optimal" conditions.
(c) Ending the reaction. The transition from "optimal" conditions to safe discharge conditions.

This defines three zones (or regimes) along the life span of a reaction. It is generally desired to proceed, all along the reaction path, at the fastest rate that does not compromise the "well being" of the product or that of the reactor and its environment. The initial "start-up" zone is usually the most critical in terms of the impact of the path followed on the overall performance of the reactor, as well as in terms of the difficulty in controlling this path. This transition usually spans a relatively wide range of

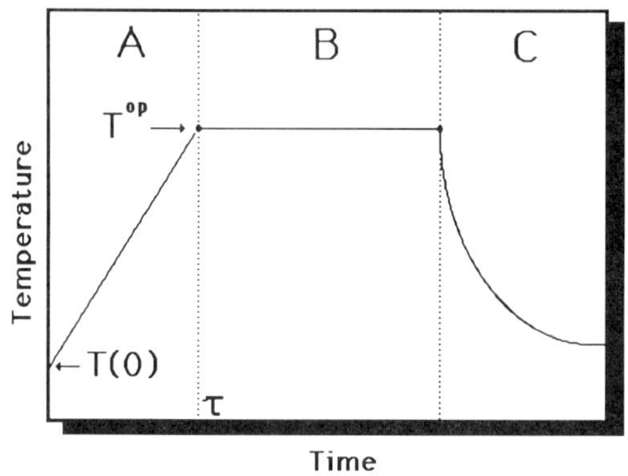

Figure 1. Schematic diagram showing the three stages of a batch reactor temperature trajectory: A, start-up, B, nominal operating conditions, and C, termination.

temperature and pressure. This makes it very difficult, if not impossible, to guide it by means of conventional, fixed-parameter, servo feedback control except with heavy manual participation of a human operator. Accordingly, much of our discussion will deal with the problems associated with the control of the reaction within this initial zone.

4. An Expert System Approach

An expert system for the control of a batch reactor should include

(a) Automated, deterministic, logic activation of all fixed preprogramable actions, such as activation of pumps and valve motors, alarms, reports, etc. This belongs to conventional automation as presently widely practiced, and is not dealt with here in any detail.

(b) Control of the process over spans of time where operating conditions are maintained within relatively narrow bands by means of conventional feedback control, possibly enhanced by adaptive and robust methods. This is briefly discussed later.

(c) "Intelligent" guidance of the process along a path yielding desired performance by mimicking manual operation by a skilled operator. This requires both a basic tactical scheme handling actions within the time scale of a single batch, and a strategic "learning" mechanism tailored to improve control from one batch to the next. This is the subject of this chapter.

In manual operation, the operator would intervene with heating or cooling actions following preset operating instructions, further modified in the field on the basis of qualitative and quantitative reasoning using intuition and experience. Concepts such as "ignition point," "run-away conditions" and "reaction completion" (which could well be qualified if a full and accurate model of the process and exact measurements of all variables and parameters were available at every instant), are sensed and judged by the operator mostly through indirect signs and measurements. The automatic expert control system should mimic this manual operation procedure while also enhancing reasoning, where possible, by

(a) qualitative analysis of trends,
(b) learning about changes in parameters,
(c) optimizing reaction trajectory.

The control strategy used is based on a set of heuristic rules, corrected in time through learning. The rule-base objective is to force a trajectory to rapidly attain a desired "optimal" temperature set point while safely avoiding runaway conditions and excessive over-shoot. These objectives are defined in crude terms capturing the essential desired path characteristics in terms of events and conditions. Some events can be quantified while others may be qualitatively and vagely defined. For example an overshoot in temperature can clearly be expressed as a percentage of a range. On the other hand a term like "reaction ignition" can be used loosely to indicate that the rate of chemical reaction at the particular prevailing conditions is such that it is expected to lead to a continuation of self-sustained reaction without any further control intervention.

Some path characteristics can be approximately quantified within bounds. Take, for example, the approach path from initial start-up to the desired operating level. Considering temperature as the most critical variable to be watched and controlled, a "family" of potential feasible paths in the temperature-time plane can be used to define a desirable path characteristic in terms of an "envelope" (see Figure 2). The definition of this envelope relies on some, at least qualitative, understanding of the process. First, heat is applies to raise the temperature of the reacting mixture to, or close to, "ignition" at the approximate temperature T_i. True, the exact path followed to this point may depend on composition, but for a given reactor the initial composition will mostly be fixed and unequivocal application of the full heating capacity at this stage can be taken to mean that the initial section of the temperature path will, to this point, always be "the same." From here on, the path followed will depend on the control actions taken. In Lewin and Lavie (1989) it is shown that constraints on the one hand, and stability considerations on the other, delineate bounds on the temperature path during start-up. This maps a zone within which one can expect the desired start-up path to be. Having characterized the "temperature set

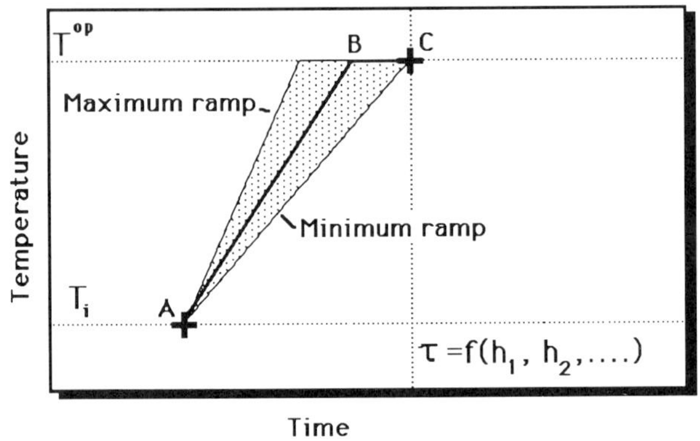

Figure 2. The start-up trajectory.

point" as a zone, one may use it as it is in formulation of the control rules by means of fuzzy logic, or, alternatively, one may arbitrarily assign a central, sharply defined path to be used in the rules. For example, one may base the rules on the central path ABC in Figure 2, using the argument that this approximates best a path that reaches the final desired temperature level tangentially.

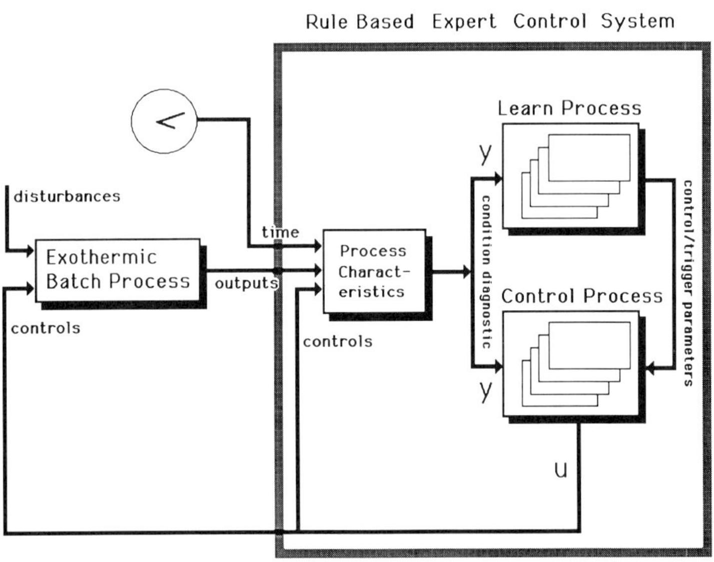

Figure 3. Automatic expert control system for batch reactor.

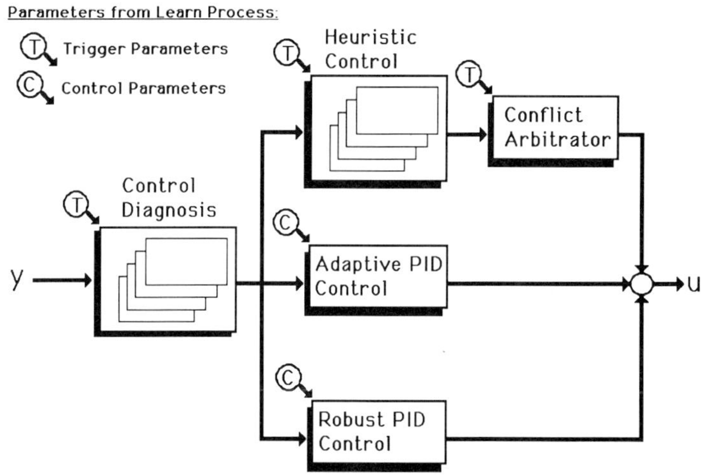

Figure 4. Control process.

4.1. Overall Expert System Description

The approach taken is illustrated by the block flow diagram in Figure 3. The process outputs (e.g., temperatures and concentrations) together with the current value of the computed control actions are used to classify the current status of the reactor on the basis of reasoning based on, among other things, deterministic process knowledge. At each sample interval, the process state is diagnosed as a *condition* (e.g., "too cold for this time," "warming too fast for this temperature and time," or "reaction rate developing too fast," etc.) by comparing a flexible grid of criteria to the output y:

$$y = \{\xi(k),\ T(k), \frac{dT}{dt}(k),\ \Delta H(k),\ t(k),\} \tag{6}$$

at each sample instant k.

This information is then used to drive two distinct processes:

(1) The information is used directly in the *control process* whose purpose is to generate the most appropriate control response to the current plant state. The structure of the *control process* is shown in Figure 4.
(2) The *condition diagnosis* is also used to feed a *learning process* where the values of control and trigger parameters are updated in the light of new information. This is shown in detail in Figure 5.

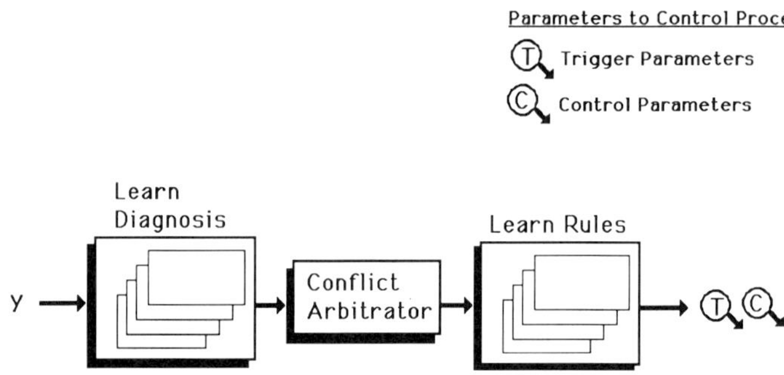

Figure 5. Learn process.

4.1.1. The Control Process (Figure 4)

To each *condition diagnosis* corresponds a predefined control action. For example, the rule controlling the temperature rise of the reactor contents to the desired operating level (Rule 2) defines controller action according to the location of the measured output relative to the grid shown in Figure 6 as well as estimated rate of heat release (see Section 6). The grid

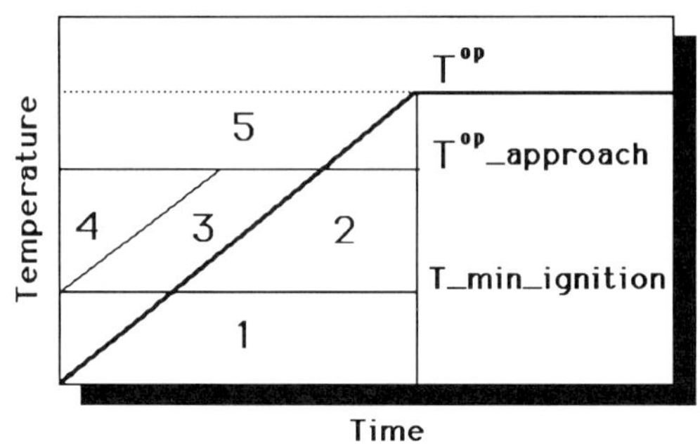

Zone	Classification	Action
1	No reaction	U_h= high_heat
2	Slow reaction	U_h= low_heat
3	Reaction ignited	U_h= U_c= off
4	Ignited but T low	U_c= low_cool
5	Ignited and close to T^{op}	U_c= high_cool
all	Excessive heat release	U_c= high_cool

Figure 6. Grid of conditions for control Rule 2.

Table 1. Sample control rules (simplified).

Rule	Description	When Activated
2	Guide start-up by temperature and ΔH	$t < \tau$
3	Watch for "potential runaway condition"	$(dT/dt > \dot{T}_3)$.and.$(T > T_3^1)$.or.$(T > T_3^2)$
4	PID Control	$(t > \tau)$.and. "no emergency condition"
5	Watch for "potential blow out condition"	$(dT/dt < \dot{T}_5)$.and.$(T < T_5^1)$.or.$(T < T_5^2)$

lines are delineated following trigger parameters whose values can be modified by learning.

Various control strategies that can be involved by the expert system are given in Table 1. We see that Rule 2 is the unique control for the initial heating-up period to time $t = \tau$, but during normal operating conditions, the remaining three rules are alternatives implemented according to the observed state of the reactor. If necessary, multiple firing of rules is checked by a *conflict arbitrator*, which will permit the most relevant rule to act. In order to remove the need for robustly tuning the PID control algorithm from batch to batch, either a robustly tuned controller or an adaptive scheme should be implemented; these alternatives are discussed in detail in Section 5. The values of the triggering parameters actuating these rules ($\dot{T}_3$, T_3^1, T_3^2, $\dot{T}_5^1$, and T_5^2) and the control action intensities applied are modified by learning in the light of *condition diagnosis*. The *learning process* is discussed next.

4.1.2. Learning Process (Figure 5)

The expert system incrementally infers desired values for the control rule triggering and intensity parameters (e.g., T_3, low-cool etc.) following the performance of the results of the current and previous batches resulting from previous actions. After waiting for a prescribed time interval in order to give the system a chance to respond to an applied control action, the process history is then labeled according to its characteristics, e.g., "response is too slow and exhibits offset," "response has excessive overshoot," or "response is too close to runaway," etc. Table 2 summarizes the complete family of *condition diagnoses*. Each history characteristic triggers an adjustment of the control rule parameters, calculated to improve performance in the next run. Thus the control scheme handles changing process conditions (such as fouling) through their effect on performance (*problem domain*).

Table 2. Sample family of condition diagnosis.

When learning occurs	Symptoms	Condition diagnosis
at $t = 1.5\tau$	Maximum temperature up to this time 10% greater than T^{op} and trajectory still not returned to T^{op}	"Excessive overshoot" (Condition L1.1)
	Maximum temperature up to this time 20% greater than T^{op} and trajectory still not returned to T^{op}	"Temperature runaway" (Condition L1.2)
	Neither Condition 1.1 nor 1.2 triggered at $t = 1.5\tau$	"Performance acceptable but could be improved"
at $t = 1.7\tau$	Maximum temperature up to this time less than 80% of T^{op}	"Insufficient heating" (Condition L2.1)
	Maximum temperature up to this time 10% greater than T^{op} and trajectory still not returned to T^{op}	"Excessive overshoot" (Condition L2.2)
	Maximum temperature up to this time 20% greater than T^{op} and trajectory still not returned to T^{op}	"Temperature runaway" (Condition L2.3)
after $t = 1.7\tau$	Maximum temperature up to this time 5% greater than T^{op}	"Slight overshoot" (Condition L9.1)
	Maximum temperature up to this time 10% greater than T^{op}	"Late excessive overshoot" (Condition L9.2)
	Maximum temperature up to this time 30% greater than T^{op}	"Late temperature runaway" (Condition L9.3)

4.1.3. Advisory Functions

The control system provides the operator with advisory information on the state of the reactor (identifies *ignition*, probability of future runaway conditions, etc.) as well major events as they occur (which rule has been triggered and when learning has modified the rule-base).

5. Associated Conventional Control Strategies

In this chapter we propose the use of an expert system that will "learn" the "best" temperature trajectory though experience with on-line reactor operation. This desired trajectory is then maintained by a hybrid control scheme in which conventional feedback control maintains the set trajectory in normal operating conditions with the expert system taking over with "shots" of either coolant or heating medium in the event of upsets (e.g., severe overshoots, reaction extinction, etc.). One of the reasons for unacceptable trajectory tracking could be a badly tuned controller. Two alternatives that can improve feedback tuning are robust fixed parameter controllers or the application of adaptive methods. The current status of conventional control strategies has been presented in Section 2; we shall now present a preliminary report concerning the implementation of the methods in the expert system.

5.1. Robust Feedback Control

Control design is a model-based activity. The type of models used span the complete spectrum from partial differential equations solved on a real-time simulator in parallel with the process to "rule-of-thumb" tuning rules for PI/PID controllers, which imply first- or second-order lag approximations of the process. Even the most sophisticated model is still only an approximation of the process to be controlled, so the designer should ensure that the controllers he or she designs be as insensitive as possible to modelling error (i.e. the controller must be *robust*). The relationships between designed controller type, model uncertainty, and achievable closed-loop performance can be elucidated by using the concept of Internal Model Control (Morari and Zafiriou, 1989).

The IMC design procedure consists of two steps. First, the controller is designed to satisfy given nominal performance specifications, thus tacitly assuming that the nominal linear model proposed for the process is accurate. The second step is to append a low-pass filter, in order to attenuate the effect of model uncertainty, which usually increases with frequencies. This has the effect of *detuning* the controller, or in other words, reducing the controller gain. In open-loop stable systems, this procedure always leads to stable performance for at worst, for large modeling uncertainties, the controller gain could be reduced to insignificance. However, in exothermic batch reactors, which are open-loop unstable processes, there will always exist some lower limit to the controller gain that guarantees stability. Thus the controller gain will have some lower limit dictated by the uncertainty of the dominant process pole (which can in principle be either negative or positive since the reactor will be stable during the portion of the batch below the kinetic ignition point) and some higher limit bounded by the need for robust stability in the face of unmodeled dynamics. The robust tuning of feedback controllers continues to be an active area of research in our group.

5.2. *Adaptive Control*

A partial review of progress in the use of adaptive methods to the control of batch reactors was given in Section 2. In our study we focus on self-tuning PID controllers by way of a comparison with the results obtained from robust tuning methods. The preliminary scheme tested was a PID algorithm featuring a gain scheduling strategy. The use of a PID controller assumes that our process can be well approximated by a second-order model. This will certainly hold for batch reactors described by our generalized reactor model but may be also true for more complicated heat removal mechanisms if the additional characteristic time constants (e.g., for heat accumulation in the metal walls) are negligible compared to that for the reaction and for the cooling/heating fluid in the jacket. In previous work (Lewin and Lavie, 1989), good control was achieved using a gain scheduling adaptive scheme for the example given in the section below assuming a quasi-linear process model with two dominant poles.

Figure 7 shows the typical performance of the adaptive scheme on the base case example, with all parameters at their nominal values. In Figure 7(a) we plot the designed and implemented temperature trajectory. Figure 7(b) shows the absolute value of the rate of temperature change with can be used as a measure of the inaccuracy of the linear model, together with the computed value for the steady-state process gain. The value of $|dT/dt| = 6°$ F/min was used as the arbitrary upper limit for gain estimation, meaning that the parameter estimation was deactivated whenever $|dT/dt| > 6°$ F/min. This high value was allowed in order to permit the adaptive algorithm to act over most of the batch cycle (the upper limit corresponds to the gradient of the desired start-up ramp). However, because of this high value, the estimated gain for batch times of under 30 min is computed on the basis of excessive linearization error and is thus unreliable. Even so, the value of the steady-state gain appears to change by almost an order of magnitude, which is the justification for the employment of some sort of adaptive scheme. In Figure 7(c), we show the computed controller gain K_c. We see the reduction in its value as the process gain increases. To complete the picture, the controller output is depicted in Figure 7(d). We continue to research the use of adaptive schemes as alternative strategies within the expert system framework for batch reactor control.

6. Testing by Simulation and Follow-up of Control Logic

The overall control strategy can now be tested on a typical example in which the consecutive non-reversible exothermic reaction: $A \rightarrow B \rightarrow C$ is carried out in a jacketed stirred tank reactor illustrated schematically in

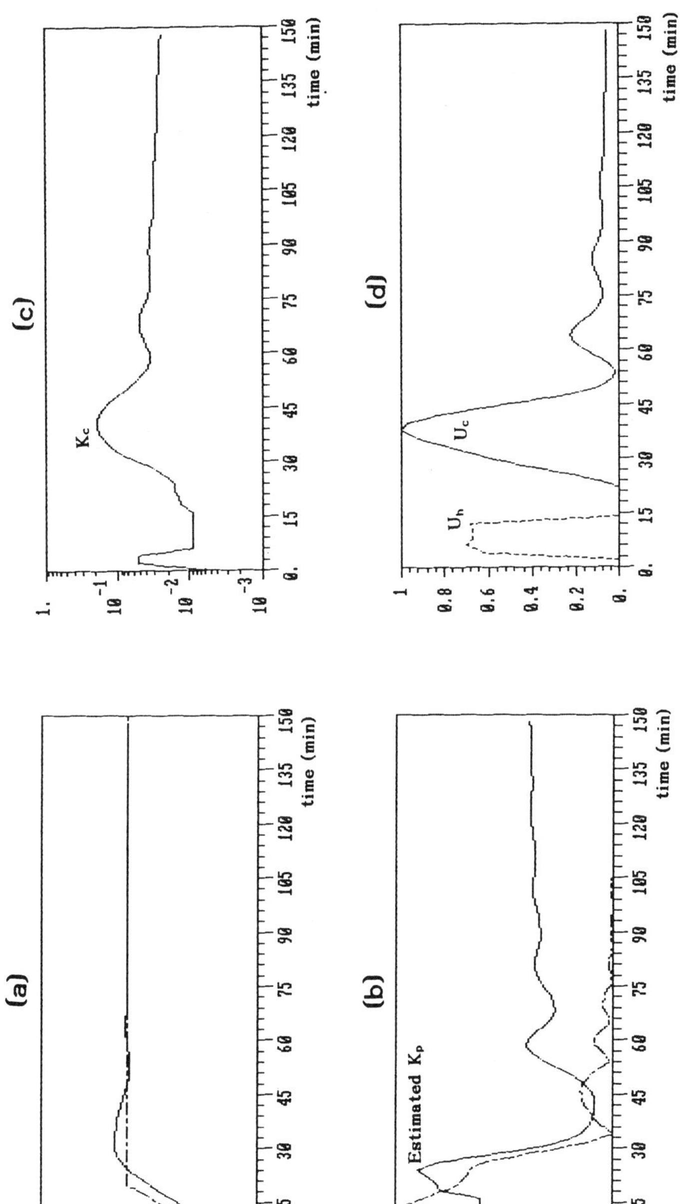

Figure 7. The performance of the adaptive scheme on the nominal base case example. (a) Designed and implemented temperature trajectory. (b) Estimated process gain and transient error. (c) Computed proportional gain. (d) Controller output.

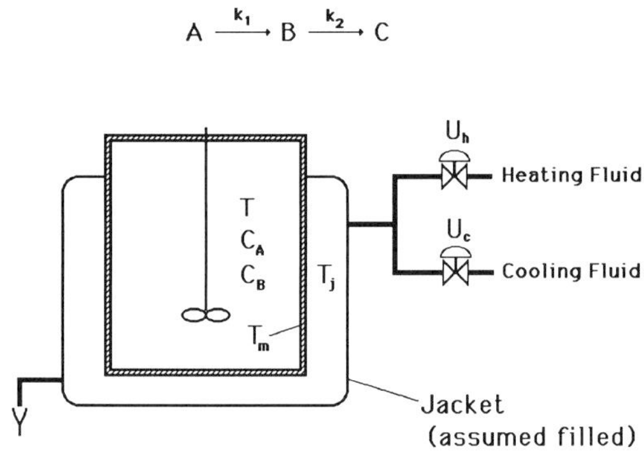

Figure 8. Simplified batch reactor showing control and state variables.

Figure 8. The temperature of the reactor inventory can be controlled by manipulation of either of two valves, which control the flow of cooling and heating fluids to the jacket. The mathematical model consists of five mass and energy balances (for A and B species concentration and for reactor, metal wall and jacket fluid temperatures) is modified from the original model presented by Luyben (1973):

$$\frac{dC_{\mathrm{A}}}{dt} = -k_1 C_{\mathrm{A}} \tag{7}$$

$$\frac{dC_{\mathrm{B}}}{dt} = k_1 C_{\mathrm{A}} - k_2 C_{\mathrm{B}} \tag{8}$$

$$\frac{dT}{dt} = \Delta H(C_{\mathrm{A}}, C_{\mathrm{B}}, T) + Q_{\mathrm{m}},$$

$$\Delta H(C_{\mathrm{A}}, C_{\mathrm{B}}, T) = \frac{k_1 C_{\mathrm{A}} \lambda_1 + k_2 C_{\mathrm{B}} \lambda_2}{\rho C_{\mathrm{p}}} \tag{9}$$

$$\frac{dT_{\mathrm{m}}}{dt} = \frac{Q_{\mathrm{j}} - Q_{\mathrm{m}}}{\theta_1} \tag{10}$$

$$\frac{dT_{\mathrm{j}}}{dt} = \phi(U_{\mathrm{h}}(T_{\mathrm{h}} - T_{\mathrm{j}}) + U_{\mathrm{c}}(T_{\mathrm{c}} - T_{\mathrm{j}})) - \frac{Q_{\mathrm{j}}}{\theta_2} \tag{11}$$

$$Q_{\mathrm{m}} = h_1(T_{\mathrm{m}} - T) \tag{12}$$

$$Q_{\mathrm{j}} = h_2(T_{\mathrm{j}} - T_{\mathrm{m}}). \tag{13}$$

Here we differentiate between system parameters that can be easily and accurately calculated (θ_1 and θ_2 [min^{-1}], which are the ratios of the heat

Table 3. Design parameters.

Variable	Formulation	Nominal value (example)
θ_1	$\frac{V_m \rho_m C_m}{V \rho C_p}$	0.272
θ_2	$\frac{V_j \rho_j C_j}{V \rho C_p}$	0.552
h_1	$\frac{h_i A_i}{V \rho C_p}$	0.071 min^{-1}
h_2	$\frac{h_{ow} A_o}{V \rho C_p}$	0.177 min^{-1}
ϕ	$\frac{F_{wo}}{V_j}$	1.058 min^{-1}

Table 4. Initial loading conditions

Variable	Value (example)
T	80°F
T_m	80°F
T_j	80°F
C_A	0.8 mole A/ft^3
C_B	0.0 mole B/ft^3
T_c	80°F
T_h	260°F

Table 5. Kinetic parameters $k_1 = eE_{a_1}/RT \quad k_2 = eE_{a_1}/RT$

Variable	Nominal value (example)
α_1	729.55 min^{-1}
α_2	6567.6 min^{-1}
Ea_1	-15000 Btu mole^{-1}
Ea_2	-20000 Btu mole^{-1}
λ_1	-40000 Btu mole^{-1}
λ_2	-50000 Btu mole^{-1}

capacities of the reactor wall to the reactor fluid and of the jacket fluid to reactor fluid, respectively, and ϕ [min^{-1}], which is the jacket space velocity) and those that may be unknown or time varying (the rate constants $k_1(T)$ and $k_2(T)$ [min^{-1}], the heat generation rate; $\Delta H(C_A, C_B, T)$ [°F/min], and the heat transfer conductances, h_1 and h_2 [min^{-1}]). A complete list of the model parameters is given in Table 3. Table 4 lists the initial loading conditions assumed at the start of each run, while Table 5 lists the actual nominal kinetic parameters used in the simulation; the control scheme of course made no direct use of information on reaction kinetics.

As discussed in the operation strategy, an estimate of the heat generation is an appropriate measure of the system reactivity, which can be derived with the help of a crude mathematical model describing the main energy balances known to prevail even if the exact numerical values of the parameters are unknown and/or likely to change in time. For the example in question, the equations below represent such energy balances:

$$\frac{dT}{dt} = \Delta H + Q_{\mathrm{m}}, \tag{14}$$

$$\frac{dT_{\mathrm{m}}}{dt} = \frac{Q_{\mathrm{j}} - Q_{\mathrm{m}}}{\theta_1}, \tag{10}$$

$$\frac{dT_{\mathrm{j}}}{dt} = \phi(U_{\mathrm{h}}(T_{\mathrm{h}} - T_{\mathrm{j}}) + U_{\mathrm{c}}(T_{\mathrm{c}} - T_{\mathrm{j}})) - \frac{Q_{\mathrm{j}}}{\theta_2}, \tag{11}$$

$$\Delta H = \frac{dT}{dt} + \theta_2 \frac{dT_{\mathrm{j}}}{dt} - \phi\theta_2\alpha - \theta_1 \frac{dT_{\mathrm{m}}}{dt}, \tag{15}$$

where

$$\alpha = U_{\mathrm{h}}(T_{\mathrm{h}} - T_{\mathrm{j}}) + U_{\mathrm{c}}(T_{\mathrm{c}} - T_{\mathrm{j}}). \tag{16}$$

At each sampling interval, an estimate of the current heat generation rate, ΔH, can be estimated using equation (15) based on system variables that are easily and accurately measured or estimated, namely the current control action (U_{c}, U_{h}), temperature measurements T_{j}, T_{m} and T, and their gradients. The dependence on temperature gradients makes the implementation of this estimate of ΔH sensitive to measurement noise. When this is significant, the measured data must be appropriately treated.

The performance of the control system was studied by means of a real-time batch reactor simulator based on the model just presented. Manual operation of the simulator provided initial operational experience allowing the initial setting of the rule-base parameters. Then the expert control system described above was tied to the simulator, allowing simulation of the controlled plant.

Figures 9–11 show the temperature trajectories of three successive batches, together with histories of the controller action taken by the expert system. The batches were run under conditions in which both the heating and cooling media are held at temperatures above their nominal values. Marked on the figures are the instants when the control system triggered new control rule action. Table 6 summarizes a sample sequence of runs as controlled by the expert system and indicates the value of trigger and controller intensity parameters that were modified from batch to batch in an attempt to improve subsequent performance. This table does not include changes effected during the batches themselves (i.e., Rule 3 and Rule 5 pulse amplitudes and durations and their triggering values, which are updated during the course of the batch.

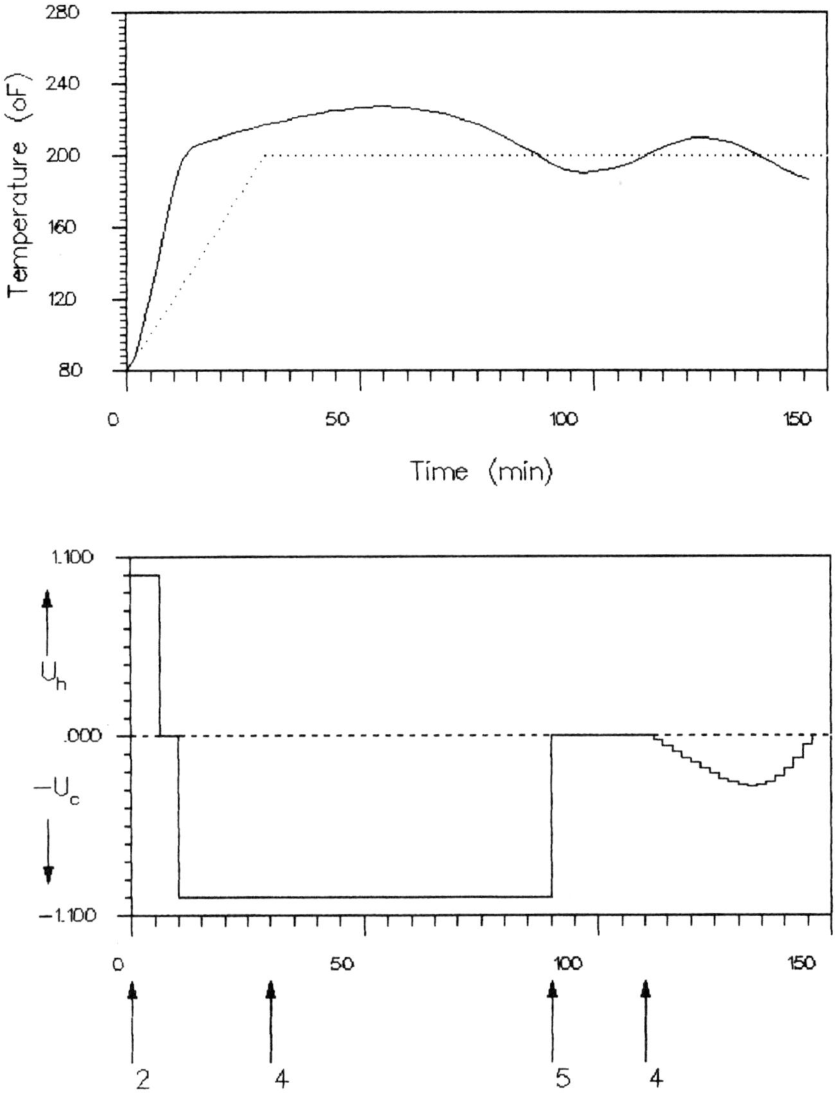

Figure 9. Development of a "safe" trajectory by the expert system: First Batch. (a) Temperature trajectories. (b) Control action.

The initial control parameters, estimated on the basis of nominal operating conditions, is shown in the first column (Batch 1) of Table 6. We note in particular that the start-up time, τ, is set to 30 minutes while the control action will be set to maximum heating for $T < 120°$ F and to maximum

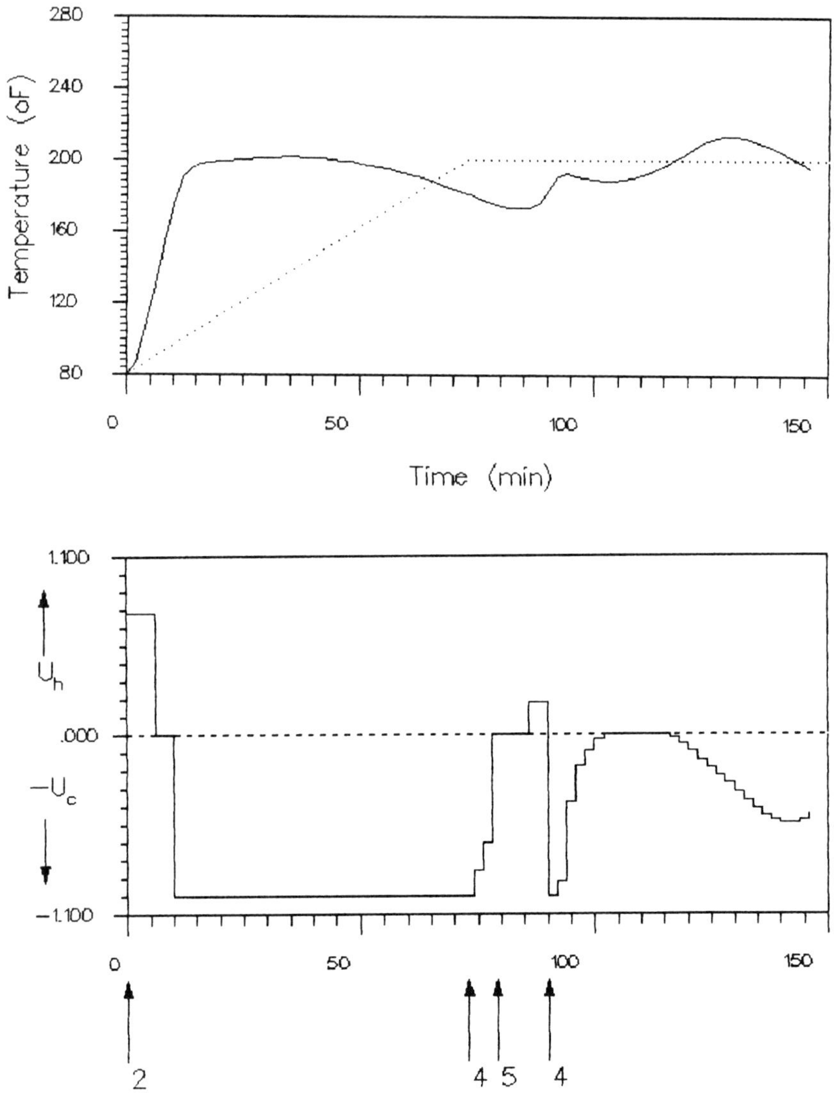

Figure 10. Development of a "safe" trajectory by the expert system: Second Batch. (a) Temperature trajectories. (b) Control action.

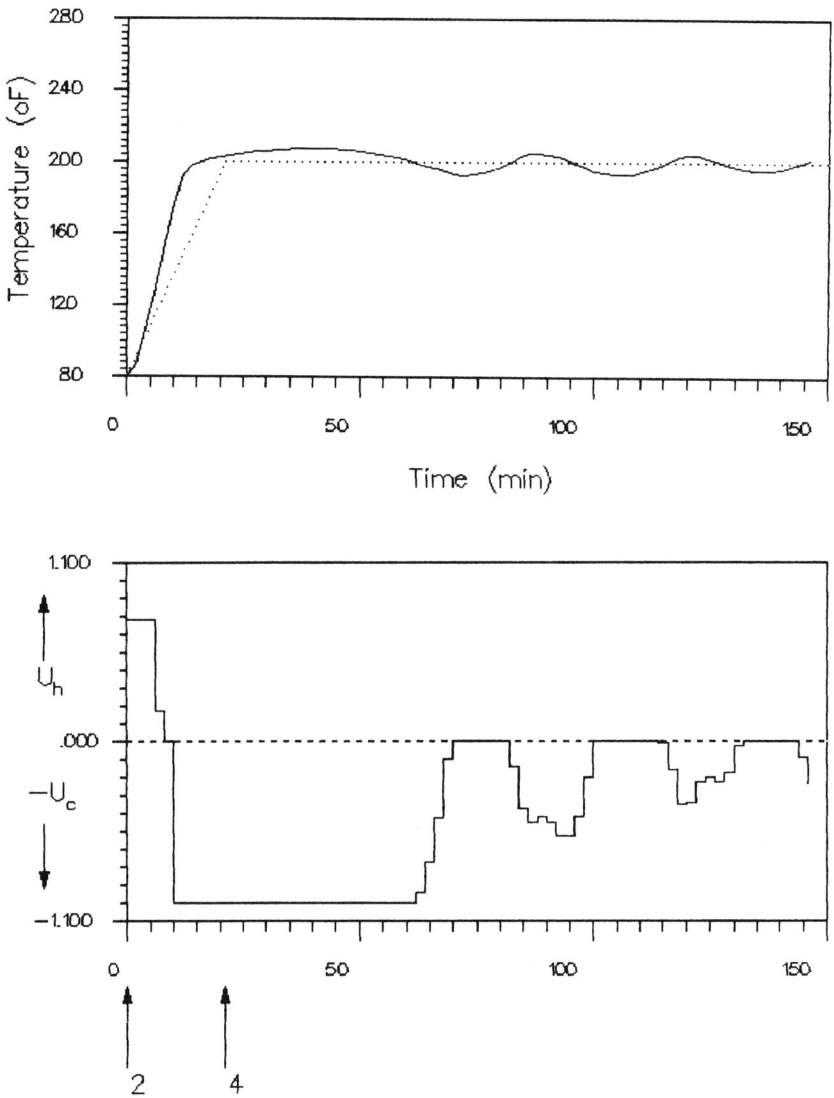

Figure 11. Development of a "safe" trajectory by the expert system: This Batch. (a) Temperature trajectories. (b) Control action.

Table 6. Values of trigger and control action intensity parameters.

Parameter name	Description	Batch 1	Batch 2	Batch 3
τ	Target time to reach nominal operating temperature (200°F)	30 min	73 min	21 min
high-cool	Amplitude of cooling pulse for Rule 2	1.00	1.00	1.00
high-heat	Amplitude of heating pulse for Rule 2	1.00	0.75	0.75
ramp-dead-band	Offset of zone 3 above ramp in Figure 6.	30°F	28°F	28°F
T^{op}-approach	Temperature above which high-cool must be applied Limit of zone 5 in Figure 6.	180°F	160°F	160°F
T-min-ignition	Temperature below which high-heat can used safely Upper limit of zone 1 in Figure 6.	120°F	110°F	110°F
ΔH-critical	Rate of heat release at which high-cool must be used	4°F/min	3°F/min	3°F/min

cooling for $T > 180°$ F. The resulting reactor temperature and control action trajectories for the first batch are illustrated in Figure 9. Then the expert system diagnoses this temperature history as "sustained excessive overshoot." It then adjusts the trigger control parameters to the values shown in the second column (Batch 2) of Table 6. The principle adjustments are

- τ is increased to 73 minutes because of apparent cooling limitations,
- High-heat is reduced by 25% to reduce heat input to the system,
- T-min-ignition is reduced to 110° G.

These changes cause the second batch to behave as shown in Figure 10. While this trajectory has considerably reduced temperature overshoot, there is now excessive cooling as indicated by the severe undershoot causing the need for additional heat input later on; this condition is diagnosed as "temperature undershoot." This is corrected immediately by overiding the PID control of Rule 4 by triggering Rule 5 instead to apply some corrective heating. As batch 2 is completed, its history is analyzed by the expert system which now diagnoses availability of sufficient heat removal capacity and hence infers that the start-up time can be reduced. Thus the only change made from Batch 2 to Batch 3 is to reduce τ to a value of 21 minutes. The resulting trajectory is shown in Figure 11, which shows a favorable response with minimal overshoot (less than 10%).

The results indicate good adaptation of the control scheme to changes in process parameters such as heat transfer coefficients, cooling jacket space velocity and kinetics parameters, even for parameter changes of an order of magnitude, after a small number of runs.

The implementation of Rule 4 is currently a fixed-PID controller. The tuning of the PID was performed in such a way as to give a good response at the operating temperature T^{op}. Alternative approaches, discussed in Section 5 and currently under development, involve either an identification scheme in parallel with the expert system, which will yield a linear model approximation to the process dynamics from which a self-tuning PID controller could be derived, or to tune robustly the PID controller to account for uncertainty both in the unstable pole position and due to unmodeled dynamics.

7. Future Goals and Directions

By definition, an expert system approach to the control of a plant can only apply to a well defined and narrow problem. Therefore, one should not expect to be able to formulate a control system of universal applicability. A case-by-case solution is unavoidable. Still, much remains to be done in terms of generalization of the elements that are common to all batch chemical reactors to enrich the domain of intuitive understanding that is so crucial to the heuristic approach. In particular, two complementing topics related to information scarcity merit attention:

(a) Developing a methodology for partial information representation that would facilitate or improve the chances of correct control action in the context of a transient process where some crucial variables (namely chemical compositions) are not measured.

(b) Systematic cross referencing of event symptoms to enrich and sharpen the induction tools as specifically applicable to the control of batch chemical reactors.

All said, it should be kept in mind that this whole field is at an extremely early stage of evolution and much progress in yet unidentified directions can be expected in the next few years.

Acknowledgments

We wish to acknowledge the participation of our students Oren Regev and Willy Rotstein in the research that led to the results presented in this chapter.

Figures 6, 9, 10, and 11, Tables 1 and 6, and portions of the text reprinted with permission from Dechema (Monograph No. 116, pp. 71-80).

Nomenclature

Symbol	**Description**
C_A, C_B	Concentrations of species A and B respectively (mole/ft^3).
C_j, C_m, C_p	Specific heats for jacket fluid, reactor walls and reactor fluid respectively (Btu/mole° F).
E_{a1}, E_{a2}	Activation energies for reactions A → B and B → C (Btu mole^{-1}).
F_{wo}	Maximum flow rate of cooling or heating fluid (ft^3/min).
h_iA_i	Reactor-wall heat transfer coefficient $i = 1$ or 2 (Btu/min° F).
$h_{ow}A_o$	Jacket-wall heat transfer coefficient (Btu/min° F).
h_1, h_2	Heat transfer conductances for reactor-wall and wall-jacket (min^{-1}).
ΔH	Heat generation rate (Btu/ft^3 min).
T, T_m, T_j	Reactor, metal wall and jacket temperatures respectively (° F except where stated).
T^{op}	Maximum operating temperature (° F).
T_h, T_c	Hot and cold supply temperatures, respectively (° F)
ΔT	Exothermic temperature rise (° F/min).
U_h, U_c	Fractions of total heating and cooling flow respectively ($0 \leq U_i \leq 1$).
V_j, V_m, V	Volume of jacket, reactor walls and reactor, respectively (ft^3).
α_1, α_2	Frequency factors for reactions A → B and B → C (min^{-1}).
θ_1, θ_2	Ratios of heat capacities of reactor wall to reactor fluid and jacket fluid to reactor fluid respectively.
λ_1, λ_2	Heats of reaction for reactions A → B and B → C (Btu mole^{-1}).
ρ_j, ρ_m, ρ	Density of jacket, reactor walls and reactor, respectively (mole/ft^3).
ξ	Reaction extent indicator.
τ	Start-up time, i.e., time taken to reach T^{op} (min).
ϕ	Jacket space velocity (min^{-1}).

References

Åkesson, I. Adaptive automatic control of reaction speed in exothermal batch processes. In *Adaptive Control of Chemical Processes*, (H. ed. Unbehauen,), pp. 99–102 (1985).

Åström, K. J. Theory and applications of adaptive control—A survey. *Automatica*, 19, pp. 471–487 (1983).

Åström, K. J., and Wittenmark, B. *Computer-Controlled Systems*, Prentice-Hall, pp. 180–181 (1984).

Bailey S. J. Artificial intelligence in industry. *Control Engng.*, 33 (14), pp 45–48 (1986).

Blickley G. Designing control systems with an expert system. *Control Engng.*, 34 (1987).

Bonvin, D., and Saner, U. On Line Procedures for Supervising the Operation of Batch Reactors. *Comp. and Chem. Engng.*, 12, pp. 371–376 (1988).

Cram, R. S., and Clarke, B. R. Expert System Monitoring and Control of a Polymer Plant. *IEE Colloquium on Expert Systems in Process Control*, Digest No 1988/47, IEE Savoy Place, London (1988).

Gygax, R. Chemical Reaction Engineering for Safety. *Chem. Eng. Sci.*, 43, pp. 1759–1771 (1988).

Hugo, P., Steinbach, J., and Stoessel, F. Calculation of the Maximum Temperature in Stirred Tank Reactors in Case of a Breakdown of Cooling. *Chem. Eng. Sci.*, 43, pp. 2147–2152 (1988).

Juba, M. R., and Hamer, J. W. Progress and Challenges in Batch Process Control. In *Chemical Process Control, CPC III* M. Morari and T. J. McAvoy, eds. pp 139–183 (1986).

Jutan, A. and Uppal, A. Combined feedforward-feedback servo control scheme for an exothermic batch reactor. *I. & E. C. Proc. Des. Dev.*, 23, pp. 597–602 (1984).

Kane L. A. AI and MAP in processing industries. *Hydro. Proc.*, 65 (6), pp. 55–59 (1986).

King, P. J., and Mamdani, E. H. The application of fuzzy control systems to industrial processes. *Automatica*, 13, pp. 235–242 (1977).

Kiparissides, C., and Shah, S. L. Self-tuning and stable adaptive control of a batch polymerization reactor. *Automatica*, 19, pp. 225–235 (1983).

Lake, R. J. Practical simulation and scale up methods for assesing the cooling of exothermic batch reactions. *Chem. and Ind.*, 2, pp. 255–258 (1977).

Lewin, D. R., and Lavie, R. Designing and Implementing Trajectories for an Exothermic Batch Chemical Reactor, accepted by *I & E. C. Res.* (1989).

Luyben, W. L. *Process Modeling, Simulation and Control for Chemical Engineers*, McGraw-Hill, pp. 160–167 (1973).

Luyben, W. L. Batch Reactor Control. *Inst. Tech.*, August, pp. 25–34 (1975).

Morari, M., and Zafiriou, E. *Robust Process Control*, Prentice-Hall, Englewood Cliffs, NJ (1989).

Morroquin, G., and Luyben, W. L. Practical control studies of batch reactors using realistic mathematical models. *Chem. Eng. Sci.*, 28, pp. 993–1003 (1973).

Niederlinski, A. *et al.* The application of self tuning control to an industrial PVC batch reactor. In *Adaptive Control of Chemical Processes*, (H. ed. Unbehauen,), pp. 193–196 (1985).

Nisenfeld E., and Turk M. Batch reactor control; could expert system help? *Intech*, April (1986).

Phillips, S. F., Seborg, D. E., and Legal, K. J. Adaptive control strategies for achieving desired temperature profiles during process start-up. 1. Model development and simulation studies. *I. & Eng. Chem. Res.*, 27, pp. 1434–1443 (1988a).

Phillips, S. F., Seborg, D. E., and Legal, K. J. Adaptive control strategies for achieving desired temperature profiles during process start-up. 2. Experimental application *I. & Eng. Chem. Res.*, 27, pp. 1443–1449 (1988b).

Regev. O., Lewin, D. R., and Lavie R. An Expert System Approach to the Control of a Batch Chemical Reactor. *AAAI-88 Workshop on Artificial Intelligence in Process Engineering*, St. Paul, Minnesota (1988).

Sripada N., Fisher D., and Morris A. An AI application for process regulation and servo control. The Univ. of Alberta, Technical report No. 86-07 (1986).

Takamatsu, T., Shioya, S., and Okada, Y. Adaptive internal model control and its application to a batch polymerization reactor. In *Adaptive Control of Chemical Processes*, (H. ed. Unbehauen,), pp. 109–114 (1985).

Tong, R. M. A control engineering review of fuzzy systems. *Automatica*, 13, pp. 559–569 (1977).

Tzouanas, V., and Shah, S. L. Adaptive pole-placement control of a batch polymer reactor. In *Adaptive Control of Chemical Processes*, (H. ed. Unbehauen,), pp. 103–108 (1985).

Umeda T., and Niida K. Process control system synthesis by an expert system. *Control-Theory and Advanced Technology*, 2 (3), pp. 385–398 (1986).

Zadeh, L. A. Making computers think like people. *IEEE Spectrum*, August, pp. 26–32 (1984).

9 Design of Protein Purification Processes by Heuristic Search

CHARLES A. SILETTI

*Massachusetts Institute of Technology**
Cambridge, Massachusetts

Abstract

Designing a commercial-scale protein purification process is a knowledge-intensive process engineering task that can be facilitated by computer-aided design. The task is sufficiently different from traditional process engineering that none of the available process design techniques are appropriate. BioSep Designer is a program that synthesizes protein purification processes. It employs physical and chemical data, mathematical models and heuristics to generate and evaluate alternative designs. A search procedure finds the optimal designs in an efficient manner.

1. Introduction

Computer-aided design for process engineering has been in existence for a number of years, and commercial programs now abound for various process engineering tasks, including process analysis (Lipowicz, 1987; Bullock, 1987), graphical flowsheet and diagram generation (Canfield, 1985), and heat exchanger network synthesis (Lipowicz, 1987). Process synthesis has also been a well established field of research since the early 1970s.

However, biotechnological processes, especially those that produce high-value products, present a number of unique challenges to the process designer. Many of the processes are not well modeled, traditional design objectives such as minimum operating cost are often irrelevant, and less easily quantified objectives, such as product quality, are more important. These aspects of protein purification are not well handled by existing process synthesis methods.

* Current address: Mobil Technical Center, Princeton, New Jersey.

ISBN 0-12-480575-2

The objective behind the development of BioSep Designer is to build a process synthesis program to address these issues. While the details of the design synthesis methodology and the computer program have been described elsewhere (Siletti, 1988), this article is a description of the design methodology used in BioSep Designer.

1.1. Large-Scale Protein Purification

Protein purification is the process by which a specific protein is recovered from a fermentation or cell culture medium. Large-scale protein purification processes produce enough product to be sold commercially. The annual production of high-value protein products, e.g., hormones or anticancer compounds, varies from less than a hundred to several thousand kilograms. A typical process, illustrated in Figure 1, involves cell rupture, solid-liquid separation, extraction of proteins from nonproteins, and the isolation of the product from other proteins in the solution. Occasionally, additional steps are required to maintain the biological activity of the product.

The biological activity of a protein is easily degraded by heat, extremes of pH, exposure to organic solvents, high shear, and exposure to gas-liquid interfaces. Consequently, most common large-scale chemical separation processes, e.g., distillation, are not applicable to protein purification. Separation of proteins from nonprotein compounds is generally accomplished by adsorption, precipitation, or, more recently, two-phase aqueous extraction. Fractionation of proteins is usually accomplished by chromatography or adsorption, although fractional precipitation or multistage extraction have been used under some circumstances. Solid-liquid separations are most commonly performed by centrifugation or filtration.

1.2. The Design Problem

Biotechnologists face the following problems when developing a commercial-scale process for producing a protein product:

- The protein purification process is expensive and accounts for at least half of the total production cost of the protein product, so the design of the purification process is an important consideration.
- New protein products and separation techniques are continually being developed, making the selection of the appropriate equipment an increasingly difficult problem.
- When the product is a pharmaceutical, legal constraints hinder the modification and improvement of an existing process. Ideally, therefore, the initial process should be nearly optimal.

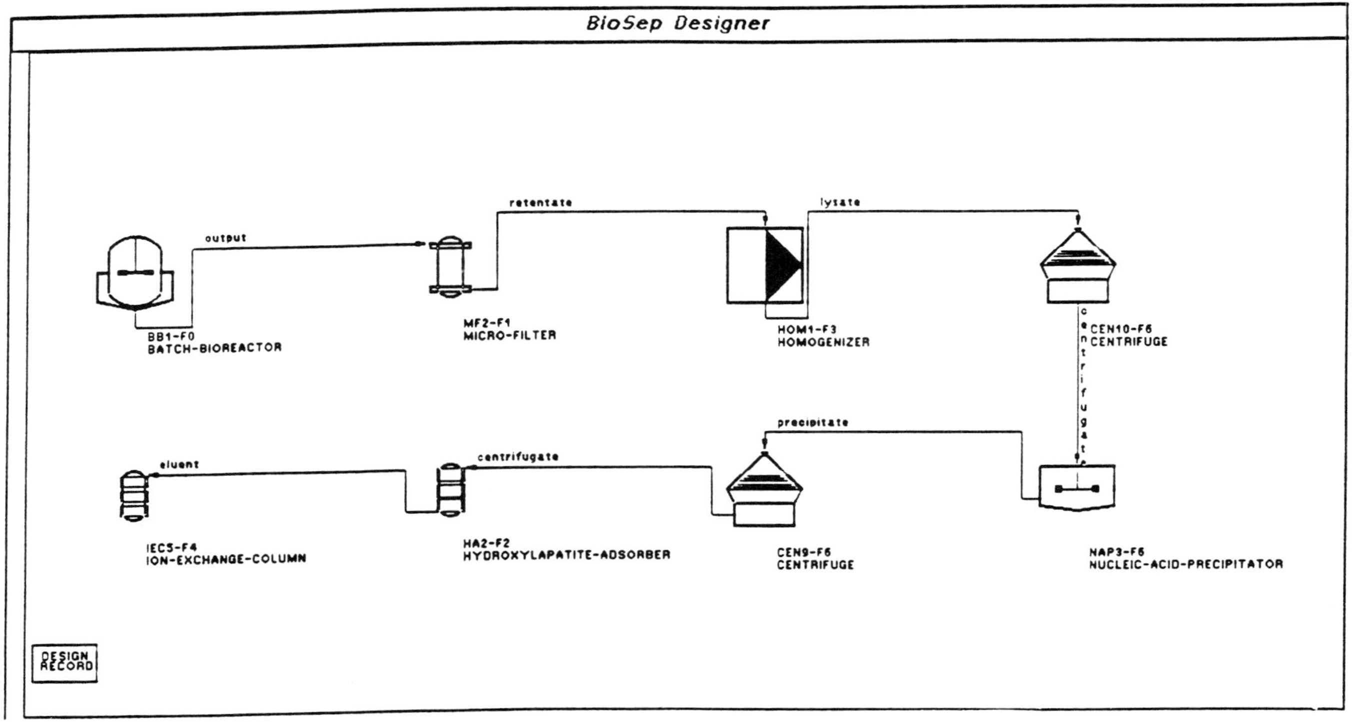

Figure 1. A typical protein purification process.

Although there are a number of papers and texts on the design of protein production processes (Asenjo and Hong, 1986; Marston, 1986; Scopes, 1984; Wheelwright, 1986), the majority of them are descriptions of the unit operation involved and provide no systematic methodology for equipment selection and specification.

1.3. The Nature of the Problem

A design—machine-generated or otherwise—is a representation of some physical entity that must carry out a specific function subject to constraints on performance, cost, and available resources (Mostow, 1985). For the purification of a biologically active protein, the physical entity is the set of equipment and their associated sizes and operating conditions which will process a feed stream that has a product concentration of 5–10% to produce an exiting stream having a product concentration of 1–5% and 1–0% contaminant concentration.

While the specific function of the protein purification process is to increase purity of the active product, the designer would also like a process that provides the following:

(1) minimum capital and operating cost,
(2) maximum product yield,
(3) maximum purity of the product,
(4) maximum reliability, i.e., the certainty that the desired purity will be achieved.

While some of these objectives, e.g., (1) and (3), are mutually exclusive, others, e.g., (1) and (2) may be achieved together.

The following information is necessary to design a protein recovery process:

(1) The desired protein and its physical properties.
(2) The biological source of the protein, its initial concentration, and the contaminants—protein and nonprotein—with which it is mixed.
(3) The required purity for the final product.
(4) The available separation equipment.
(5) The outside or legal constraints that might be placed on the problem.

1.4. Overview

The remainder of this article will describe (1) how the design of protein recovery processes relates to other types of design problems, (2) some of the initial approaches that were taken in the development of BioSep Designer, (3) the formulation of the design problem as a heuristic search, and (4) conclusions and the current status of BioSep Designer.

2. Approaches to Design

2.1. *Classification of Design Problems*

Sriram (1986) has presented a method for classifying design problems. In a *routine design*, the components of a design are known, and there is a specific procedure or plan for assembling these components. In *innovative design*, the components of the design are known, but there is no straightforward procedure for assembling the components in a way that meets the constraints and objectives for a given problem. Finally, in a *creative design* problem, not even the components of the design are completely known.

2.2. *Previous Process Synthesizers*

The synthesis of separation processes, for both proteins and other substances, is an innovative design problem. The materials and processes are known, but there is no clearly defined procedure for their selection and configuration. Previous research in the synthesis of general chemical separation sequences and of chemical processing plants (Siirola *et al*, 1971; Seader and Westerberg, 1977) investigated some of the issues of innovative design. The major differences that distinguish BioSep Designer are the specific synthesis strategy and the supporting design facilities in the computer implementation.

The chemical process synthesizers developed in the 1970s use hierarchical design to develop an initial flowsheet for a chemical process. In the approach taken by Lu and Motard (1987), for example, the synthesizer first considers the overall feed and product streams, second the reaction path-ways and reactor configurations, third the preliminary flowsheet with separation, and finally, the fully specified flowsheet. The actual designing is done by searching for operations that reduce the differences between entering and exiting streams. The separation sequence synthesizer of Seader and Westerberg uses a similar scheme that applies a set of heuristics for developing an initial process design. In both cases, the hierarchy precludes any conflicts. For example, finding a solution in one step of the hierarchy does not interfere with the solutions of other steps. Also, in both cases only a single preliminary design is generated.

While the general methodologies of hierarchical design can be applied to the design of protein recovery processes, none of the existing synthesis procedures is directly applicable. Because both the technology and the objectives in protein purification differ from those of other chemical separations, the typical heuristics, e.g., "consider distillation," used for chemical process synthesis cannot be applied.

2.3. Hierarchical Design for Protein Purification

In many of the biotechnology handbooks and texts (e.g., Aiba *et al.*, 1971; Atkinson, 1983), protein purification processes are typically presented first as a general scheme consisting of a series of stages as shown in Figure 2, and details for each stage are described one by one. This organization suggests that the design of a protein purification process might be approached in a hierarchical or top-down manner, beginning with a general outline of the process and proceeding to the design of the equipment.

Two general approaches to hierarchical design can be taken. One approach is to begin making decisions at the top level and proceed to more detailed levels. In this approach, the decisions made at the top levels may place constraints on the decisions to be made at the more detailed levels. In a second approach, exemplified by MOLGEN (Stefik, 1981), constraints are posted and checked before decisions are made. The former approach is useful in the design of protein purification processes because few con-

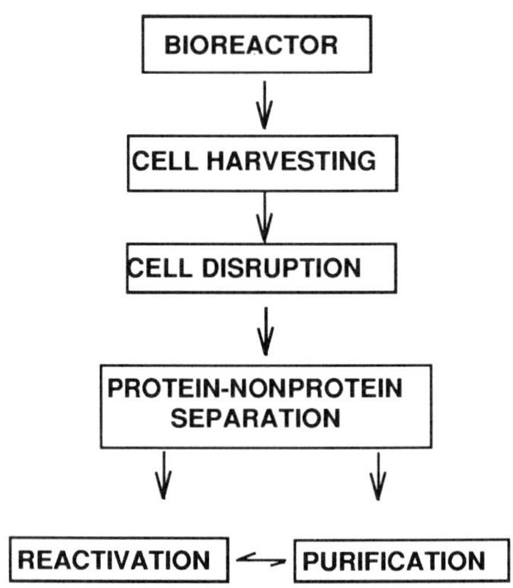

Figure 2. Stages in Protein Purification.

straints can be found without actually generating and simulating parts of the design.

In BioSep Designer, a protein purification process evolves through the following stages:

Step 1, Input: The design at each point consists of a set of problem specifications which include the protein product and its physical characteristics, the biological sources of the product, the contaminants, and the required purity and condition of the final product. Additional information, such as the physical and chemical properties of the protein, is either looked up on the system's database or estimated. The user is questioned only when the necessary information cannot be otherwise obtained.

Step 2, Basic Operations (Abstract Design): A purification scheme consisting of the general operations is developed based on the statement of the problem. Specific equipment is not yet specified.

Step 3, Ordering: The order in which to select equipment for each of the operations defined in the previous step is determined. This step is necessary because equipment selections made for one operation can place constraints on subsequent operations. For example, ion exchange cannot be used for protein fractionation if a previous step, e.g., extraction, has left a high salt content in the solution. This problem is overcome by ordering the steps so that the most constrained and most important operations are specified first, so constraints fall on less critical operations.

Step 4, Make Selections: In the order specified in Step 3, equipment is selected for the operations in the abstract design. For a given operation, all the pertinent equipment is screened, and equipment that would destroy the product or introduce incompatibilities with previously selected equipment is omitted from further consideration. For example, heat treatment would be omitted from consideration as a means of rupturing cells containing a thermolabile product.

Step 5, Analysis: The sizes and specification of the equipment are chosen to satisfy the original problem specifications. If sufficient information is available, the flowsheet is simulated and analyzed in terms of performance and cost.

These steps provide a reasonable way of organizing the design of a protein purification process, but they do not provide any means to select an alternative for any of the operations. In BioSep Designer, a rule-based approach is used to detemine the feasible alternatives for the operations in the abstract design and for the equipment in each operation in an abstract design (Steps 2 and 4). In each case, an objective function is then applied to select one of the available alternatives. This strategy, shown in

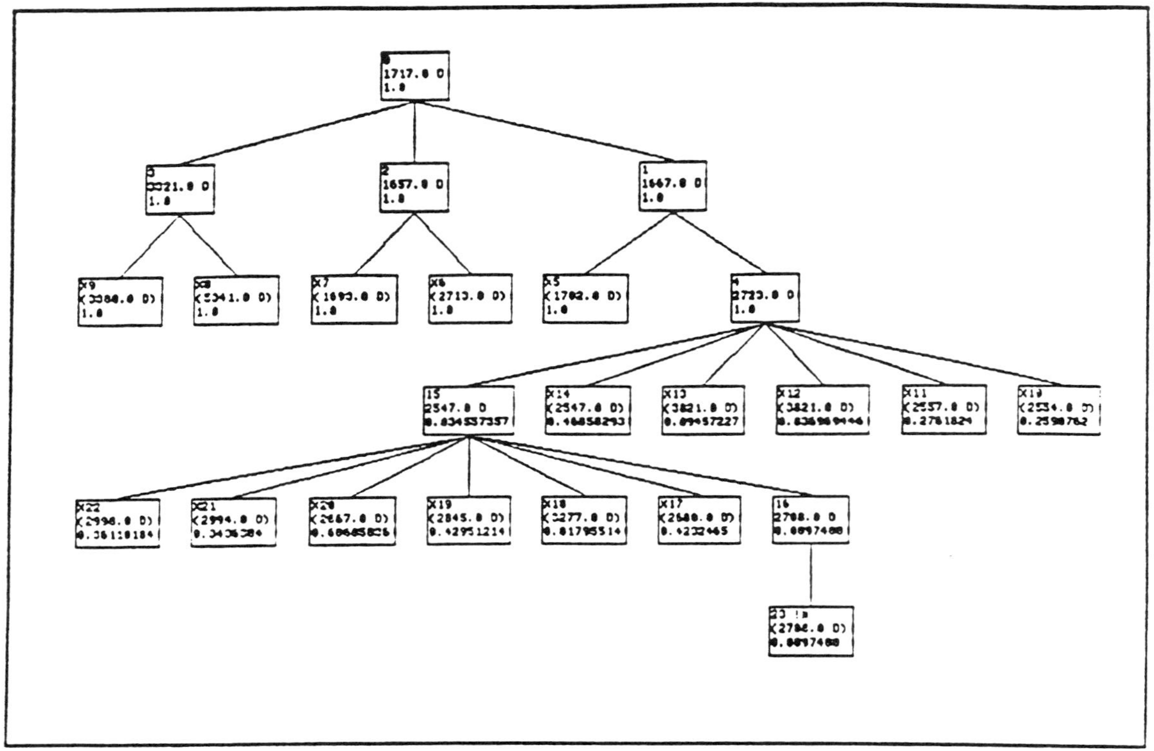

Figure 3. Tree of alternative processes for interferon purification numbers in the boxes show partial design number (upper left), predicted processing cost (center line), and predicted yield (bottom line).

Figure 3, is a best first search through the space of possible equipment configurations.

When the protein to be recovered is a pharmaceutical, the nature of the objective function can be elusive. Unlike most chemical processes, operating and capital costs are less important than product quality and realiability. According to industry experts, product purity and the degree of confidence in the purity are primary issues in the design of purification processes; cost is considered secondarily. Because product quality objectives are difficult to formulate in economic terms, BioSep Designer compromises by allowing the user to choose one or more different objectives appropriate for the given design task. The objectives include maximum confidence in the purity, minimum operating cost, and minimum total (operating and capital) cost. When more than one objective is specified, the first one is applied, and secondary objectives are used if the primary objective fails to eliminate any alternatives. For example, if equipment for cell disruption were being considered, an objective function dealing with product purity would not be able to distinguish among the different types of cell disruption devices, so a secondary objective, e.g., operating cost, would be invoked.

All objective functions are formulated in the predictive manner shown in equation 1.

$$CT = (Ck + Cu + Cs)/(\text{yield}),$$

where Ck is the cost associated with the known equipment, Cu is the predicted cost of equipment needed to perform operations for which equipment has not yet been selected, and Cs is the costs of known supplement steps. This is necessary because if each decision were made independently, the resulting process would not be optimal. If, for example, the total cost of the process were used as the objective and decisions were made nonpredictively, the resulting design methodology would generate processes consisting of many inexpensive but less effective processing steps. The resulting design, however, would have a high total cost and a low yield.

2.4. *Design as Search*

The major drawback to the preceding approach is that one alternative is selected at any given decision step without regard to the accuracy of the estimated cost function. At any point in the hierarchy, there may be several alternatives that cannot be resolved by the cost or objective function. The accuracy of the predictive objective function, however, improves as the design becomes more detailed.

A heuristic search strategy was developed to find the smallest set of designs that includes the optimum without generating unnecessary alternatives. Using this strategy, BioSep Designer follows the hierarchy outlined above. At each level in the hierarchy, the system examines all the design alternatives by calculating, for each alternative, upper and lower limits for the value of the objective function for approximately 90% confidence. The design with the lowest base value for its objective function (assuming that the problem is to minimize the function) is designated as the current best design. All the designs that have lower bounds greater than the upper bound of the best design are discarded as suboptimal. The result is the set of completed designs, all of which are optimal within the accuracy allowed by the system.

BioSep Designer does not attempt to optimize the sizes or operating conditions of the purification equipment. In the design of protein purification processes, the common practice for sizing equipment is to use design relations to determine an approximate size as a guideline for selecting among the available sizes. Approximate operating conditions can be determined by applying a number of guidelines, but the true operating conditions are determined experimentally. Optimization, therefore, would do little to improve the utility of the program.

2.5. Limitations to the Approach

There are two limitations to this design strategy. It is limited to innovative but not creative design, it cannot predict unusual protein properties that may affect the design, and optimality cannot be guaranteed.

The approach to design taken in BioSep Designer is not capable of creative design. Processes that deviate from the general scheme shown in Figure 1, cannot be found by the system unless the rules are modified. Furthermore, the correctness of a design for a given protein is based on the assumption that the protein does not display unusual behavior, e.g., interactions or low solubility, that cannot be predicted by the program.

The approach would guarantee an optimal design if true bounds on the accuracy of the estimated objective function were known. In reality, these bounds can only be estimated, and they serve as an adjustment to the breadth of the search.

3. BioSep Designer

3.1. Architecture and Implementation

BioSep Designer is implemented as a Lisp program and uses an object-oriented representation. The construction of a design centers about four types of objects: the design object, the abstract design, equipment objects,

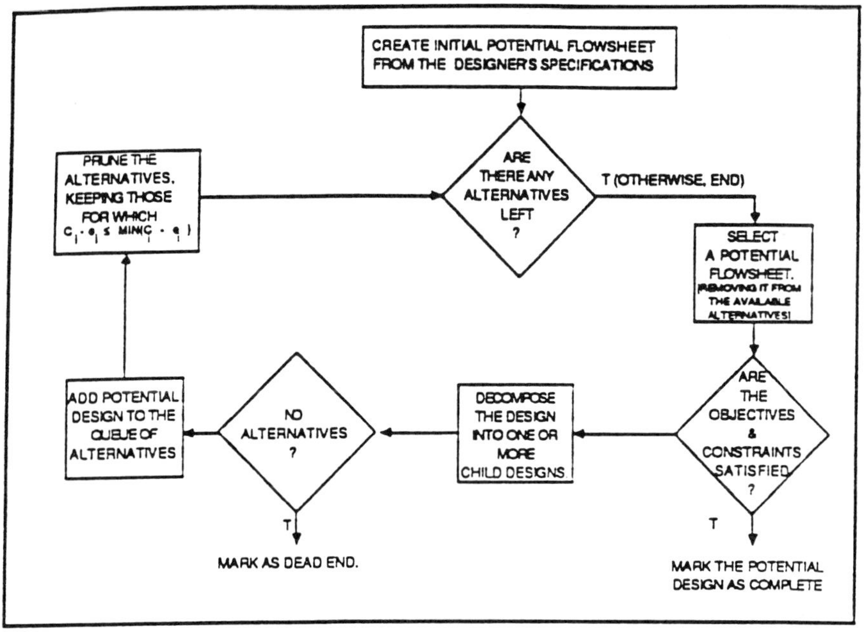

Figure 4. Process for interferon recovery devised by BioSep Designer.

and proteins. Rules for the selection of design alternatives are processed by a forward chaining inference engine developed for BioSep Designer. The synthesis routine is executed by a series of tasks placed on an agenda that is continually updated as the design proceeds.

At the outset of a design session, the user specifies a protein from a library of protein objects. If the physical properties of the protein, e.g., molecular weight, isoelectric point, and hydrophobicity, are not known, they are estimated from the protein's amino acid composition—an abbreviation of the molecular structure. A set of rules is then used to determine the feasible abstract designs. Each abstract design consists of a series of processing steps (abstract processes) that specify an operation to be carried out but not the actual equipment. Each abstract process is associated with set of rules that are used to determine the feasible types of equipment. The objective function is applied to eliminate poor designs at both the abstract design and equipment selection stages. Supplemental design facilities in BioSep Designer allow a user to modify and simulate completed designs. A diagram of the entire procedure is shown in Figure 4.

3.2. An Example

The large-scale production of recombinant human gamma interferon provides a reasonable test case for BioSep Designer. Currently produced for research purposes, gamma interferon is a potential pharmaceutical with antiviral and possibly anticancer activity. The protein has been successfully cloned in a biologically active form in both cultured mammalian cells and the bacteria *E. coli* (Gray *et al*,). *E. coli*, when genetically modified to overproduce a protein, will often manufacture at least a portion of the protein in an insoluble and biologically inactive form called an inclusion body. The biological activity of inclusion proteins can be restored by exposing them to compounds that disrupt their folded conformation (Jones *et al*, 1985).

The design begins with a statement of the problem that provides information about the product, method of production and final specifications. In this example, recombinant interferon is to be produced by *E. coli* grown in a batch fermentation process. Approximately 60% of the product is assumed to be in the form of inclusion bodies which account for about 20% of the total cellular protein. An annual production of 300 kg of 95% pure active protein is used as a basis, but there is no established market value.

The objective in this case is to find process designs with the least total cost per unit of product produced. The specifications of the design problem are summarized in Table 1.

BioSep Designer commences by determining the size of the bioreactor and its approximate cycle time. In this example, there are two possible designs for the recovery process at the abstract level. (Figure 3). In the first, the inclusion proteins are recovered and their biological activity restored. In the second, only the active cytoplasmic proteins are recovered.

Table 1. Specifications for interferon (IFN) purification.

organism	*E. coli*
product location	cytoplasm/inclusion body
product composition	10%
protein composition	65%
protein distribution	20% in inclusion bodies
bioreactor type	batch fermentor
nitrogen source	yeast extract
carbon source	glucose
final cell density	15 g/l
required purity	at least 95%
product use	pharmaceutical
operating hours/day	24
days/year	250
expected yield	50%

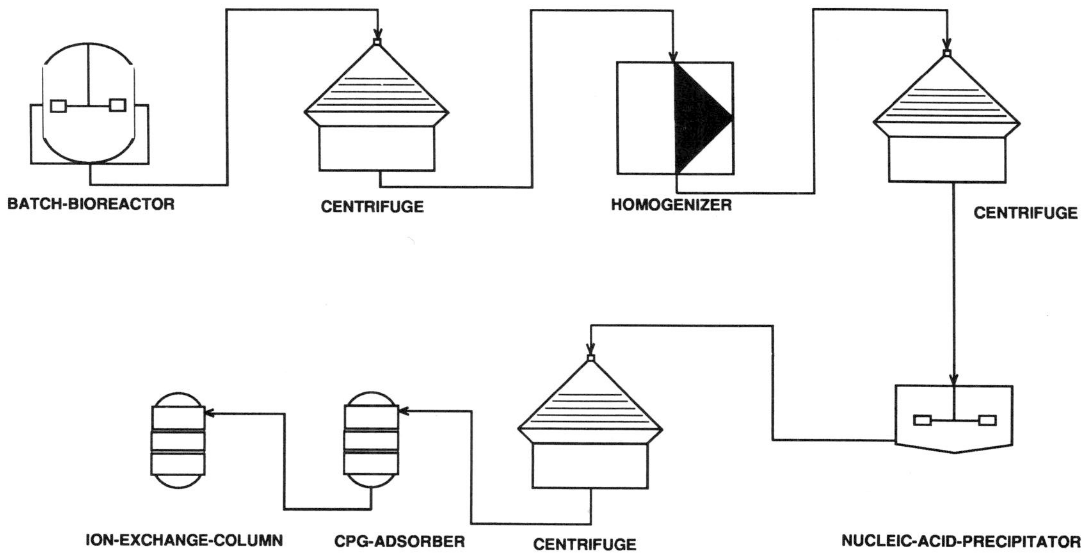

Figure 5. Interferon purification process.

These alternatives are assigned to potential flowsheets 0 and 1, respectively. An early economic analysis shows that the cost and low yield of the renaturation route render flowsheet 0 at least an order of magnitude more costly than flowsheet 1, so flowsheet 0 is discarded immediately.

The abstract design represented by flowsheet 1 comprises the operations of cell harvesting, cell disruption, protein extraction, and final purification. As discussed previously, the selection of process equipment for each of these abstract processes requires that the ordering be found that allows the equipment selection to proceed with a minimum of dead-end solutions. In this case, the protein extraction step is the most constrained as well as the most critical.

The high-resolution step is affected by the protein extraction step and is therefore selected afterward. The remaining steps of cell disruption and cell harvesting are independent of the rest of the process and may be selected at any time. BioSep Designer decides to select equipment for the basic operations in the following order:

(1) Protein extraction,
(2) Cell disruption,
(3) Cell harvesting,
(4) Protein purification.

For the protein extraction process, the available choices at the equipment level include controlled pore glass and hydroxyl apatite adsorption columns, bioaffinity columns, or ammonium sulfate precipitation. The alternatives retained are adsorption on controlled pore glass and precipitation with ammonium sulfate corresponding to potential flowsheets 2 and 3 respectively. For each of these alternatives, cell disruption by osmotic shock, high-pressure homogenization, and bead milling are considered. In each case, osmotic shock is discarded as too costly, while the other two alternatives—milling and homogenization—are considered further.

At the protein purification stage, the alternatives are cation exchange, anion exchange, hydrophobic interaction chromatography, size exclusion, and affinity or bioaffinity chromatography. The latter two are not considered for this problem because no good general affinity media are known for the nonglycosylated form of gamma interferon, and immunospecific media are, for this case, not considered.

For each of the potential flowsheets under consideration, the alternatives of anion exchange, cation exchange, hydrophobic interaction, and size exclusion are considered. If the task of selecting a protein purification step fails to yield a design that meets the purity requirements, the selection is repeated until the required purity is achieved or until there are no feasible purification steps left.

Having achieved the required purification and annual production, a design carries the implicit requirement that is must be feasible, which further implies that, among other requirements, all the adjacent processing steps must be compatible with each other. Therefore each of the the remaining potential flowsheets are checked to ensure that the processing steps can be arranged in a way that allows the product of a given unit to be a feed stream for the next unit. If discrepancies are detected, unit operations, such as the removal of cell lysate for the resuspension of precipitated protein, are inserted to overcome such problems.

The result for this example is that after exploring on the order of 100 intermediate design states, a set of 28 potential designs is retained. The optimal design employs microfiltration for cell harvesting, homogenization for cell disruption, centrifugation for clarification, controlled pore glass adsorption for protein extraction, and ion exchange for protein purification. The estimated purity of the final product is 98%, and the estimated yield is 60%.

Upon completion of a design, the user may obtain explanations for any item in the design. The explanation provided by the system gives the rules that added that item as a feasible alternative or explains the reason that a particular item was ruled out. The system can also describe other alternative that could have been chosen. In this example, the process is reasonable because it has chosen the least costly equipment that takes advantage of the unusually high isoelectric point and hydrophobicity of gamma interferon.

The designs suggested by BioSep Designer bear a close resemblance to documented processes for purifying the protein. Both ion exchange and controlled pore glass adsorption have been used for the purification of mammalian immune interferon, but most of the published methods for the purification of the recombinant protein have concentrated on the recovery of the inclusion protein (Jones *et al.*, 1985).

4. Conclusion

The design of protein purification processes can be viewed as the selection and sequencing of a set of process units that will recover a protein product from a bioreactor effluent. Guidelines and heuristics based on physical reasoning experience, and common sense must be used to determine which basic operations must be carried out, while a search procedure is used to select the best abstract designs and ultimately the best processes. A heuristic search strategy is employed to find efficiently the best design altenatives.

The design procedure is limited by the heuristics and will not generate creative designs. Hence its ability to resolve among different design alternatives is limited by the accuracy of its estimation of the objective function.

The program BioSep Designer, however, was developed with the idea that it should be modified by its users, so facilities are available for adding new types of protein purification equipment and new rules for generating design alternatives.

References

Aiba, S., *et al. Biochemical Engineering*, Academic Press (1971).

Asenjo, J., and Hong, J. *Separation, Recovery, and Purification in Biotechnology*. ACS Symposium Series (1986).

Atkinson, P. Development of Downstream Processing. *J. Chem. Tech. and Biotech.*, 32, p. 100 (1982).

Bullock, R. H. Computer Aided Design on Personal Computers. *Chem. Eng.*, 94, p. 143 (1987).

Canfield, F. B. Computer Aided Design in the Process Industry. *CEP*, 81, p. 31 (1985).

Gray, P., Leung, D., Pennica, D., Yelverton, E., Najarian, R., Simonsen, C., Derynkck, R., Sherwood, P., Wallace, D., Berger, S., Levinson, A., and Goeddel, D. Expression of Human Immune Interferon cDNA in *E. coli* and Monkey Cells. *Nature*, 295 (11) (1982).

Jones, A., Olson, K., and Shire, S. Purification and Activity Assurance of Precipitated Heterologous Proteins. U.S. Patent No. 4,512,922 (1985).

Lipowicz, M. Heat Exchanger Software. *Chem. Eng.*, 93 (15), p. 73 (1987).

Lu, D., and Motard, R. L. Computer Aided Total Flowsheet Synthesis. *Comp. and Chem. Eng.*, 9, p. 431 (1987).

Marston, F. The Purification of Eucaryotic Polypeptides Synthesized in *Eshcerichia coli. Biochemical J.*, 240, p. 1 (1986).

Mostow, J. Toward Better Models of the Design Process. *AI Magazine*, 6 p. 44, (1985).

Scopes, R. *Protein Purification Principles and Practice.* Springer-Verlag (1984).

Seader, J. and Westerberg A. A Combined Heuristic and Evolutionary Strategy for Synthesis of simple Separation Sequences. *AlChE J.*, 23 (6), p. 951 (1977).

Siirola, J., Powers, G., and Rudd, D. The Synthesis of System Designs III. Toward a Process Concept Generator. *AIChE J.*, 17 (3), p. 343 (1971).

Siletti, C. Computer Aided Design of Protein Recovery Processes. PhD Thesis, Massachusetts Institute of Technology, Department of Chemical Engineering (1988).

Sriram, D. DESTINY: A Model for Integrated Structural Design, *Journal for AI and Engineering*, October 1986.

Stefik, M. Planning with Constraints, (MOLGEN: Parts I and II), *Artificial Intelligence*, 16, p. 111 (1981).

Wheelwright, S. Designing Downstream Processes for Large Scale ProteinPurification. *Biotechnology*, 5, p. 789 (1986).

10 An Adaptive Heuristic-Based System for Synthesis of Complex Separation Sequences

Y. W. HUANG AND L. T. FAN

Department of Chemical Engineering
Kansas State University
Manhattan, Kansas

Abstract

This chapter presents a framework for developing an intelligent computer software environment, specifically an adaptive knowledge-based system, for general separation scheme synthesis without heat integration. The adaptivity of the system manifests itself in the self-adjustment of the weights of the heuristics employed in the knowledge base. The separation problems under consideration are complex in the sense that each can have multiple input and output streams. Moreover, the separation schemes generated are non-sharp, in which some or all product streams may be impure, and any component is allowed to be distributed between two or more product streams.

1. Introduction

Design is the central function of essentially every engineering discipline. In process engineering, process design is a vital activity which generates a schema for a process system converting raw materials into useful products (Nishida *et al.*, 1981). The initial, most creative step of process design is process synthesis; in this step, component parts or processing units are selected and interconnections among them are identified for yielding a preliminary system structure or flowsheet. The next step is analysis of the

ISBN 0-12-480575-2

resultant system structure. This step yields a set of design parameters, such as equipment sizes and operating conditions. Searching for an optimal design involves iterative execution of synthesis and analysis.

Synthesis is the step controlling the quality of process design; it is also the step hindering successful process design automation because of its extreme complexity, which can be ascribed to the multitude of creative activities involved in the step. Note, however, that although not in an algorithmic format, sufficient domain knowledge does exist for process synthesis. This knowledge manifests itself as engineering heuristics, pointers or rules of thumb. To effect a systematic synthesis based on available nonalgorithmic knowledge, advanced reasoning strategies, e.g., blackboard reasoning enhanced by some non-conventional logic, are highly desirable. Moreover, computer software environments in general, and their underlying host languages in particular, that are more versatile than FORTRAN and BASIC need be exploited to represent and utilize the domain knowledge for process synthesis.

In light of the observations made above, it appears that process design automation is possible only if the synthesis step can be effectively computerized. In studying chemical process synthesis, at least five types of problems have been identified: 1) separation scheme synthesis, 2) heat exchanger network synthesis, 3) reaction path synthesis, 4) control system synthesis, and 5) total flowsheet synthesis (Hendry *et al.*, 1973; Westerberg, 1980). In dealing with any synthesis, we usually resort to an approach that is either algorithmic oriented or heuristic oriented. The former states exactly the synthesis problem to be solved and then insures that the best solution is identified among the possible alternatives. The latter relies on rules of thumb to search among the alternatives; this approach resembles more closely the way a human expert tackles the problem. The synthesis may begin with or without a feasible initial flowsheet (Hendry *et al.*, 1973; Westerberg, 1980).

The framework of an approach that can facilitate "total" process design automation is delineated in this chapter. The focus is on computerizing the synthesis step to attain a "superior" process structure. One type of synthesis problems namely, the separation scheme synthesis, will be solved to demonstrate the effectiveness of the approach. The approach, heuristic oriented, is rooted in the notions of object-oriented database, fuzzy logic, learning mechanism in neural networks, and object-oriented LISP programming. The first, object-oriented database (Huang and Fan, 1988a), is concerned with efficient and correct representation of engineering information on a computer. The second, fuzzy logic (Zimmermann, 1985; Zadeh, 1988), offers an effective vehicle to handle the subjective and often ambiguous knowledge encountered in process synthesis. The third, learning mechanism in neural networks (Rumelhart *et al.*, 1986), provides a systematic way for tuning individual weights of the rules in a rule set by representing the rule set in a network format. The last, object-oriented LISP programming (Stefik and Babrow, 1985), enables complex information and reasoning strategies to be systematically implemented on a computer.

2. Problem Specifications

Separation scheme synthesis can be defined generally as follows (Westerberg, 1980; Lu and Motard, 1982):

> Given a set of source mixtures, with amount specified, create a cost-optimal structure of separation units and simple stream dividers and mixers to produce a set of specified products.

The task can be visualized as consisting of three phases, namely, 1) selection of separation units, 2) generation of an initial separation scheme, and 3) evolution of the initial scheme. Although each phase has its own significance, the second phase appears to play a decisive role in yielding a superior final flowsheet; it is also the most difficult phase to perform owing to the combinatorial nature of the problem (Thompson and King, 1972; Hendry *et al.*, 1973; Shoaei and Sommerfeld, 1986). In the phase, the general sequence of separation units is determined; the result greatly restrains the quality and cost of the final flowsheet. A broad spectrum of techniques, ranging from those being heuristic oriented (Lockhart, 1947; Harbert, 1957; Heaven, 1969; King, 1971; Nishimura and Hiraizumi, 1971; Thompson and King, 1972), algorithmic oriented (Hendry and Hughes, 1972; Hendry *et al.*, 1973; Rathore *et al.*, 1974; Westerberg and Stephanopoulos, 1975; Biess *et al.*, 1986; Pibouleau and Domenech, 1986; Floudas, 1987, 1988), or evolutionary oriented (Stephanopoulos and Westerberg, 1976; Mahalec and Motard, 1977a, 1977b), to those incorporating AI concepts (Fan *et al.*, 1987; Huang and Fan, 1988b), have been proposed to achieve the synthesis of a separation scheme in general and that of a separation sequence in particular. The development of techniques for separation scheme synthesis has been comprehensively reviewed by Huang (1988). It is also schematically summarized in Figure 1.

Mocsny (1986) has classified problems of the separation scheme synthesis with increasing degrees of complexity into the following five types.

Type 1. Problems, each with a single feed stream, no heat integration, pure product streams, a single separation method, and sharp separators.

Type 2. The same as Type 1 except that heat integration is involved.

Type 3. The same as Type 2 except that any product stream can contain more than one component; however, no component is allowed to distribute among the product stream (sharp product streams).

Type 4. The same as Type 3 except that instead of a single feed, multiple feeds are present.

Type 5. The same as Type 4 except that the product streams have one or more distributing components (non-sharp product streams), and that the separators may yield non-sharp product streams (non-sharp separators).

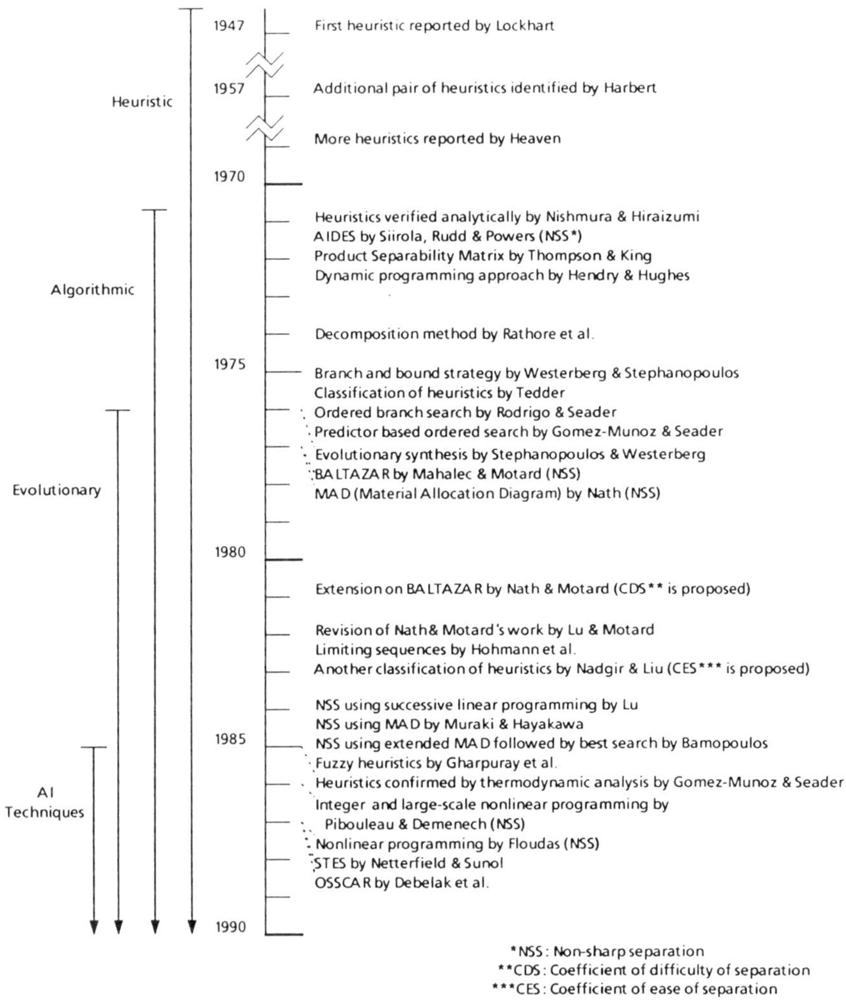

Figure 1. Progress of separation scheme synthesis techniques.

Separation schemes presented in this chapter do not involve heat integration and non-sharp separators; otherwise, they possess the highest degree of complexity. Heat integration and non-sharp separators are excluded mainly due to the following reasons.

(a) Not all separation schemes require heat integration. In fact, heat integration is useful only for highly energy-intensive separation schemes, e.g., distillation.
(b) Even for an energy-intensive separation scheme, almost all the heuristics for synthesizing it have been conceived by those experienced in total flowsheet synthesis; various thermodynamic laws and economic factors are taken into account in these heuristics. The resultant separation scheme almost always leads to an optimal or near-optimal total flowsheet from the energy point of view.
(c) Numerous references are available on the heat-integrated separation scheme synthesis (Rathore *et al.*, 1974; Umeda *et al.*, 1979; Muraki and Hayakawa, 1981; Naka *et al.*, 1982; Andrecovich and Westerberg, 1985; Westerberg and Andrecovich, 1985; Mocsny, 1986); however, the present work will be devoted to solving separation scheme synthesis problems with higher levels of complexity.
(d) Not all non-sharp separations can be achieved by non-sharp separators. For example, if A, B, and C (ranked in separation property) are three components contained in a source stream, then the product sets, AC and BC, can never be obtained with non-sharp separators alone.
(e) On the contrary, all non-sharp separations can be achieved by the integrated use of sharp separators, dividers, and mixers.

3. Knowledge Representation Strategy

A knowledge base for complex separation scheme synthesis, as for other applications, must contain two types of knowledge, namely, declarative and procedural. The former is knowledge of knowing "what," and the latter knowing "how." In the present approach, knowledge of both types is represented in the form of frames (objects).

3.1. Declarative Knowledge

Knowledge manifesting itself as a set of facts is recognized to be declarative (Barr and Feigenbaum, 1981). Here, at least three groups of declarative knowledge have been identified and deemed useful for the synthesis of a separation scheme. They include information concerning the results of previously solved problems, the properties of chemicals to be separated, and the status of a current solution.

3.1.1. Representation of Previously Solved Problems

An intelligent knowledge base should be equipped with the ability to preserve experiences, i.e., to recognize previously solved problems. For separation scheme synthesis, each candidate problem is specified by a set of input streams and a set of output streams, with their respective compositions predetermined for individual streams. It is rather common that even a very small variation in the composition of one stream can affect significantly the structure of the resultant optimum separation scheme. This has rendered the storage of every flowsheet of a solved problem in a knowledge base or database impractical. Alternatively, only essential information contained in each case of a solved problem should be preserved.

Two of the essential pieces of information contained in any separation flowsheet are the types of separators (separation methods) employed and the relative locations or sequence of these separators. The latter is invariably determined after the former. Given a problem specification, the sequence can almost always be reproduced following a set of rules; the knowledge involved is procedural in nature. This will be elaborated later. On the other hand, although attempts have been made to derive general rules for the selection of separation methods (King, 1971; Ulrich, 1984; Netterfield and Sunol, 1987), they appear to have relatively little success in real-world applications. This can be attributed to the multicomponent and multiphase characteristics of a separation system, the specifications of which necessitate a vast number of parameters. This renders an induction of general rules almost impossible. Consequently, the designers often prefer to select a separation method based on experience (King, 1971). Thus, the scattered knowledge learned from individual cases can best be represented declaratively in a knowledge base.

As an example, consider the separation of xylene isomers (King, 1976). Crystallization is accepted as the best industrial separation method when the desired product set has p-xylene in one stream and o-xylene and m-xylene in another stream. (Freezing points for these three isomers are 13.3° C, −25.2° C, and −47.9° C, respectively.) However, distillation is a common method of choice in the industry if the production of o-xylene in one stream and m-xylene and p-xylene in another stream is desirable. (Boiling points for the isomers are 144.4° C, 139.1° C, and 138.3° C, respectively.) Nevertheless, neither approach is adopted for laboratory analyses of xylene isomers; instead, we often resort to chemical reaction (with $KMnO_4$) or gas-liquid chromotography. Obviously, it is almost impossible to encode scattered knowledge like this into a set of general rules.

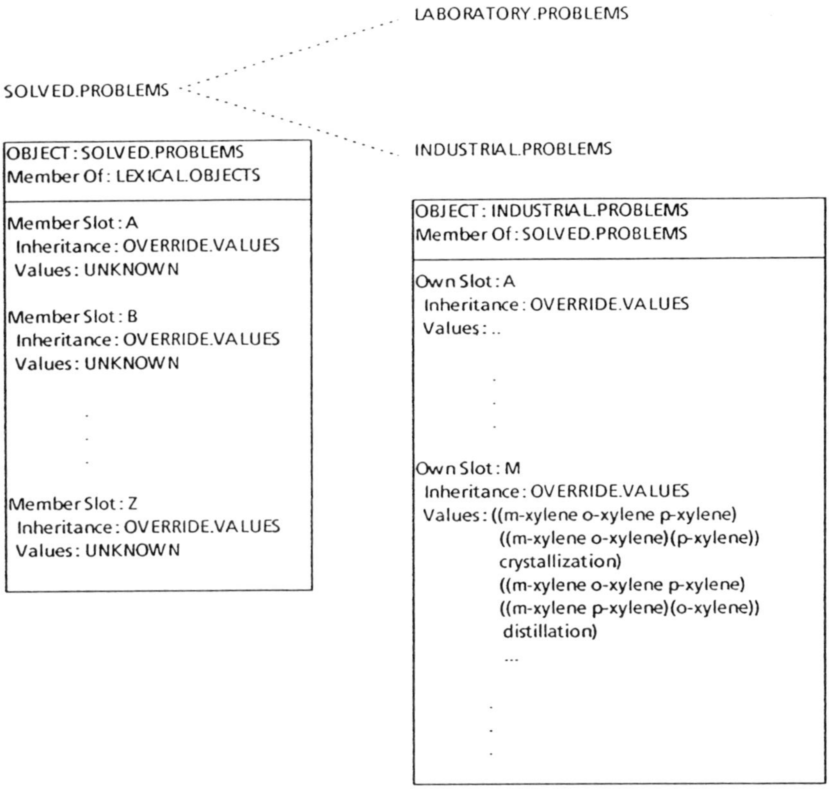

Figure 2. Data structure of the known-case knowledge base. Each known case is stored declaratively as a list (INSTREAM OUTSTREAMS METHOD).

Separation methods identified in various case studies, such as those described in the preceding paragraph, are stored declaratively in the form of objects in the approach presented in this chapter. The data structure is depicted in Figure 2, in which each known separation case is expressed by a list

```
(INSTREAM OUTSTREAMS METHOD)
```

where

```
INSTREAM   = (IN.COMPONENT1 IN.COMPONENT2 · · ·),
OUTSTREAMS = (OUTSTREAM1 OUTSTREAM2 · · ·)
           = ((OUT.COMPONENT11 OUTCOMPONENT12 · · ·)
              (OUT.COMPONENT21 OUTCOMPONENT22 · · ·) · · ·),
METHOD     = a separation method.
```

When the INSTREAM and OUTSTREAMS of a new problem match exactly those of an entry in the known-case database, the METHOD of that entry is retrieved and employed for the new problem. If the retrieval of the entries with input and output specifications "similar" to those of the new problem is also desired, we can resort to fuzzy set theory to modify the retrieval mechanism. However, such a modification will not be discussed in this chapter.

The atoms (the simplest building blocks of a list, either numeric, e.g., 24, or symbolic, e.g., propane) in INSTREAM and in each sublist of OUTSTREAMS are sorted alphanumerically. Various industrial and laboratory separation methods are then placed lexicographically in the slots of two objects, INDUSTRIAL.PROBLEMS and LABORATORY .PROBLEMS (Figure 3). For example, all available industrial separation

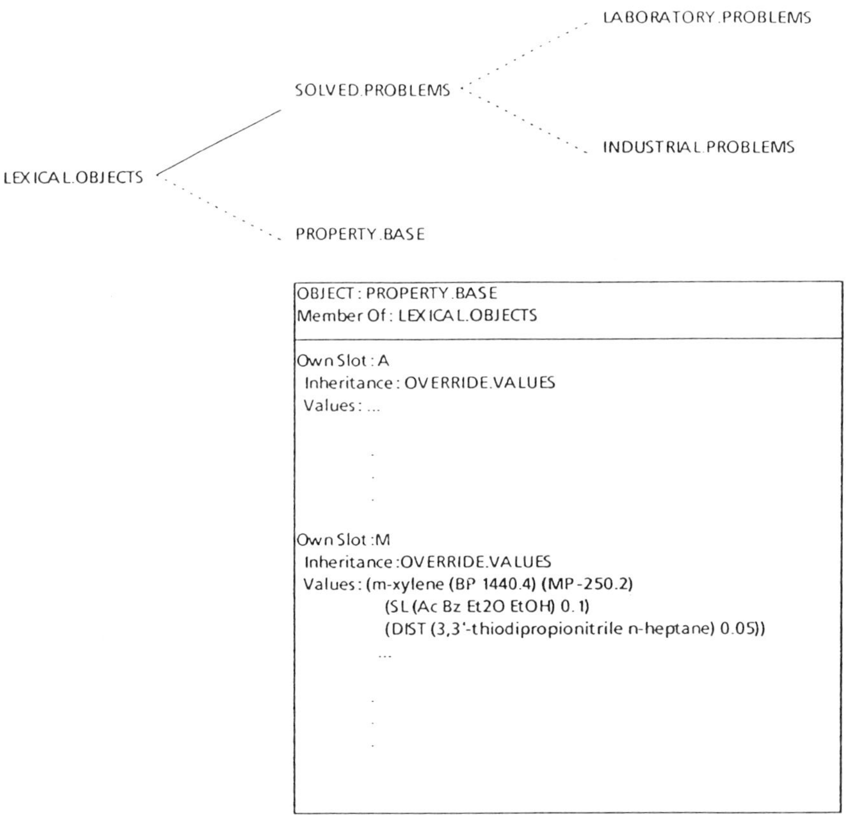

Figure 3. Data structure of the property knowledge base. Each property of a chemical is stored declaratively as a sublist (PROP [MED] VAL).

cases in which the names of the first components of the input streams starting with an "m" (or "M") will be placed in slot M of object INDUSTRIAL. Although a separation process represented in this fashion contains only a single input stream, the resultant knowledge base is ready to aid the synthesis of separation processes with multiple input streams—the separation method for each of the multiple streams should be determined independently.

3.1.2. Representation of the Properties of Chemicals

When the specifications of a new separation case do not match those of any known cases, decisions concerning which separators to select and, subsequently, what arrangement should be made for each separator are usually made in the light of the chemical and/or physical properties of the species to be separated. A partial list of those properties that are most influential in selecting a separation method is given in Table 1 (Netterfield and Sunol, 1987; Null, 1987; Muralidhara, 1988). In practice, boiling point (usually reflecting the vapor pressure), melting point, and molecular weight (usually reflecting the molecular size) are three of the most important. They have been measured for a myriad of substances and are available in numerous references (Perry, 1973; Weast and Astle, 1985). The remaining properties listed in Table 1, however, are unique properties of specific compounds. Their values are often medium-dependent. For example, a

Table 1. Some industrial separation processes based on different properties

Separation property	Technology	Remark
Vapor pressure (B.P.)	Distillation	E
Melting point	Crystallization	E
Molecular size (M.W.)	Ultrafiltration/microfiltration	T
Molecular size +shape	Gel filtration/chromatography	T
Solubility	Crystallization/extraction	E/M
Preferential solubility	Leaching	M
Chemical reaction equilibrium	Ion exchange/chromatography	T
Equilibrium sorption	Adsorption/chromatography	M
Ionic mobility	Electrophoresis/electrodialysis	T
Diffusivity	Dialysis	T
Diffusivity +solubility	Reverse osmosis/gas membrane	T
Diffusivity +chem. rxn. equil.	Liquid membrane	T

Remarks:

E: Energy-based separation (thermodynamically least efficient, and yet technically and commercially developed to the maximum extent possible currently).

M: Mass-based separation.

T: Transport-based separation (least overall energy efficiency).

substance's solubility or diffusion coefficient (except its self-diffusion coefficient) can only be measured in a mixture; the ionic mobility of a species is determined only after an electrolytic solution is chosen. Such dependence has rendered an enumeration of the values for these properties ineffectual. Estimation methods based on thermodynamic properties of a substance and the medium in which it exists serve conventionally as standard techniques to evaluate these properties (Perry, 1973; Reid *et al.*, 1977; Shieh and Fan, 1983). It should be noted that both the estimation of various thermodynamic properties of a substance and the estimation of a selected property of the substance in a certain medium based on the resultant thermodynamic properties are tedious tasks. Each is usually considered an independent research topic in its own right.

The database in the approach of this chapter is not intended to encompass all properties (of different chemicals), especially not those medium-dependent ones. Alternatively, known separation processes, based on these medium-dependent properties, could be stored in the known-case knowledge base. Nevertheless, the data structure of the database is sufficiently flexible so as to accommodate these medium-dependent properties and their values whenever needed. Selected properties of each chemical substance are represented explicitly in the form of

```
(NAME (PROP1 [MED1] VAL1) (PROP2 [MED2] VAL2) · · · ·)
```

where VAL*n* denotes the value of the *n*th property (PROP*n*) measured in medium "*n*" (MED*n*) for substance NAME. The square brackets in the representation signal that the existence of a medium is optional. For example, an entry for m-xylene is

```
(m-xylene  (BP 144.4)
           (MP -25.2)
           (MW 106.17)
           (SL (Ac Bz EtOH) 0.1)
           (DIST (3,3'-thiodipropionitrile n-
           heptane) 0.05)).
```

In the example, m-xylene is recorded with a boiling point (BP) of 144.4° C, a melting point (MP) of −25° C, and a molecular weight (MW) of 106.17 g/mole. These are the properties actively exploited in the approach of this chapter. Additionally, the example portrays m-xylene possessing a solubility (SL) of at least 10% (weight basis) in acid (Ac), benzene (Bz), ethyl ether (Et2O), or ethanol (EtOH), and a distribution coefficient (DIST) of 0.05 (weight fraction ratio) in a ternary system of m-xylene(B)/3,3'-thiodipropionitrile(S)/n-heptane(A) ($X_{BA}/X_{BS} = 0.05$ for the tie line of the lowest B concentration reported). The additional information is reserved for future use.

The above representation scheme has given rise to a collection of lists, each recording pertinent properties of a chemical substance. The lists are stored similar to previously solved problems; they are placed lexicographically in object PROPERTY.BASE (see Figure 3). The object shares the same superclass as that of SOLVED.PROBLEMS, i.e., LEXICAL.OBJECTS.

3.1.3. Representation of the Status of a Current Solution

The development status of a separation scheme at any moment of the synthesis must be preserved to respond to a query of how the scheme is reached. The status can be delineated by the following pieces of information.

a. The compositions of source stream expressed as

```
  (INSTREAM1 INSTREAM2 · · ·)
= (((INAME11 IAMOUNT11) (INAME12 IAMOUNT12) · · ·)
   ((INAME21 IAMOUNT21) (INAME22 IAMOUNT22) · · ·)
   · · ·)
```

where source stream *i* (INSTREAM*i*) is a list consisting of a number of sublists

```
(INAMEij IAMOUNTij), 1 ≤ j ≤ total number of
                             components in INSTREAMi
```

in which INAME*ij* and IAMOUNT*ij* designate, respectively, the name and amount of component *j* in INSTREAM*i*. The list of source streams is identical to that of the input streams at the outset of the development. It becomes a list containing all output streams of all dividers and separators at each subsequent step of synthesis. Note that for a specific INSTREAM*i*, (INAME*ij* IAMOUNT*ij*)'s are sorted so that INAME*ij*'s are in an ascending alphanumeric order.

b. The compositions of target streams expressed as

```
  (OUTSTREAM1 OUTSTREAM2 · · · )
= (((ONAME11 OAMOUNT11) (ONAME12 OAMOUNT12) · · ·)
   ((ONAME21 OAMOUNT21) (ONAME22 OAMOUNT22) · · ·)
   · · ·).
```

The construction of the list is similar to that of the source streams. The list remains the same as that of the product streams throughout a synthesis task. For a specific OUTSTREAM*i*, (ONAME*ij* OAMOUNT*ij*)'s are sorted so that ONAME*ij*'s are in an ascending alphanumeric order.

c. The partial separation scheme, containing pertinent data on individual dividers, mixers and separators generated thus far. For each divider, mixer or separator, these data include the coordination (position in the flowsheet); the rule yielding this creation, a pointer to its immediate upstream and downstream dividers, mixers or separators if they exist, and the compositions of input and output streams.

The frame data structure provides an effective tool to integrate the above information stored in a number of objects; this is exemplified by Figure 4. Object NEW.SCHEME in Figure 4a can be interpreted as an erasable blackboard. Each unit (divider, mixer or separator), generated on the blackboard, is recorded as a variable name, e.g., DISTILLATION. UNIT.1 (not shown in Figure 4a), in slot PICTURE.INSTANCES: the name points to the respective object instance where the detailed log of the unit is kept (Figure 4b). The modularization of data facilitates the incorporation of intelligence into the knowledge base. For example, the designer can pinpoint any of the units on the blackboard to retrieve information concerning the unit or fine tune the unit at any time during

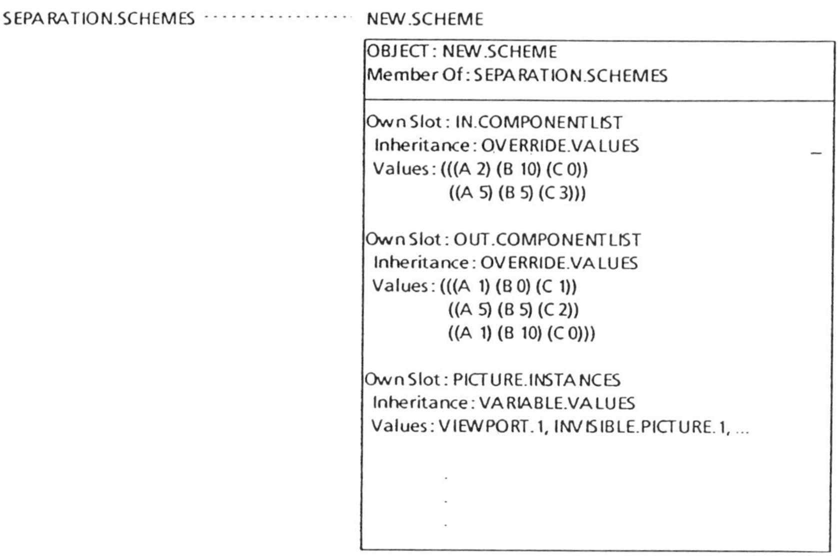

(a)

Figure 4. (a) Data structure of the "blackboard" holding the current status of a separation scheme. (b) Data structure of process units with each mixer, divider, or separator manifesting itself as a picture on the "blackboard." The data structure stores essential information pertaining to the existence of each unit, in addition to other data.

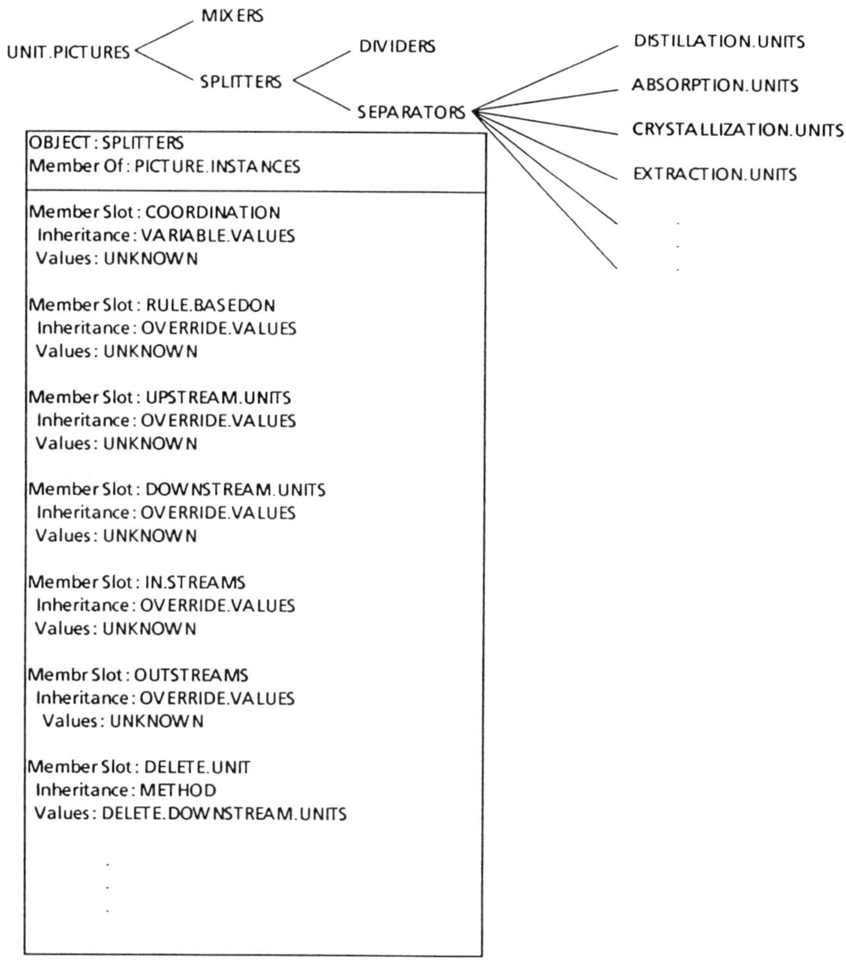

(b)

Figure 4. (*Cont.*)

the synthesis task. Note that tuning on a unit usually results in deletion of its downstream units; the synthesis task will resume from the tuning point.

3.2. *Procedural Knowledge*

A procedural representation of a piece of information is essentially a plan for the use of that information. Heuristic rules employed in this work typify

Table 2. Heuristic rules employed in the present work for separation scheme synthesis

1.	Rules for the selection of a separation unit
1.1	IF boiling points vary widely THEN use distillation.
1.2	IF melting points vary widely THEN use crystallization.
1.3	IF molecular weights vary widely THEN use ultrafiltration.
1.4	IF two or more methods are equally feasible THEN favor energy-based one to mass-based one to transport-related one.
2.	Rules for synthesizing an initial separation scheme
2.1	IF a corrosive component is found THEN remove it first.
2.2	IF a by-product is generated THEN process it first.
2.3	IF a feed stream and a product stream consist of the same components THEN match the product stream with the feed stream by using a divider.
2.4	IF a separation can yield a stream that matches completely one of the product stream THEN do it first.
2.5	IF a divider can be used although a product stream has some extraneous components with respect to a feed stream THEN use it.
2.6	IF an MSA is used THEN remove it first.
2.7	IF adjacent separation properties vary widely THEN separate at the largest adjacent property.
2.8	IF adjacent separation properties do not vary widely AND relative quantities vary widely THEN separate at the component with the largest quantity.
2.9	IF adjacent separation properties do not vary widely AND relative quantities do not vary widely AND a near 50/50 separation exists THEN separate at the near 50/50 position.
2.10	IF adjacent separation properties do not vary widely AND relative quantities do not vary widely AND no obvious 50/50 separation exists THEN separate by direct sequencing.
3.	Rules for evolution of the initial scheme
3.1	IF the designer wants to override the design THEN accept it.
3.2	IF an output from a divider is an input to another divider THEN merge the dividers.
3.3	IF an output from a mixer is an input to another mixer THEN merge the mixers.
3.4	IF two separations are parallel (performing the same task) THEN merge them.

knowledge of this type. Obtained through a comprehensive literature review (Huang, 1988) and listed in Table 2 in an IF-THEN format, these rules are classified into three groups, of which the second group will be emphasized in this chapter. It is believed that the creation of the best possible initial separation scheme is most critical in producing a good final scheme. Rules 2.3 and 2.5, originally proposed in this work, encourage the use of dividers. Rules 2.4 and 2.6 to 2.10, determine the separation position when a separator is to be employed. It should be pointed out that a mixer will be automatically generated whenever a product stream cannot be matched completely in a single synthesis step.

At this juncture, it is worth clarifying two points of procedural knowledge. First, the procedural knowledge, when described, often appears to be declarative. This is especially true when the knowledge is expressed in a high-level language where the reasoning mechanisms are hidden. As evidence, each rule in Table 2 appears to be a declaration of a certain fact. However, inference based on the rule yields stepwise directions to the generation of a separation scheme. This argument implies that no clear-cut distinction exists between declarative and procedural knowledge. The distinction lies in the nature of the knowledge and the mode of its usage. Second, procedural knowledge need not be presented in a sequential manner. In fact, step-by-step instructions do not normally exist for creative types of tasks, e.g., process synthesis.

The rules (procedural knowledge) in the present approach are represented in the form of objects as mentioned earlier. Figure 5 illustrates the representation. The rules are organized into three classes (sets), namely METHOD.RULES, SEQUENCE.RULES, and EVOLUTION. RULES. The firing of a rule within a rule set is controlled by the instructions described in slot MONITOR. Each member rule in a rule set is characterized by its name and the values stored in the five slots, including WEIGHT, ANTECEDENTS, CONSEQUENCES, WEIGHTED.TRUTHVALUE, and VERBAL.FORM.

Slot WEIGHT keeps the associated weight of each rule. The initial weights of individual rules are tabulated in Table 3. These values are assigned subjectively after the relative importance of the rules is evaluated in the light of the comprehensive literature review (Huang, 1988); they represent the domain experts' general belief in weighing and balancing the rules and are independent of the separation method. This belief, nevertheless, can be fine-tuned; the evolution of these values will be discussed in Section 5.

Slot ANTECEDENTS holds an algorithm to calculate the truth value of each rule; some highlights of the algorithms are found in Table 3. Considering Rule 1.1 as an example, the degree of truth of "boiling points vary widely (B)" can be determined by the membership function, $\mu_B(^*)$ (see Appendix A),

$$\mu_B(^*) = \left[\frac{(BP_2 - BP_1)}{1 + (BP_2 - BP_1)}\right], \tag{1}$$

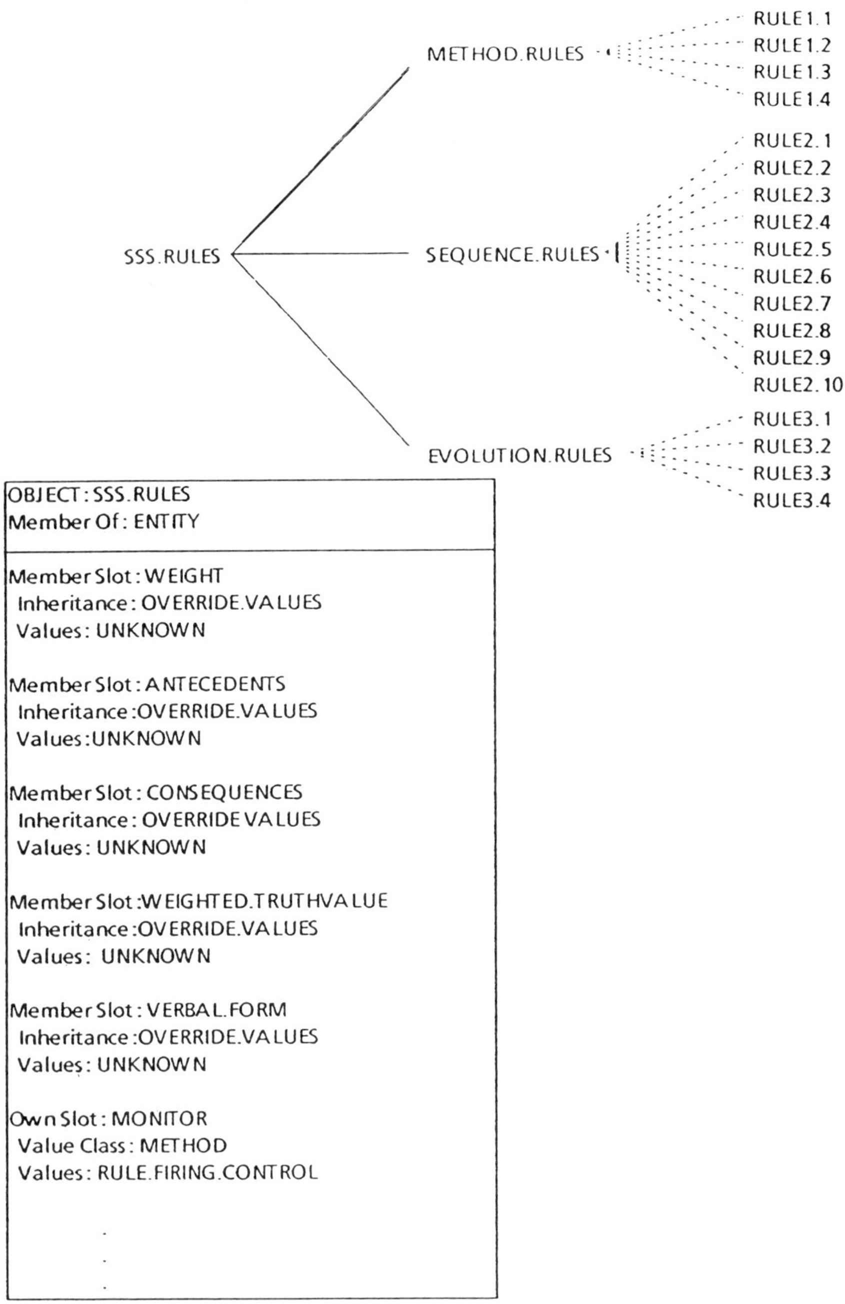

Figure 5. Data structure of the rules in the present system.

Table 3. Initial weights and highlights of the algorithms to calculate the truth value of the antecedents of each rule

Rule No.	Initial Weight	Algorithm Highlight
1.1	0.9	$(BP_{\max} - BP_{\min})/(1 + BP_{\max} - BP_{\min})$.
1.2	0.5	$(MP_{\max} - MP_{\min})/(1 + MP_{\max} - MP_{\min})$.
1.3	0.2	$(MW_{max} - MW_{\min})/(1 + MW_{\max} - MW_{\min})$.
1.4	1.0	Detect the existence of multiple methods.
2.1	1.0	Ask user.
2.2	0.98	Detect a by-product system.
2.3	0.96	Examine the source/target stream patterns.
2.4	0.94	Examine the source/target stream patterns.
2.5	0.92	Examine the source/target stream patterns.
2.6	0.90	Detect an MSA.
2.7	0.85	$\max\limits_{i=1,n}((SP_{i+1} - SP_i)/(1 + SP_{i+1} - SP_i))$.
2.8	0.80	$\max\limits_{i=1,n}(1/(1 + SP_{i+1} - SP_i)) \wedge \max\limits_{i=1,n}((Q_{i+1} - Q_i)/(1 + Q_{i+1} - Q_i))$.
2.9	0.70	$\max\limits_{i=1,n}(1/(1 + SP_{i+1} - SP_i)) \wedge \max\limits_{i=1,n}(1/(1 + Q_{i+1} - Q_i)) \wedge \left(\max\limits_{i=1,(n-1)} \left[\left\Vert \dfrac{\sum_{j=1}^{i} Q_j}{\sum_{j=i+1}^{n} Q_j} \right\Vert\right]\right)$.
2.10	0.60	$\max\limits_{i=1,n}(1/(1 + SP_{i+1} - SP_i)) \wedge \max\limits_{i=1,n}(1/(1 + Q_{i+1} - Q_i)) \wedge \left(1 - \max\limits_{i=1,(n-1)} \left[\left\Vert \dfrac{\sum_{j=1}^{i} Q_j}{\sum_{j=i+1}^{n} Q_j} \right\Vert\right]\right)$.
3.1	1.0	Ask user.
3.2	0.95	Ask user.
3.3	0.90	Ask user.
3.4	0.85	Ask user.

for two-component separation, or in general,

$$\mu_{\mathrm{B}}(^*) = \max_{i=1,n}\left[\frac{(BP_{i+1} - BP_i)}{1 + (BP_{i+1} - BP_i)}\right], \tag{2}$$

where BP_i and BP_{i+1} are boiling points of two successive components i and $i+1$ (in terms of ascending boiling point ranking) in a source stream of n components, and the "max" operation quantifies the term "boiling points vary widely" based on the largest variation in boiling points of two successive components in the source stream of n components. Expression (2) equals 0 when $BP_{i+1} = BP_i$ and approaches 1 when $BP_{i+1} \gg BP_i$. In other words, the larger the value of expression (2), the higher the degree of truth in the claim "boiling points vary widely (B)". By the same token, the truth value of the antecedent "adjacent separation properties do not vary widely (¬S) AND relatively quantities do not vary widely (¬Q) AND a near 50/50 split is available (F)" of Rule 2.9 can be reflected by the membership function, $\mu_{\neg\mathrm{S},\neg\mathrm{Q},\mathrm{F}}(^*)$ (see Appendix B).

$$\begin{aligned}\mu_{\neg\mathrm{S},\neg\mathrm{Q},\mathrm{F}}(^*) &= \mu_{\neg\mathrm{S}}(^*) \wedge \mu_{\neg\mathrm{Q}}(^*) \wedge \mu_{\mathrm{F}}(^*)\\ &= \max_{i=1,n}\left[\frac{1}{1 + (SP_{i+1} - SP_i)}\right] \wedge \max_{i=1,n}\left[\frac{1}{1 + (Q_{i+1} - Q_i)}\right]\\ &\quad \wedge \max_{i=1,(n-1)}\left[\left\|\frac{\sum_{j=1}^{i} Q_j}{\sum_{j=i+1}^{n} Q_j}\right\|\right],\end{aligned} \tag{3}$$

where $\wedge$ denotes the fuzzy "AND" operator (Zimmermann, 1985), SP an adjacent separation property, and Q a quantity. Again, the greater the value of expression (3), the higher the degree of truth in the antecedent of Rule 2.8.

The truth value of each rule is then combined with the weight of the corresponding rule to yield a weighted truth value (Fan *et al.*, 1987); the result is deposited in slot WEIGHTED.TRUTHVALUE. When a rule is selected to be fired, the action to be taken is coded in slot CONSEQUENCES. Slot VERBAL.FORM saves a verbal description of a rule.

4. Reasoning Strategy

The problem-solving strategy adopted in the present system loosely follows the blackboard model (Nii, 1986). The model can be described by elaborating on its three major components.

a. *The knowledge sources.* Each heuristic rule in a rule set, defined as an object, represents a knowledge source. The three sets of rules constitute knowledge sources of three hierarchical levels, namely method identification, sequence determination, and flowsheet evolution. In reasoning, individual knowledge sources (rules) at any level are kept separately and independently. The knowledge sources, whose sole function is to modify the blackboard, change it stepwise until a solution is found. Note that only the knowledge sources can modify the blackboard.

b. *The blackboard data structure.* The framelike object NEW.SCHEME, described earlier, serves as the blackboard (see Figure 4a); it holds information from the solution space. This comprises essentially the partial solution and the eventual final solution of a synthesis. The progress of synthesis revealed on the blackboard is hierarchically organized into levels of analysis, from method identification to sequence determination followed by flowsheet evolution. Information associated with a solution at one level becomes the input to the set of knowledge sources at this level, which, in turn, generate new information at the same or other levels.

c. *Control.* The knowledge sources respond opportunistically to changes in the blackboard. To avoid conflicts that may arise among the knowledge sources, an initial weight is assigned subjectively to each heuristic in the light of its popularity among domain experts (see the second column of Table 3). No order is established *a priori* for applying the heuristics: the selection of a heuristic at each step will depend on its weight and truth value. The weight will be self-adjusted at the end of each application based on the result of the synthesis, if desired.

A schematic summary of the reasoning strategies of the proposed system is depicted in Figure 6. One component in the figure which has not been mentioned so far is the Mathematical toolbox. It contains supportive operations such as fuzzy algebra and linear programming. The former facilitates the calculation of truth values and weighted truth values. The latter performs stream splitting as a standard resource allocation problem (Lu and Motard, 1985).

5. Adaptation Mechanism

Two major concepts have been adopted by cognitive scientists in emulating the mechanism of adaptation (learning). The first concept assumes that the goal of learning is to formulate explicit rules (propositions, productions, etc.). TEIRESIAS and subsequently EMYCIN are two classical systems implementing this concept (Buchanan and Shortliffe, 1984). These systems acquire new knowledge, i.e., learn, by comparing particular cases with

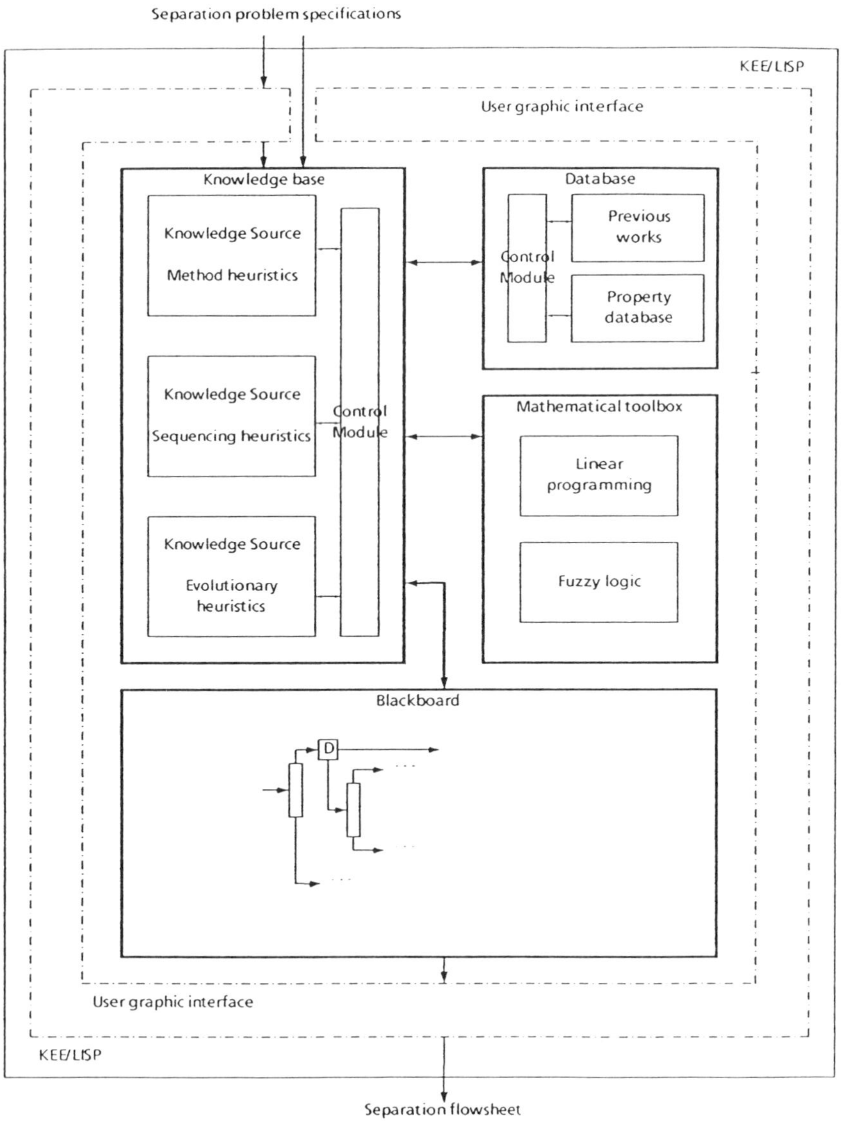

Figure 6. Schematic of the reasoning strategy for separation scheme synthesis.

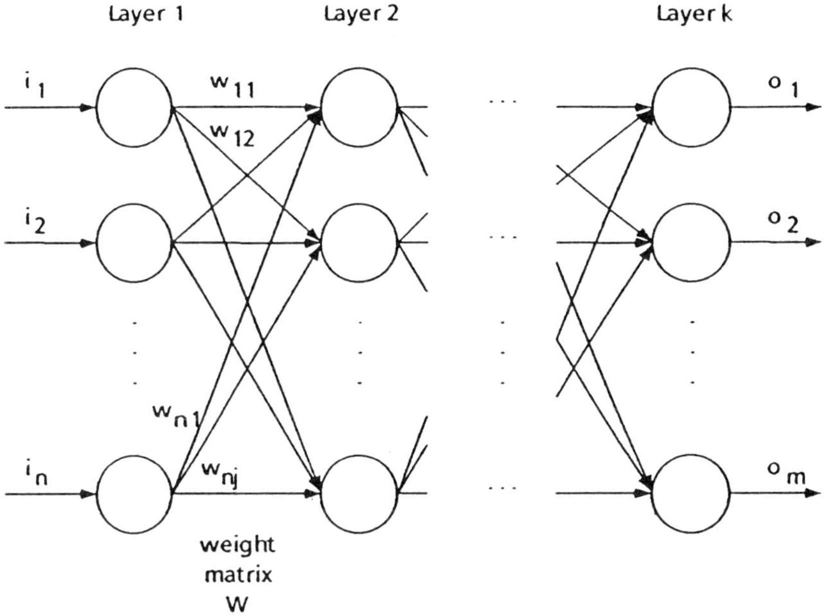

Figure 7. General multi-layer parallel distributed model (neural network). Note that the number of processing units at different layers need not be identical.

innate domain knowledge and a certain starting set of rules; this gives rise to the formulation of new hypothetical general rules.

The second concept concerning the learning mechanism differs drastically from the first. This concept is popular among researchers in parallel distributed processing (PDP). It assumes that an intelligent computing system can be constructed by connecting a number of simple processing units. The resultant system is known as a neurocomputer, neural network, artificial neural system or connectionist system (Figure 7). The power of such a system lies in the massive interconnections among the processing elements, each of which has an adjustable strength. The goal of learning by a PDP model is not to formulate explicit rules; rather, it is the acquisition of connection strengths allowing a network of simple units to act as if it knew the rules (McClelland *et al.*, 1986).

In spite of the strong rule-base orientation, the learning mechanism incorporated into the present system is based on the concept rooted in PDP-based modeling. Explicitly, learning in the present approach is achieved by tuning the weights of various rules. The justification for this seemingly unorthodox approach is as follows:

First, the sets of rules listed in Table 2 result from an exhaustive review of previous works. The table contains the majority of significant rules

IF A AND B AND C THEN X OR Y
(= IF A AND B AND C THEN X
IF A AND B AND C THEN Y)
IF B AND D THEN Y
IF C THEN Z

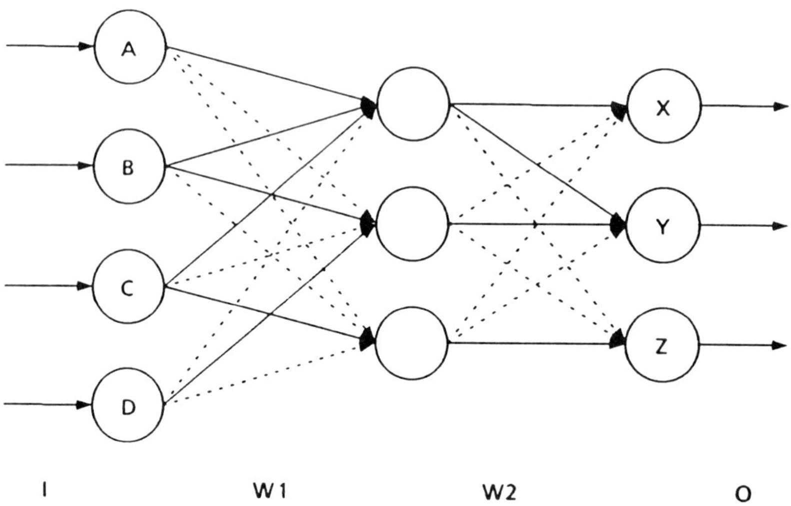

I = truth values of the propositions in the antecedents
W1 = weighting factors of the antecedents
W2 = weighting factors of the rules
O = weighted truth values of the rules

Figure 8. Representation of a rule set with a parallel distributed processing (PDP) model. (a) A rule set. (b) A PDP network representing rule set (a); dotted lines are connections with zero weight.

applicable to separation scheme syntheses. An attempt to formulate additional rules for each particular case merely degrades the generality and efficacy of the rule sets. What could and should be done to improve the quality of the rule sets, thus, is to assign properly the weights to the rules.

Second, it appears that a rule set can readily be presented in a network format. Each proposition or group of propositions in a rule set is a node (unit), each logical relationship between two nodes (two propositions or two groups of propositions) a connection, and each weight of a logical relationship the strength of the connection. Figure 8 presents such a network; it resembles a fuzzy causal network (Kosko, 1988). The activation and threshold functions of the units need be defined to satisfy the nature of the rule set. Strikingly, the network shares the same topology as a layered neural network (Rumelhart *et al.*, 1986). In light of the premises that the weights are the parameters to be adjusted (in enhancing the quality of a rule set) and that a rule set bears strong similarities to a neural network, it is only natural that the learning concept in the neural network be applied to tuning the weights of a rule set.

The general scheme to map a rule set into a neural net-like network suggested by Figure 8 can often be simplified. Specifically, the three-layer network depicted in the figure can be readily reduced to a two-layer network when individual propositions are assumed to possess equal weight for every rule in the rule set. This reduction has been attained in the present implementation to map the rule sets summarized in Table 2 into a two-layer neural network displayed in Figure 9, where each unit in the input layer represents the antecedent of a rule, and each unit in the output layer the consequence of a rule.

Virtually all learning rules for PDP models can be considered a variation of the Hebbian learning rule (Hebb, 1949; also see Rumelhart *et al.*, 1986).

$$\Delta w_{ij} = g(a_i(t), t_i(t)) \cdot h(o_j(t), w_{ij}), \tag{4}$$

where $a_i(t)$ signifies the activation of unit u_i at time t, $t_i(t)$ the teaching input to u_i, and $o_j(t)$ the output signal from unit u_j. The simplest version of Hebbian learning does not involve the teacher, and the functions, g and h, are proportional to their first arguments; this yields

$$\Delta w_{ij} = \eta \cdot a_i \cdot o_j, \tag{5}$$

where η is the constant of proportionality representing the learning rate. In other words, the equation states that the adjustment of the strength between two units in a network is proportional to the products of their simultaneous activation. Learning of this type is unsupervised and competitive in nature; its major application is regularity detection. In this

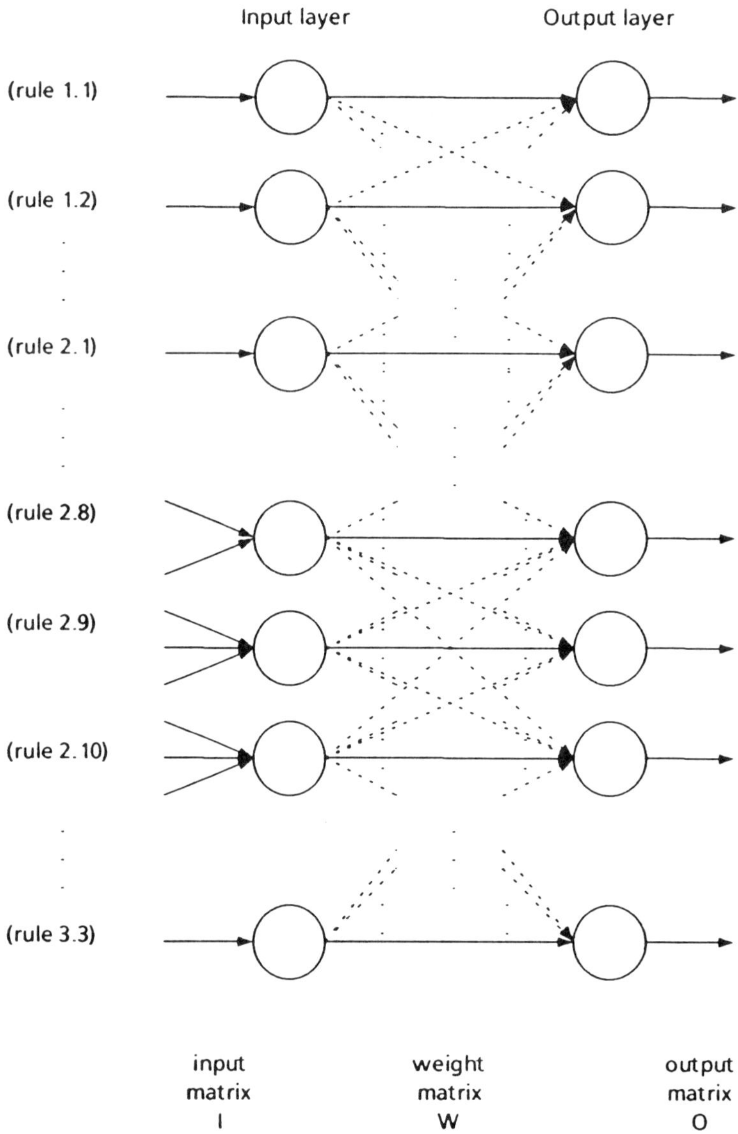

I = truth values of the propositions in the antecedents.
W = weighting factors of the rules.
O = weighted truth values of the rules.

Figure 9. PDP model for representing the rule sets in the present work. The seemingly simple two-layer "non-interconnected" network is sufficient when individual propositions are equally weighted and connected only with logical AND.

learning paradigm, a network is trained to develop its own feature representation of input stimuli to capture the most salient features of the population of input patterns. Apart from regularity detection, the other major learning paradigm developed for PDP models is known as associative learning, whose goal is to store patterns so that they can be evoked in the future. In certain cases, the two modes of learning blend into each other (Rumelhart *et al.*, 1986).

While variations of Hebbian learning are available for both regularity detection and associative learning (Wasserman and Schwartz, 1988), Equation (5) with its proportionality constant, η, equaling 1 is chosen for the approach of this chapter to tune the weights of rules. The reason is twofold. First, tuning the weights to force a set of rules to respond optimally to arbitrary inputs is equivalent to finding salient features of the inputs. This calls for regularity detection. Second, as a test of concept, Equation (5) is appealing from the standpoints of logical implication and mathematical simplicity. Accordingly, we define

$$\Delta w_{ij} = a_i \cdot o_j \tag{6}$$

to govern the learning mechanism of the current system. We further define the activation functions and the output functions for the units (nodes in Figure 9) in the input layer to be

$$a_i = \min\,(I_i), \qquad o_i = a_i, \tag{7}$$

and those for the units (nodes in Figure 9) in the output layer to be

$$a_j = \max\,(O_i), \qquad o_j = a_j \tag{8}$$

In these two equations, I_i's are the inputs to unit i of the input layer, and O_i's are outputs from different units in the input layer, which become the inputs to unit j of the output layer. The activation functions, a_i's, in Equation (7) realize a version of fuzzy "AND" operation, and the activation functions, a_j's, in Equation (8) realize a version of fuzzy "OR" operation (Williams, 1986).

The learning process can be turned on or off at any stage of synthesis. When it is on, the input to each unit in the input layer is the combined truth value of the antecedent (see Figure 9). Each output from the output layer denotes the weighted truth value of rules connoted by the non-null connections between the input and output layers. The synthesis process will pause and seek the user's opinion subsequent to each creation of a processing unit. The changes in the weighting factors, Δw_{ij}'s, will be in the plus direction if the user confirms positively, i.e., if the result of the synthesis is deemed correct by the user; the changes will be in the minus direction if the user responds negatively. Note that because of the user feedback, the present learning mechanism is sometimes categorized as a "supervised" one although the learning process is governed by the "nonsupervised" Hebbian rule, Equation (5).[1] While the exact categorization of learning mechanisms is beyond the scope of this chapter, it is worth mentioning that the human monitoring described above achieves the "gating" effect performed by sigma-pi units in a neural network (Williams, 1986). The "gating" effect can be described as the control of the output value from one unit by multiplying it by a certain quantity before the value enters another unit.

The new weighting factors will be normalized to values between 0 and 1.[2] Two rules govern the normalization to enforce or relax the relative weights among the rules in a rule set.

a. If all rules have been activated during the preceding synthesis cycle, then 1) add a constant to every weighting factor, if necessary, so that the resultant smallest weighting factor becomes positive and at least 0.1, and 2) divide every weighting factor by a constant, if necessary, so that the resultant largest weighting factor does not exceed 1.
b. If only part of the rules has been activated, then 1) maintain the weighting factors of those rules that were not activated; 2) record the original value, κ (the value before learning), of the weighting factor which is the greatest among those that have been adjusted; 3) add a constant to every weighting factor that has been changed, if necessary, so that the resultant smallest weighting factor becomes positive and at least 0.1; and 4) divide every weighting factor that has been changed by a constant so that the resultant largest weighting factor restores its value κ.

A scenario of normalization according to the rules delineated above will be demonstrated with an example.

6. Example

This example consists of two multicomponent feed streams that are to be separated into three multicomponent product streams; the problem has been solved previously (Mahalec and Motard, 1977a; Floudas, 1987). The

[1] This type of learning, falling between supervised and unsupervised learning, is sometimes referred to as reinforcement learning where an external "teacher" merely indicates whether the response to an input is good or bad.

[2] The weight normalization is rarely mentioned in any neural network learning mechanism. Qualitatively, the significance of individual weights, and thereby the performance of a neural network, will not change with or without weight normalization. The reason for defining weight normalization in this work is to simplify the mergence of the concept of fuzzy membership value with that of weight in neural networks.

Table 4. Data for example 1 (Mahalec and Motard, 1977a; Floudas 1987)

	Species			
	A %	B %	C %	Quantity
Feed stream 1	0.168	0.833	0.000	12
Feed stream 2	0.385	0.385	0.231	13
Product stream 1	0.500	0.000	0.500	2
Product stream 2	0.417	0.417	0.167	12
Product stream 3	0.091	0.909	0.000	11

Note: A, B, and C are ranked and the degrees of difficulty of separation between any two adjacent species are assumed to be identical. (Reproduced by permission of the American Institute of Chemical Engineers.)

composition of each stream along with the degree of difficulty of separation is reproduced in Table 4. Some operational details of the current system will be elaborated in this example.

The system commences by acquiring the numbers of feed and product streams and the names of the components (Figure 10a). This gives rise to a spreadsheet to receive quantitatively the problem specifications (Figure 10b). The specifications are validated automatically by the spreadsheet for mass balance before acceptance. Once accepted, the specifications are stored in slots IN.COMPONENTLIST and OUT.COMPONENTLIST of object NEW.SCHEME as illustrated in Figure 4a. The control of the system is then passed to the blackboard for the synthesis.

In the quest for a separation method at the first level of synthesis, the system either resorts to its own (known-case and property) databases or lets the user specify the method. The choice is up to the user through manipulation of a menu (Figure 11). In the default mode, the system searches through the known-case databases first. A separation method will be recommended if it is found fitting or appropriate for the input and output streams specified. This is followed by a search in the property database for the relevant separation factors. If no method is identified in the first database, the system will still proceed with the search in the second. This is to discover a property whose values vary significantly among the specified components. If such a property is determined, a separation method will be selected based on the first set of knowledge sources, i.e., Rules 1.1 to 1.4 in Table 2. If no complete information is obtained after searching through both databases, the user will be prompted to specify a method, or certain separation factors, or both, depending on the missing information. The procedure is summarized in Figure 12. Since no information is available for the fictitious components, A, B, and C, in

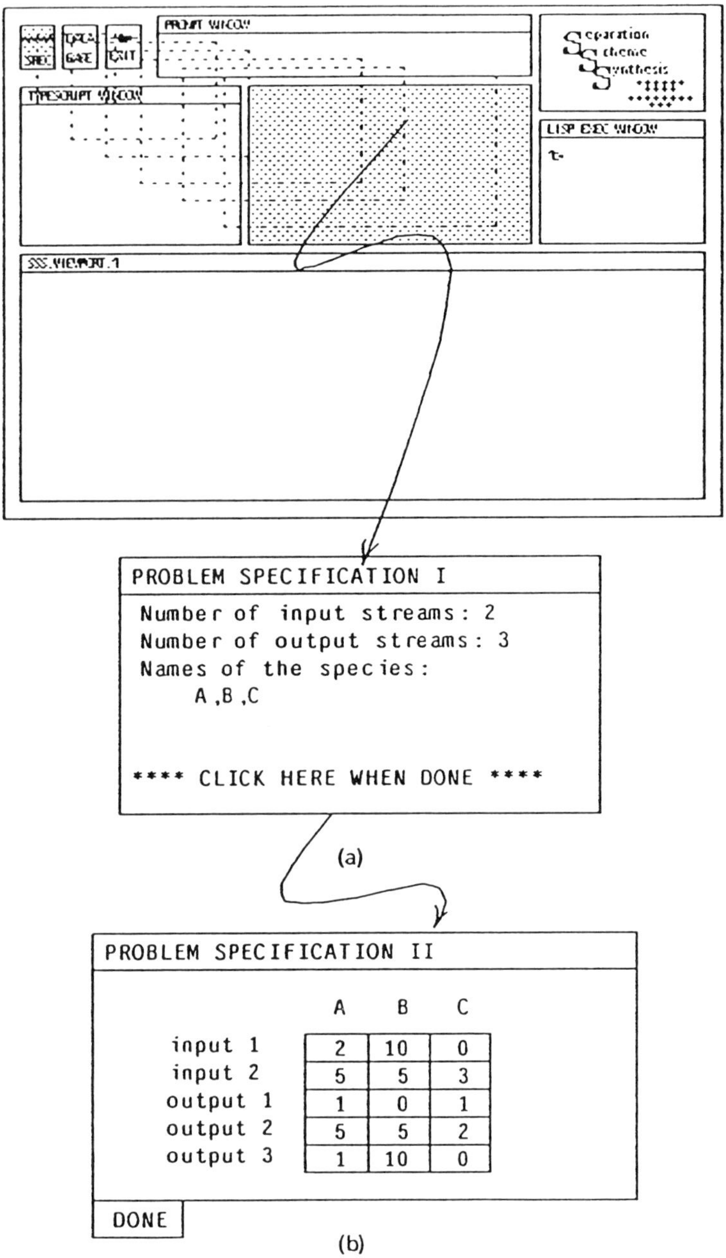

Figure 10. Problem specification in the present system. The sektch on the top portrays the actual screen layout.

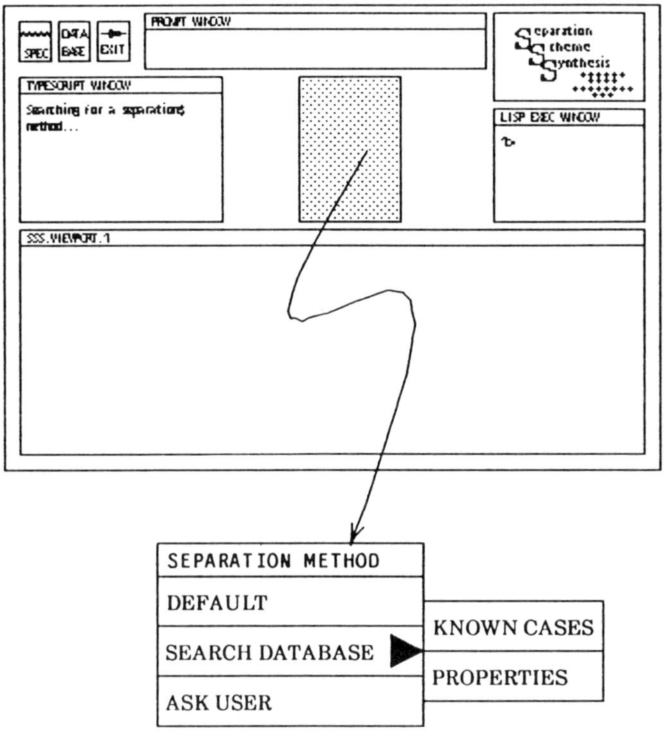

Figure 11. Search for a separation method.

this example, the user is prompted to specify both a separation method and the relevant separation factors. The responses, conforming to the data employed by the original workers (Mahalec and Motard, 1977a; Floudas, 1987), are distillation for the separation method and an equal degree of difficulty for separation between A/B and B/C.

The system engages in the second level of synthesis after a separator type is specified. Separators or dividers will be generated to form a separation train at this level. The decision is made within the discretion of the second set of knowledge sources (Rules 2.1 to 2.10 in Table 2). At this stage, the weighted truth values of the knowledge sources 1 through 10, i.e., Rules 2.1 through 2.10, are 0.0, 0.0, 0.96, 0.0, 0.92, 0.0, 0.0, 0.712, 0.438, and 0.225, respectively; they are briefly explained below.

Knowledge source 1. It is assumed that none of the fictitious components is corrosive.

Knowledge source 2. No by-product exists at this stage.

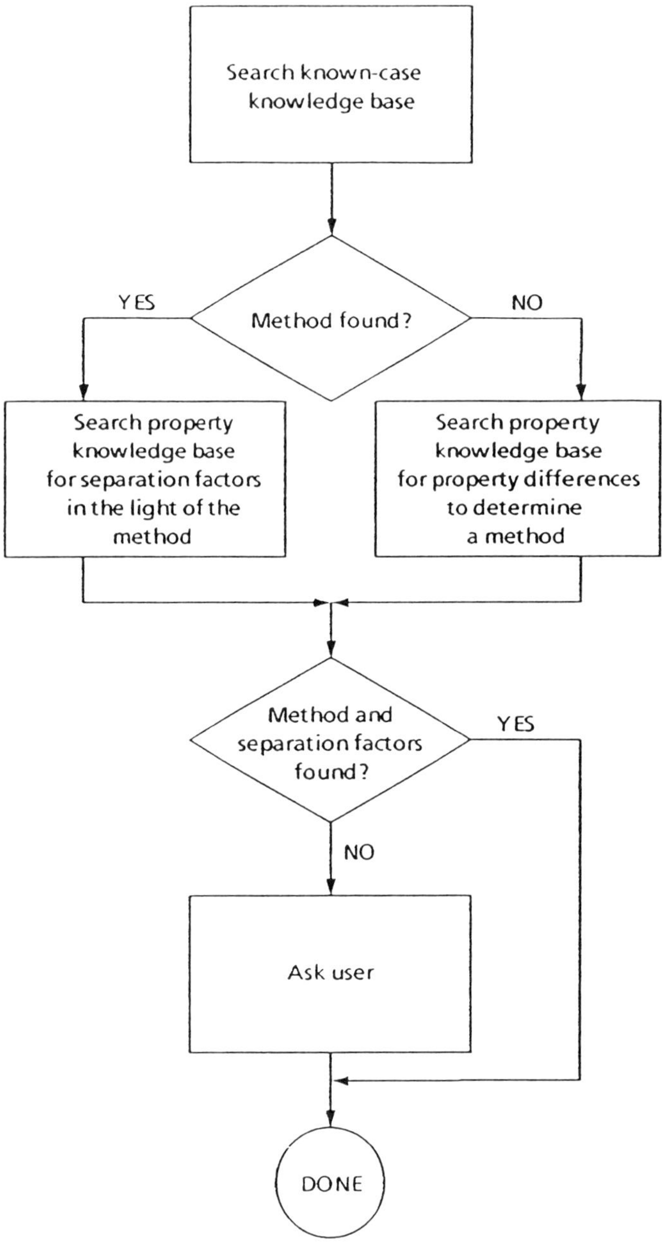

Figure 12. Procedure to search for a separation method.

Knowledge source 3. The feed/product stream pattern reveals that feed stream 1 and product stream 3 consist of the same components and so do feed stream 2 and product stream 1. On the first found basis, feed stream 1 is divided so that one of the by-product streams matches at least partially product stream 3 (see Appendix C). The weighted truth value hence equals

(The truth value of the antecedent of Rule 2.3)

× (The weighting factor of Rule 2.3)

$= 1.0 \times 0.96$

$= 0.96.$

Knowledge source 4. A pattern, indicating that a separation will lead to a by-product stream matching one of the product streams, does not exist (see Appendix C).

Knowledge source 5. The feed/product stream pattern reveals that although product stream 2 has an extraneous component, C, with respect to feed stream 1, the use of a divider is feasible. Specifically, feed stream 1 can be divided so that one of the by-product streams matches partially product stream 2 (see Appendix C). The weighted truth value hence equals

(The truth value of the antecedent of Rule 2.5)

× (The weighting factor of Rule 2.5)

$= 1.0 \times 0.92$

$= 0.92.$

Knowledge source 6. No MSA (mass separating agent) is in use.

Knowledge source 7. The adjacent separation factors are identical.

Knowledge source 8. The truth value of this knowledge source is determined by (see Rule 2.8 in Table 3)

$$\max_{i=1,n}\left[\frac{1}{1+(SP_{i+1}-SP_i)}\right] \wedge \max_{i=1,n}\left[\frac{(Q_{i+1}-Q_i)}{1+(Q_{i+1}-Q_i)}\right]$$

$$= \begin{cases} 1 \wedge \max\left[\dfrac{10-2}{1+10-2}\right] = 0.89 & \text{for feed stream 1} \\ 1 \wedge \max\left[\dfrac{5-5}{1+5-5}, \dfrac{5-3}{1+5-3}\right] = 0.67 & \text{for feed stream 2.} \end{cases}$$

The most feasible separation is determined by the rule with the highest truth value. Thus, the overall truth value for this knowledge source will be based on feed stream 1; it equals

$$\begin{aligned}&(\text{The truth value of the antecedent of Rule 2.8})\\&\times(\text{The weighting factor of Rule 2.8})\\&=0.89\times 0.8\\&=0.712.\end{aligned}$$

Knowledge source 9. The truth value of this knowledge source is determined by (see Rule 2.9 in Table 3)

$$\begin{aligned}&\max_{i=1,n}\left[\frac{1}{1+(SP_{i+1}-SP_i)}\right]\wedge\max_{i=1,n}\left[\frac{1}{1+(Q_{i+1}-Q_i)}\right]\\&\quad\wedge\max_{i=1,(n-1)}\left[\left\|\frac{\sum_{j=1}^{i}Q_j}{\sum_{j=i+1}^{n}Q_j}\right\|\right]\\&=\begin{cases}1\wedge\max\left[\dfrac{1}{1+10-2}\right]\wedge\max\left[\left\|\dfrac{2}{10}\right\|\right]\\[2ex]1\wedge\max\left[\dfrac{1}{1+5-5},\dfrac{1}{1+5-3}\right]\wedge\max\left[\left\|\dfrac{5}{5+3}\right\|,\left\|\dfrac{5+5}{3}\right\|\right]\end{cases}\\&=\begin{cases}1\wedge 0.125\wedge 0.2=0.125 & \text{for feed stream 1}\\1\wedge 1\wedge 0.625=0.625 & \text{for feed stream 2.}\end{cases}\end{aligned}$$

The overall weighted truth value for this knowledge source will be calculated based on feed stream 2. In other words, if this knowledge source wins the opportunity to modify the blackboard at this step, feed stream 2 need be separated into two streams, A and BC. The overall weighted truth value is

$$\begin{aligned}&(\text{The truth value of the antecedent of Rule 2.9})\\&\times(\text{The weighting factor of Rule 2.9})\\&=0.625\times 0.7\\&=0.438.\end{aligned}$$

Knowledge source 10. The truth value of this knowledge source is determined by (see Rules 2.10 in Table 3)

$$\max_{i=1,n}\left[\frac{1}{1+SP_{i+1}-SP_i}\right] \wedge \max_{i=1,n}\left[\frac{1}{1+Q_{i+1}-Q_i}\right]$$

$$\wedge\left(1-\max_{i=1,(n-1)}\left[\left\|\frac{\sum_{j=1}^{i} Q_j}{\sum_{j=i+1}^{n} Q_j}\right\|\right]\right)$$

$$=\begin{cases} 1 \wedge \max\left[\dfrac{1}{1+10-2}\right] \wedge 1-\max\left[\left\|\dfrac{2}{10}\right\|\right] \\ 1 \wedge \max\left[\dfrac{1}{1+5-5}, \dfrac{1}{1+5-3}\right] \wedge 1-\max\left[\left\|\dfrac{5}{5+3}\right\|, \left\|\dfrac{5+5}{3}\right\|\right] \end{cases}$$

$$=\begin{cases} 1 \wedge 0.125 \wedge 0.8 = 0.125 & \text{for feed stream 1} \\ 1 \wedge 1 \wedge 0.375 = 0.375 & \text{for feed stream 2} \end{cases}$$

The overall weighted truth value is

(The truth value of the antecedent of Rule 2.10)

× (The weighting factor of Rule 2.10)

$= 0.375 \times 0.6$

$= 0.225$

Knowledge source 3, i.e., Rule 2.3, clearly dominates at this step because its weighted truth value is 0.96, the greatest among the 10 knowledge sources. This gives rise to the creation of the first divider (solid line) illustrated in Figure 13. The division ratio is decided by the linear programming technique. Note that the labeling of the streams in the figure is for illustration only; a pop-up menu can be activated at any time to retrieve the same information or to override the design. At this point, the system behaves differently depending on whether or not the learning mechanism is activated; the default is not activated. In this mode, the synthesis continues by repeating the steps described so far with 1) a new set of source streams consisting of the unprocessed streams left from the previous step and the unmatched product streams from the divider, 2) a new set of target streams consisting of the residual target streams, and 3) the same set of weights for the rules. If the learning mechanism is activated, the system pauses to inquire of the user for an evaluation of a result. A positive evaluation, as in the present case, results in the weight adjustments portrayed in Table 5.

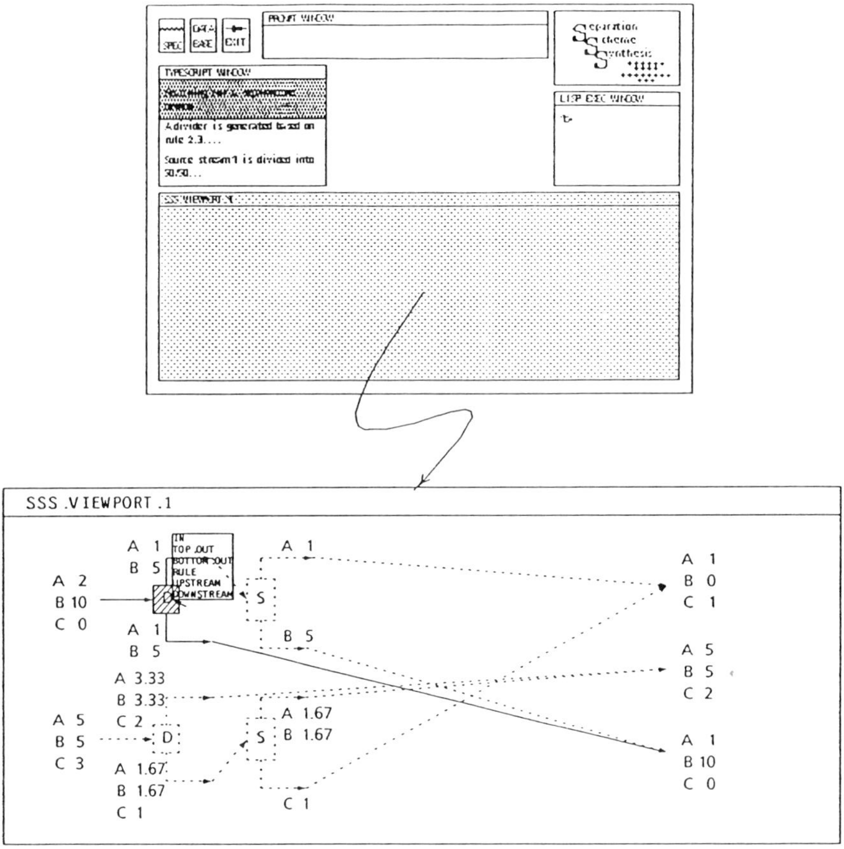

Figure 13. Solution of example 1.

Table 5. A scenario of weight adjustment

Rule No.	Initial Weights	$\Delta w = a_i o_j$	Adjusted Weights	Normalized Weights
2.1.	1.0	0	1.0	1.0
2.2.	0.98	0	0.98	0.98
2.3.	0.96	0.96	1.92	0.96
2.4.	0.94	0	0.94	0.94
2.5.	0.92	0.92	1.84	0.92
2.6.	0.90	0	0.90	0.90
2.7.	0.85	0	0.85	0.85
2.8.	0.80	0.634	1.434	0.72
2.9.	0.70	0.274	0.974	0.49
2.10.	0.60	0.084	0.684	0.34

Note that the adjustments render the important rules more important (with greater relative weights).

An initial separation scheme has been obtained after four iterations of the procedure described above. The result is depicted in Figure 13; it is the same as those obtained by Mahalec and Motard (1977a) and Floudas (1987). Hereafter, the system passes the initial scheme to the next level of (evolutionary) analysis. The scheme remains unchanged, however, after the analysis in this example.

7. Discussion and Conclusion

Several additional problems of separation scheme synthesis have been reworked to test the effectiveness of the proposed system. The examples include those nontrivial cases of Lu (1984), Lu and Motard (1985), and Floudas (1987,1988), the emphasis of which is entirely on sequencing, the most crucial step in generating a quality final separation scheme. Although the other two levels of analysis, namely, method identification and flowsheet evolution, are not fully demonstrated, their roles as integrated parts of a complete system should not be overlooked.

The specifications of the problems are tabulated in Tables 6 through 10. The results for the respective problems solved by the present system are illustrated in Figures 14 through 18. In the examples, the present system generates solutions up to 47% better than those obtained previously. The comparisons are based on the cost index (CI) defined as (Floudas, 1987)

$$CI = \sum [\,(\text{mass load})_i \times (\text{difficulty of separation})_i]^b$$

where b is a constant ($b = 0.6$ according to the well known six-tenth law).

Table 6. Data for example 2 (Lu, 1984)

	Species		
	A	B	C
Feed stream 1	200	120	80
Feed stream 2	25	200	275
Product stream 1	20	80	0
Product stream 2	205	240	355

Note: A, B, and C are ranked and the degrees of difficulty of separation between any two adjacent species are assumed to be identical.

Table 7. Data for example 3 (Lu and Motard, 1985)

	Species			
	A	B	C	D
Feed stream 1	100	100	100	100
Product stream 1	60	50	20	0
Product stream 2	40	50	0	0
Product stream 3	0	0	80	40
Product stream 4	0	0	0	60

Note: A, B, C, and D are ranked and the degrees of difficulty of separation between any two adjacent species are assumed to be identical. (Reproduced by permission of the Pergamon Press.)

Table 8. Data for example 4 (Floudas, 1987)

	Species			
	A	B	C	D
Feed stream 1	15	20	10	15
Product stream 1	5	10	4	10
Product stream 2	10	10	6	5

Note: A, B, C, and D are ranked and the degrees of difficulty of separation are 2.5, 3.0, and 1.5 for A/B, B/C, and C/D, respectively. (Reproduced by permission of the American Institute of Chemical Engineers.)

Table 9. Data for example 5 (Floudas, 1987)

	Species			
	A	B	C	D
Feed stream 1	15.0	20	10	15
Product stream 1	7.5	10	4	10
Product stream 2	7.5	10	6	5

Note: A, B, C and D are ranked and the degrees of difficulty of separation are 2.5, 3.0, and 1.2 for A/B, B/C, and C/D, respectively. (Reproduced by permission of the American Institute of Chemical Engineers.)

Table 10. Data for example 6 (Floudas, 1988)

	Species			
	A	B	C	D
Feed stream 1	0.00	136.08	226.80	181.44
Feed stream 2	45.36	136.08	226.80	181.44
Product stream 1	10.00	100.00	100.00	150.00
Product stream 2	20.00	50.00	200.00	150.00
Product stream 3	15.36	122.16	153.60	62.88

Note: A: Propane
B: Isobutane
C: n-Butane
D: Isopentane

The degrees of difficulty of separation are 2.35, 8.29, and 2.22 for A/B, B/C, and C/D, respectively. (Reproduced by permission of the Institute of Chemical Engineers.)

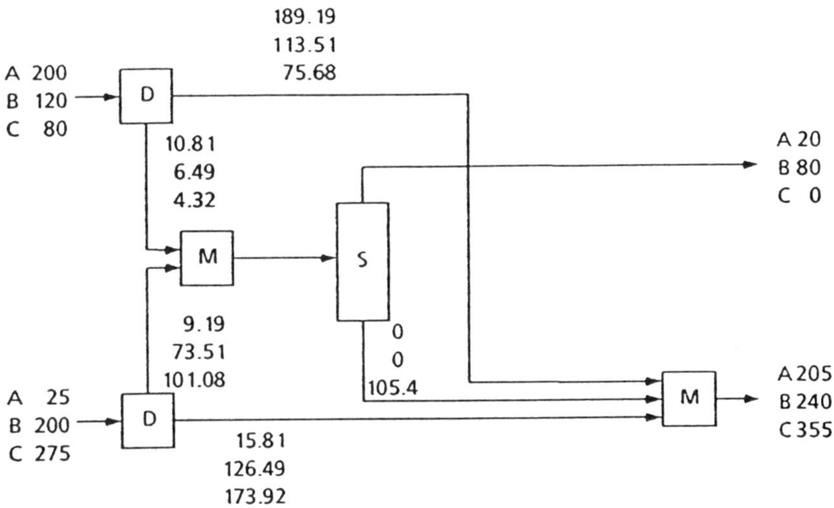

Figure 14. Solution of example 2. It is identical to that obtained by Lu (1984) with successive linear programming.

The adaptive mechanism of the present system follows the simplest nonsupervised Hebbian learning law. The network mapped from a rule set has two layers. A node in the input layer represents the antecedent of a rule; it is connected to one node in the output layer representing the consequence of a rule. This configuration is appropriate when all the propositions in the antecedents are equally weighted and connected with

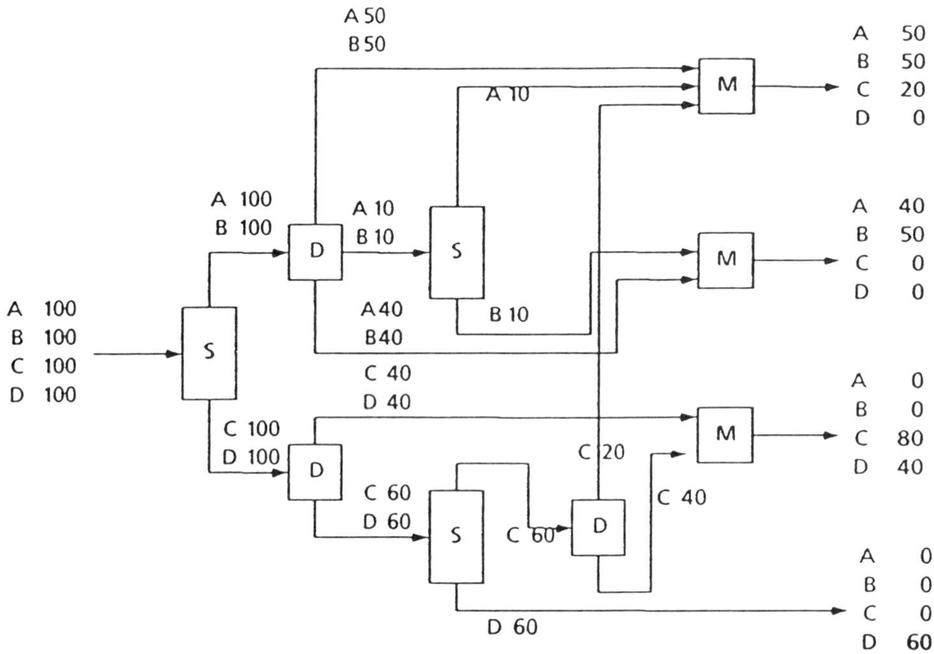

Figure 15. Solution of example 3. It is identical to that obtained by Lu and Motard (1985) after two rounds of evolution of the initial flowsheet.

logical ANDs. Nevertheless, the propositions in the antecedent of a rule may not have equal weighting factors, and they may be connected with logical ORs in some applications. In those cases, a rule set may map into a network with three or more layers. More sophisticated learning laws, such as back propagation (Rumelhart *et al.*, 1986), will be needed to tune the weights in these networks. It is worth mentioning here that although the classical layered neural network model is dubious from the perspective of neurophysiology, it offers a mathematically sound method to effect self-organization of heuristics in performing fuzzy rule-based programming.

It is expected that, as powerful as it is, the present system may not produce the best or globally optimal solutions at all times. A software program or even a human expert guaranteed always to generate the best design is yet to be born in any field. Furthermore, it is our belief that speed is not of primary importance in performing most engineering designs and syntheses on a computer, especially with higher and higher speed processors available at lower and lower costs on a wide spectrum of platforms nowadays. An ideal knowledge-based system at this stage is the one that

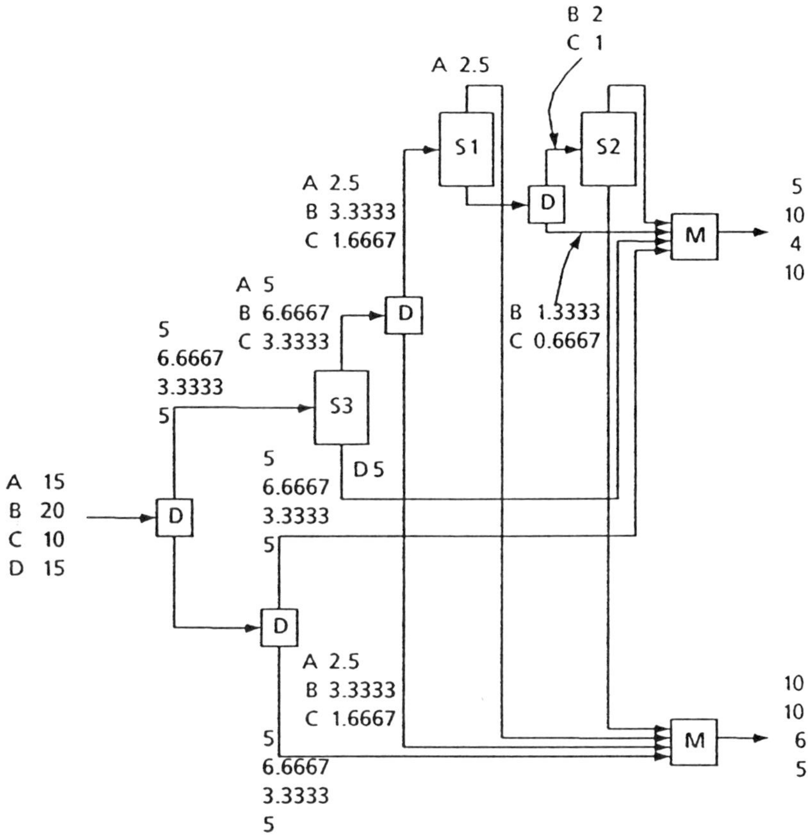

$$\text{Cost index} = (20\times 1.5)^{0.6} + (7.5\times 2.5)^{0.6} + (3.\times 3)^{0.6}$$
$$= 17.238$$

Figure 16. Solution of example 4. It is 7.8% superior to that obtained by Floudas (1987).

interacts intimately with the user throughout the course of design or synthesis and yields a near optimal initial solution most of the time. These are the qualities found in the present system.

Acknowledgment

This is contribution 89-435-B, Department of Chemical Engineering, Kansas Agricultural Experiment Station, Kansas State University, KS 66506, U.S.A.

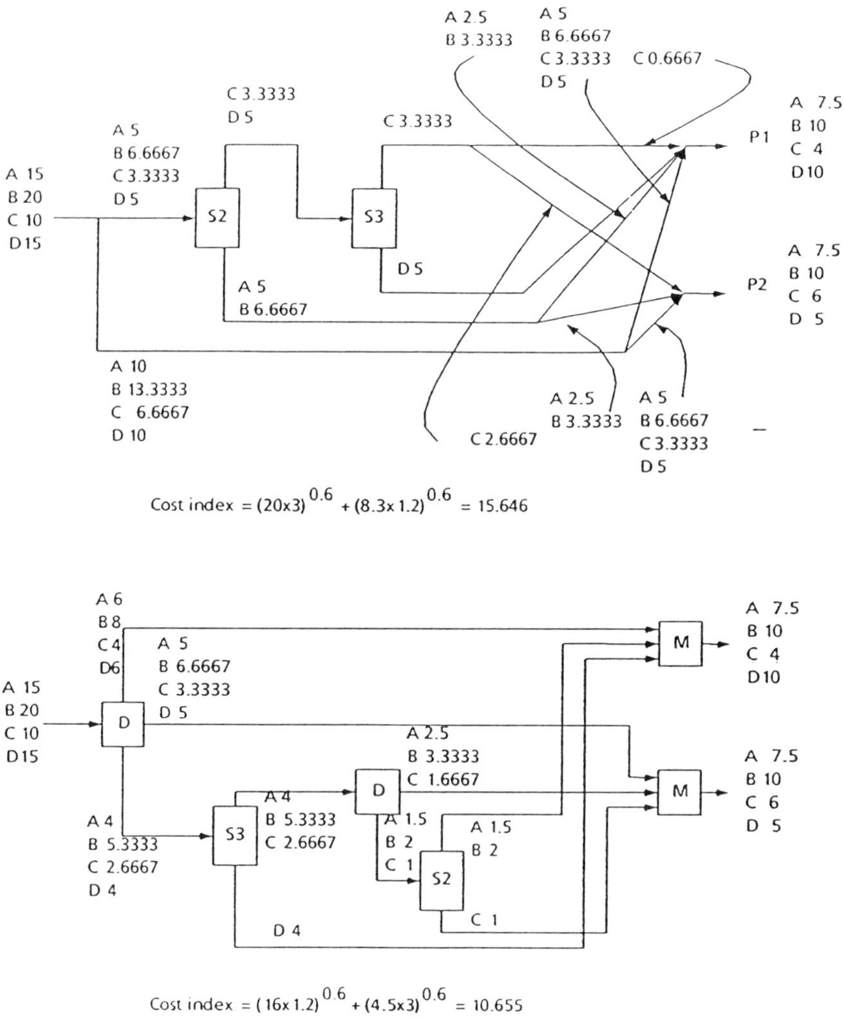

Figure 17. Solution of example 5. It is 46.8% superior to that obtained by Floudas (1987). (a) Separation scheme obtained by Floudas (1987) with nonlinear programming. (b) Separation scheme obtained by the proposed system.

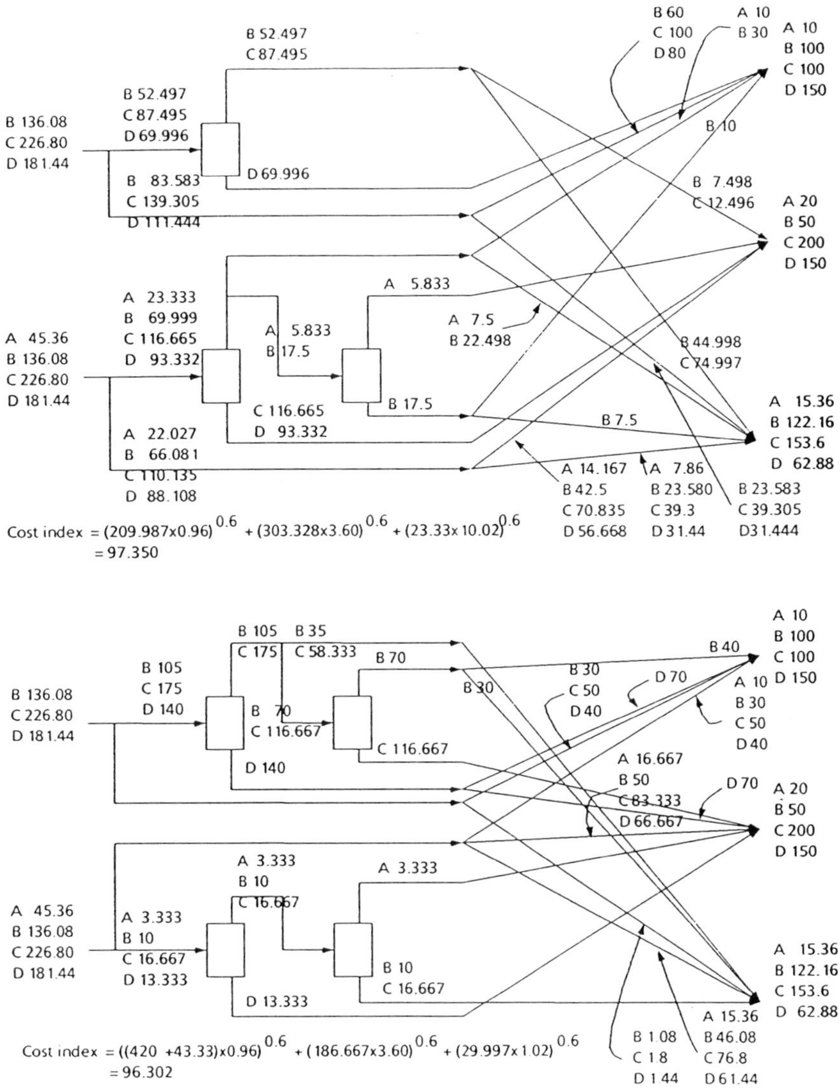

Figure 18. Solution of example 6. It is 1.0% superior to that obtained by Floudas (1988). (a) Separation scheme obtained by Floudas (1988) with MLP. (b) Separation scheme obtained in the present study with heuristics.

Appendix A: Quantitative Expression for the Term "Vary Widely"

Suppose that P is the property underlying separation method M and that P_1 and P_2 ($P_2 > P_1$) are the property values for species 1 and 2, respectively. The degree of difficulty in separating the two species based on method M is known to be a function of $(P_2 - P_1)$, i.e., the difference in the values of the property underlying the separation method (Nadgir and Liu, 1983; Fan *et al.*, 1987). Specifically, the greater the value of $(P_2 - P_1)$, i.e., the wider the property values vary, the easier the separation.

Clearly, the difference $(P_2 - P_1)$ by itself is the simplest quantitative index to the term "vary widely" in separation. However, since theorecti-cally,

$$0 \leq (P_2 - P_1) < \infty, \tag{A.1}$$

It is inconvenient to adopt the index for a formal analysis, especially one involving fuzzy logic. It is highly desirable to generate an index normalized to the range of [0, 1]. This can be accomplished in various ways (Kaufmann, 1975); one possible way is to define the index as

$$\frac{1}{1 + k(P_2 - P_1)^n} \tag{A.2}$$

Taking the complement of this normalized index, we obtain

$$0 \leq 1 - \frac{1}{1 + k(P_2 - P_1)^n} \leq 1, \tag{A.3}$$

which leads naturally to the following membership function for the linguistic term "vary widely":

$$\mu_{\text{``vary widely''}}(P_2 - P_1) = \frac{k(P_2 - P_1)^n}{1 + k(P_2 - P_1)^n}. \tag{A.4}$$

In these expressions, $k\,(> 0)$ and $n\,(> 0)$ are two parameters affecting the sensitivity of the index. The larger the k or the n, the more sensitive the index (Figure A.1).

Let us focus on k (and assume $n = 1$) as an example to demonstrate the significance of the parameters. It is obvious that in temperature measurement, a 5°C deviation is identical to a 9°F deviation. Consequently, the membership function, Equation (A.4), should have a value of $k = 9$ in describing Celsius temperatures and a value of $k = 5$ in describing Fahrenheit temperatures when comparing the temperature differences in

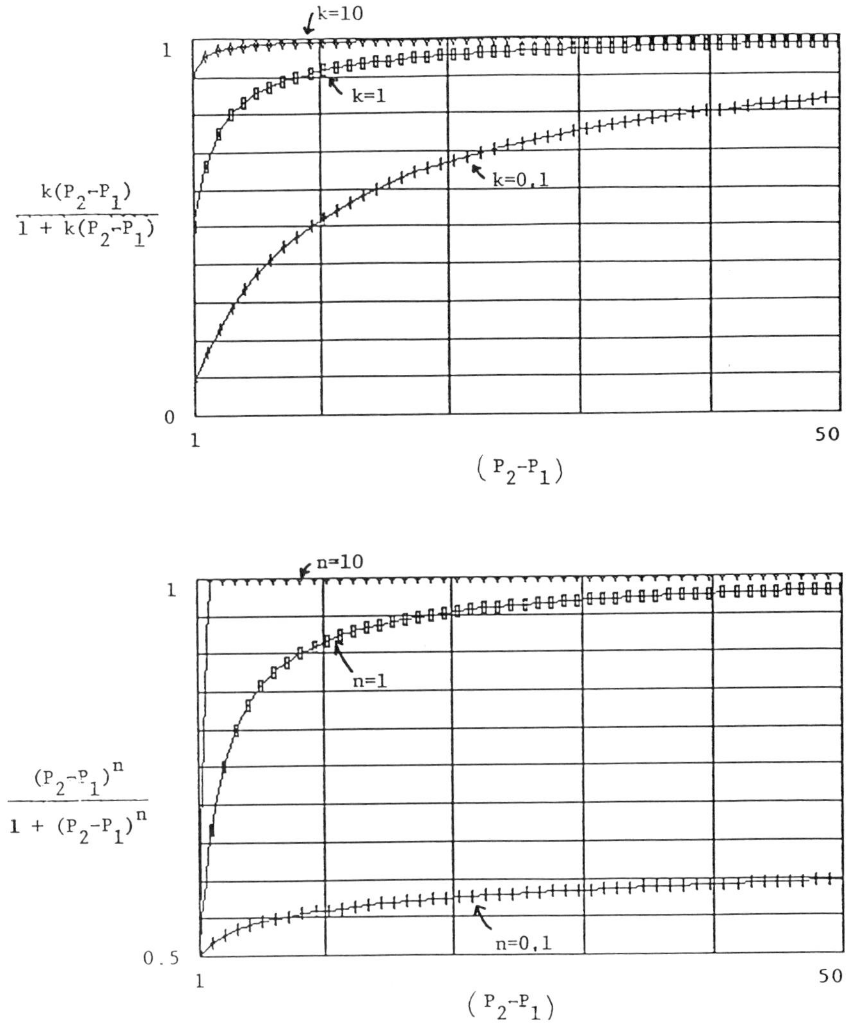

Figure A.1. Effects of k and n on the sensitivity of membership function "vary widely."

two scales. Nevertheless, in applications where properties under consideration do not correlate with one another, the values for k and n can be conveniently assigned to be 1. This leads to Equations (1) and (2) in the text.

It is worth commenting that at least two additional normalization schemes are frequently employed in yielding a membership function for linguistic terms such as "vary widely" (Kaufmann, 1975; Fan *et al.*, 1987). The resultant membershp functions are

$$\mu'_{\text{"vary widely"}}(P_2 - P_1) = 1 - e^{-k(P_2 - P_1)^n} \tag{A.5}$$

and

$$\mu''_{\text{"vary widely"}}(P_2 - P_1) = \left[\frac{(P_2 - P_1)}{P_2}\right]. \tag{A.6}$$

Nevertheless, neither equation is adopted in the present work. The first is rejected for the very steep slope generated by the exponential function; the second for the dependence on P_2.

Appendix B: Fuzzy Membership Function Representing the Antecedent of Rule 2.9

Recall that Rule 2.9 reads

```
IF   adjacent separation properties do not vary widely
     (¬S),
AND  relative quantities do not vary widely (¬Q),
AND  a near 50/50 separation exists (F)
THEN separate at the near 50/50 position.
```

The fuzzy membership function for the whole antecedent is

$$\begin{aligned}
&\mu_{\text{"adjacent separation properties do not ..."}}(*)\\
&\quad = \mu_{\neg S,\neg Q,F}(*)\\
&\quad = \mu_{\text{"adjacent separation properties do not vary widely"}}(*)\\
&\qquad\qquad \wedge \mu_{\text{"relative quantities do not vary widely"}}(*)\\
&\qquad\qquad \wedge \mu_{\text{"a near 50/50 separation exists"}}(*)\\
&\quad = \mu_{\neg S}(*) \wedge \mu_{\neg Q}(*) \wedge \mu_F(*).
\end{aligned} \tag{B.1}$$

It has been presented in Appendix A that for a two-component system, the concept of "vary widely" between two measurements of the property P can be quantified by

$$\mu_{\text{"vary widely"}}(P_2 - P_1) = \frac{(P_2 - P_1)}{1 + (P_2 - P_1)}. \tag{B.2}$$

The complement of this expression yields

$$\mu_{\text{“do not vary widely”}}(P_2 - P_1) = 1 - \frac{(P_2 - P_1)}{1 + (P_2 - P_1)}$$

$$= \frac{1}{1 + (P_2 - P_1)}. \tag{B.3}$$

This expression can be generalized for a multicomponent system as

$$\mu_{\text{“do not vary widely”}}(P_2 - P_1) = \max_{i=1,n}\left[\frac{1}{1 + (P_{i+1} - P_i)}\right]. \tag{B.4}$$

Accordingly, $\mu_{\neg S}(^*)$ and $\mu\ _Q(^*)$ in equation (B.1) are, respectively,

$$\mu_{\neg S}(^*) = \max_{i=1,n}\left[\frac{1}{1 + (SP_{i+1} - SP_i)}\right] \tag{B.5}$$

and

$$\mu_{\neg Q}(^*) = \max_{i=1,n}\left[\frac{1}{1 + (Q_{i+1} - Q_i)}\right]. \tag{B.6}$$

$\mu_F(^*)$ in equation (B.1) is quantified as

$$\max\left[\left\|\frac{Q_1}{Q_2 + \cdots + Q_n}\right\|, \left\|\frac{Q_1 + Q_2}{Q_3 + \cdots + Q_n}\right\|, \ldots, \left\|\frac{Q_1 + \cdots Q_{n-1}}{Q_n}\right\|\right]$$

$$= \max_{i=1,(n-1)}\left[\left\|\frac{\sum_{j=1}^{i} Q_j}{\sum_{j=i+1}^{n} Q_j}\right\|\right], \tag{B.7}$$

where

$$\|X\| = \begin{cases} X & \text{if } X \leq 1 \\ 1/X & \text{if } X > 1. \end{cases} \tag{B.8}$$

Clearly,

$$0 < \|X\| \leq 1.$$

The greater the value of $\|X\|$, the higher the availability of a 50/50 separation. The “max” operation in equation (B.7) identifies the position with the most obvious 50/50 separation. Combining equations (B.5) through (B.7) with fuzzy logic AND operators yields a quantitative expression for the right-hand side of equation (B.1); this leads to equation (3) in the text.

Appendix C: Pattern Recognition for Stream Division or Separation

The feasibility of performing stream division or separation can be determined by the feed/product stream relation defined by a matrix-like format

$$M_i = [\mathbf{F}_i/\mathbf{P}] = \begin{bmatrix} F_{i1}/P_{11} & F_{i2}/P_{12}\cdots \\ F_{i1}/P_{21} & F_{i2}/P_{22}\cdots \\ \vdots & \vdots \quad \vdots \end{bmatrix},$$

where $\mathbf{F}_i$ represents the i-th feed stream, $\mathbf{P}$ the product streams, F_{ij} the amount of component j in feed stream i, and P_{kj} the amount of component j in product stream k. Furthermore, we define

$$F_{ij}/P_{kj} = \begin{cases} 1 & \text{if } F_{ij} = P_{kj} \\ 1/2 & \text{if } F_{ij} > 0,\ P_{kj} > 0, \text{ and } F_{ij} \neq P_{kj} \\ 0 & \text{if } F_{ij} = 0 \text{ and } P_{kj} > 0 \\ -1 & \text{if } F_{ij} > 0 \text{ and } P_{kj} = 0. \end{cases}$$

Literally, we observe the following.

1. A combination of "1" and "1/2" in the kth row of matrix M_i signifies that (1) feed stream i and product stream k consist of the same components, and (2) feed stream i can be divided so that one of the by-product streams matches product stream k. When two or more rows exhibit this pattern, feed stream i can be divided simultaneously to yield by-product streams matching at least partially each of the product streams designated by the row numbers. The optimal ratio of division will be determined by linear programming. When the pattern is detected in more than one matrix (i.e., when more than one source stream can be divided to yield by-product streams matching one or more product streams), the stream divisions can be performed on the basis of one feed stream at a time. The order is insignificant.
2. A pattern of $[1\,1\ldots -1\,-1\ldots]$ or $[-1\,-1\ldots 1\,1\ldots]$ (a series of "1" followed by a series of -1 or vice versa) in the kth row indicates that a separation of feed stream i will yield a by-product stream matching completely product stream k. When two or more rows exhibit this pattern (i.e., when feed stream i can be separated in more than one way to match completely different product streams), it should be separated so that one of the resultant by-product streams matches one of the product streams. The selection may be made on the first found basis. If the pattern is detected in more than one matrix, i.e., if more than one feed stream can be separated

to yield by-product streams matching completely one or more product streams, then the separation can be performed one feed stream at a time. The order is insignificant.

3. A combination of "1," "1/2," and "0" with at least one "0" in the kth row also indicates that, although product stream k has some extraneous components with respect to feed stream i, the feed stream can be divided so that one of the by-product streams matches product stream k. When two or more rows exhibit this pattern, feed stream i can be divided simultaneously to yield by-product streams matching partially each of the product streams designated by the row number. The optimal ratio of division will be determined by linear programming. When the pattern is detected in more than one matrix (i.e., when more than one source stream can be divided to yield by-product streams matching one or more product streams), the stream division can be performed one feed stream at a time. The order is insignificant.

For example 1 in the text, we have

$$M_1 = \begin{bmatrix} 2/1 & 10/0 & 0/1 \\ 2/5 & 10/5 & 0/2 \\ 2/1 & 10/10 & 0/0 \end{bmatrix} = \begin{bmatrix} 1/2 & -1 & 0 \\ 1/2 & 1/2 & 0 \\ 1/2 & 1 & 1 \end{bmatrix}$$

and

$$M_2 = \begin{bmatrix} 5/1 & 5/0 & 3/1 \\ 5/5 & 5/5 & 3/2 \\ 5/1 & 5/10 & 3/0 \end{bmatrix} = \begin{bmatrix} 1/2 & -1 & 1/2 \\ 1 & 1 & 1/2 \\ 1/2 & 1/2 & -1 \end{bmatrix}.$$

Note that the third row of M_1 and the second row of M_2 have the pattern of all "1" and "1/2"; thus these matrices reveal that feed stream 1 and 2 can be divided to match product streams 3 and 2, respectively. Nevertheless, the system will perform stream division to match one product stream at a time. The ordering is on the first found basis. At this step, stream 1 will be split first to match product stream 3.

References

Andrecovich, M. J., and Westerberg, A. W. An MILP Formulation for Heat-Integrated Distillation Sequence Synthesis. *AIChE J.*, 31, pp. 1461–1474 (1985).

Barr, A., and Feigenbaum, *The Handbook of Artificial Intelligence*, Vol. 1. William Kaufmann, Los Altos, CA (1981).

Biess, G., Gruhn, G., and Janicke, W. Synthesis of Multicomponent Separation Trains using Shortest Path Algorithms. *Hungarian J. Ind. Chem.*, 14, pp. 91–97 (1986).

Buchanan, B. G., and Shortliffe, E. H. *Rule-Based Expert Systems*, pp. 302–330. Addison-Wesley, Reading, MA (1984).

Fan, L. T., Gharpuray, M. M., and Huang, Y. W. Sequencing a Separation via Fuzzy Heuristics. *Chem. Engg.*, 94 (3), pp. 57–59 (1987).

Floudas, C. A., Separation Synthesis of Multicomponent Feed Stream into Multicomponent Product Stream. *AIChE J.*, 33, pp. 540–550 (1987).

Floudas, C. A. Multicomponent Multi-Feed Multi-Product Synthesis of Distillation Sequences. *Inst. Chem. Eng. Symp. Ser.*, No. 109, pp. 61–73 (1988).

Harbert, W. D. Which Tower Goes Where. *Petroleum Refiner*, 36 (3), pp. 169–174 (1957).

Heaven, D. L. Optimum Sequencing of Distillation Columns in Multicomponent Fractionation. MS Thesis, University of California, Berkeley, CA (1969).

Hebb, D. O. *The Organization of Behavior.* Wiley, New York, NY (1949).

Hendry, J. E., and Hughes, R. R. Generating Separation Process Flowsheets. *Chem. Engg. Prog.*, 68 (6), pp. 71–76 (1972).

Hendry, J. E., Rudd, D. F., and Seader, J. D. Synthesis in the Design of Chemical Processes. *AIChE J.*, 19, pp. 1–15 (1973).

Huang, Y. W. Designing Intelligent Knowledge & Data Bases for Separation Scheme Synthesis—A Case Study of Knowledge Engineering in Process Engineering. PhD Dissertation, Chapter 2, Kansas State University, Manhattan, KS (1988).

Huang, Y. W. and Fan, L. T. Designing an Object-Relation Hybrid Database for Chemical Process Engineering. *Comput. Chem. Engg.*, 12, pp. 973–983 (1988a).

Huang, Y. W. and Fan, L. T. Fuzzy Logic Rule Based System for Separation Sequence Synthesis: An Object-Oriented Approach. *Comput. Chem. Engg.*, 12, pp. 601–607 (1988b).

Kaufmann, A. *Introduction to the Theory of Fuzzy Subsets*, Vol. 1., pp. 168–171, Academic Press, New York, NY (1975).

King, C. J. *Separation Process*, pp. 665–675, McGraw-Hill, New York, NY (1971).

King, C. J. Selection and Sequencing of Separations. In *New Developments in Separation Methods* (E. Grushka, ed.), pp 1–11, Marcel Dekker, New York, NY (1976).

Kosko, B. Fuzziness and Neural Networks. *Proceedings of International Workshop on Fuzzy System Applications*, IFSA, Kyushu Institute of Technology, pp. 5–7 (1988).

Lockhart, F. J. Multi-Column Distillation of Natural Gasoline. *Petroleum Refiner*, 26, pp. 104–108 (1947).

Lu, M. D. A Study of General Flowsheet Synthesis. Ph.D. Dissertation, Washington University, St. Louis, MO (1984).

Lu, M. D., and Motard, R. L. A Strategy for the Synthesis of Separation Sequences. *Inst. Chem. Engg. Symp. Ser.*, No. 74, pp. 141–151 (1982).

Lu, M. D., and Motard, R. L. Computer-Aided Total Flowsheet Synthesis. *Comput. Chem. Engg.*, 9, pp. 431–445 (1985).

Mahalec, V., and Motard, R. L. Procedures for the Initial Design of Chemical Process Systems. *Comput. Chem. Engg.*, 1, pp. 57–68 (1977a).

Mahalec, V., and Motard, R. L. Evolutionary Search for an Optimal Limiting Process Flowsheet. *Comput. Chem. Engg.*, 1, pp. 149–160 (1977b).

McClelland, J. L., Rumelhart, D. E., and Hinton, G. E. The Appeal of Parallel Distributed Processing. In *Parallel Distributed Processing: Explorations in the Microstructure of Cognition* (D. E. Rumelhart, *et al.*, eds.), Vol. 1, pp. 3–44, MIT Press, Cambridge, MA (1986).

Mocsny, D. J. Studies in the Synthesis of Heat Exchanger Networks and Separation Systems. Ph.D. Dissertation, University of Cincinnati, Cincinnati, OH (1986).

Muraki, M., and Hayakawa, T. Synthesis of Distillation Separation Process with Energy Integration. *J. Chem. Eng. Jpn.*, 14, pp. 233–238 (1981).

Muralidhara, H. S. The Combined-Fields Approach to Separations. *Chemtech*, 18 (4), pp. 229–235 (1988).

Nadgir, V. M., and Liu, Y. A. Studies in Chemical Process Design and Synthesis: Part V: A Simple Heuristic Method for Systematic Synthesis of Initial Sequences for Multicomponent Separations. *AIChE J.*, 29, pp. 926–934 (1983).

Naka, Y., Terashita, M., and Takamatsu, T. A Thermodynamic Approach to Multicomponent Distillation Systems Synthesis. *AIChE J.*, 28, pp. 812–820 (1982).

Netterfield, T, and Sunol, A. K. An Expert System for Separation Technology Selection. Paper presented at the AIChE Annual Meeting, New York, Nov. 15–20 (1987).

Nii, H. P. Blackboard Systems: The Blackboard Model of Problem Solving and the Evolution of Blackboard Architectures. *AI Magazine*, 7 (2), pp. 38–53 (1986).

Nishida, N., Stephanopoulos, G., and Westerberg, A. W. A Review of Process Synthesis. *AIChE J.*, 27, pp. 321–351 (1981).

Nishimura, H., and Hiraizumi, Y. Optimal System Pattern for Multicomponent Distillation Systems. *Intern. Chem. Eng.*, 11, pp. 188–193 (1971).

Null, H. R. Selection of a Separation Process. In *Handbook of Separation Process Technology* (R. W. Rousseau, ed.), Chap. 22, Wiley, New York, NY (1987).

Perry, J. H. *Chemical Engineers' Handbook*, 5th edn., pp. 3-6–3-44, McGraw Hill, New York, NY (1973).

Pibouleau, L., and Domenech, S. Discrete and Continuous Approaches to the Optimal Synthesis of Distillation Sequences. *Comput. Chem. Eng.*, 10, pp. 479–491 (1986).

Rathore, R. N. S., Wormer, K. A. V., and Power, G. J. Synthesis Strategies for Multicomponent Separation Systems with Energy Integration. *AIChE. J.*, 20, pp. 491–502 (1974).

Reid, R. C., Prausnitz, J. M., and Sherwood, T. K. *The Properties of Gases and Liquids*, 3rd ed., McGraw Hill, New York, NY (1977).

Rumelhart, D. E., McClelland, J. L., and the PDP Research Group. *Explorations in the Microstructure of Cognition, Vol. 1: Foundations*, MIT Press, Cambridge, MA (1986).

Shieh, J. H., and Fan, L. T. *Energy and Exergy Estimation Using the Group Contribution Method, ACS Symposium Ser.* No. 235, pp. 351–371, ACS, Washington, DC (1983).

Shoaei, M., and Sommerfeld, J. T. Catalan Numbers in Process Synthesis. *AIChE. J.*, 32, pp. 1931–1933 (1986).

Stephanopoulos, G., and Westerberg, A. W. Studies in Process Synthesis—II. Evolutionary Synthesis of Optimal Process Flowsheets. *Chem. Engg. Sci.*, 31, pp. 195–204 (1976).

Stefik, M., and Bobrow, D. G. Object-Oriented Programming: Theme and Variation. *AI Magazine*, 6 (4), pp. 40–62 (1985).

Thompson, R. W., and King, C. J. Systematic Synthesis of Separation Schemes. *AIChE J.*, 18, pp. 941–948 (1972).

Ulrich, G. D. *A Guide to Chemical Engineering Process Design and Economics*, Wiley, New York, NY (1984).

Umeda, T. Computer Aided Process Synthesis. *Comput. Chem. Eng.*, 7, pp. 279–309 (1983).

Umeda, T., Niida, K., and Shiroko, K. A. Thermodynamic Approach to Heat Integration in Distillation Systems, *AIChE J.*, 25, pp 423–429 (1979).

Wasserman, P. D., and Schwartz, T. Neural Network, Part 2. *IEEE Expert*, 3 (1), pp. 10–15 (1988).

Weast, R. C., and Astle, M. J. *Handbook of Data on Organic Compounds*, CRC Press, Boca Raton, FL (1985).

Westerberg, A. W. A Review of Process Synthesis, *ACS Symp. Ser.*, No. 124, pp. 53–87, ACS, Washington, DC (1980).

Westerberg, A. W., and Andrecovich, M. J. Utility Bound Nonconstant QWT for Heat-Integrated Distillation Sequence Synthesis. *AIChE J.*, 31, pp. 1475–1479 (1985).

Westerberg, A. W., and Stephanopoulos, G. Studies in Process Synthesis—I. Branch and Bound Strategy with List Techniques for the Synthesis of Separation Schemes. *Chem. Engg. Sci.*, 30, pp. 963–972 (1975).

Williams, R. J. The Logic of Activation Functions. In *Parallel Distributed Processing: Explorations in the Microstructure of Cognition* (D. E. Rumelhart, et al., eds.), Vol. 1, pp 423–443, MIT Press, Cambridge, MA (1986).

Zadeh, L. A. Fuzzy Logic. *IEEE Computer*, 21 (4), pp. 83–92 (1988).

Zimmermann, H. J. *Fuzzy Set Theory and Its Application*, pp. 121–156, Kluwer Academic, Hingham, MA (1985).

Other Suggested Readings

Bamopoulos, G. Synthesis of Non-Sharp Separation Sequences. PhD Dissertation, Washington University, St. Louis, MO (1984).

Bamopoulos, G., Nath, R., and Motard, R. L. Heuristic Synthesis of Nonsharp Separation Sequences. *AIChE J.*, 34, pp. 763–780 (1988).

Bheemaiah, P., Husain, A., Devi, G. R., and Gangiah, K. Synthesis of Separation Sequences based on Heuristics. *J. Sci. Ind. Res.*, 37 (4), 175–183 (1978).

Chen, M. R. Process Synthesis of Distillation Sequences—A Multiple Objective Approach. PhD Dissertation, Columbia University, New York, NY (1985).

Cheng, S. H. Systematic Synthesis of Sloppy Multicomponent Separation Sequences. PhD Dissertation, Virginia Polytechnic Institute and State University, Blacksburg, VA (1987).

Cheng, S. H., and Liu, Y. A. Heuristic Synthesis of Sloppy Multicomponent Separation Sequences. Paper no. 91a, AIChE Annual Meeting, New York, NY, Nov. 15–20 (1987).

Debelak, K. A., Leuze, M. R., Bourne, J. R., Whitlow, J. E., Antao, B. A., Patino-Siliceo, O., and Pruett, D. J. A Prototype Expert System for Separation Science. Paper presented at the AIChE Summer National Meeting, Minneapolis, MN Aug. 16 (1987).

Floudas, C. A., and Anastasiadis, S. H. Synthesis of Distillation Sequences with Several Multicomponent Feed and Product Streams. *Chem. Engg. Sci.*, 43, pp. 2407–2419 (1988).

Floudas, C. A., and Paules, G. E. A Mixed-Integer Nonlinear Programming Formulation for the Synthesis of Heat-Integrated Distillation Sequence. *Comput. Chem. Engg.*, 12, pp. 531–546 (1988).

Gadkari, P. B. Analysis and Synthesis of Distillation Column Sequences Using Total Interstage Flow. PhD Dissertation, University of Cincinnati, Cincinnati, OH (1987).

Gharpuray, M. M. Studies in Computer-Aided Process Design Using Fuzzy Set Theory. PhD Dissertation, Kansas State University, Manhattan, KS (1985).

Gharpuray, M. M., Tazaki, E., and Fan, L. T. A Fuzzy Heuristic Approach for Synthesis of Separation Sequences. In *Analysis of Fuzzy Information*, Vol. 2, (J. C Bezdek, ed.), pp. 89–106, CRC Press, Boca Raton, FL (1987).

Gomez- Munoz, A., and Seader, J. D., Separation Sequence Synthesis by a Predictor Based Ordered Search. *AIChE J.*, 22, pp. 970–979 (1976).

Gomez-Munoz, A., and Seader, J. D. Synthesis of Distillation Train by Thermodynamic Methods. *Comput. Chem. Eng.*, 9, pp. 311–341 (1985).

Hendry, J. E. Computer-Aided Synthesis of Optimal Multicomponent Separation Sequences. PhD Dissertation, University of Wisconsin, Madison, WI (1972).

Hohmann, E. C., Sander, M. T., and Dunford, H. New Approach to the Synthesis of Multicomponent Separation Schemes. *Chem. Engg. Comm.*, 17, pp. 273–284 (1982).

Johns, W. R., and Romero, D. The Automated Generation and Evaluation of Process Flowsheets. *CACE'79*, Montreux, Switzerland, Apr. 8–11 (1979).

Kadhem, A, Meszaros, I., and Fonyo, Z. Nonsharp Distillation Systems: Parametric Study for Distributing Recovery Ratios. *Hungarian J. Ind. Chem.*, 15, pp. 197–209 (1987).

Keller, G. E. Separation: New Directions for an Old Field. *AIChE Monograph Series*, Vol. 83 (1987).

Liu, Y. A. Process Synthesis: Some Simple and Practical Developments. In *Recent Developments in Chemical Process and Plant Design* (Y. A. Liu, Jr. McGee, H. A., and W. R. Epperly eds.), pp. 147–260, Wiley, New York (1987).

Lu, M. D., and Motard, R. L. An Expert System for Computer-Aided Flowsheet Synthesis, *Inst. Chem. Engg. Symp. Ser.*, 92, pp. 517–528 (1985).

Mocsny, D., and Govind, R. Feasibility of Partitioning Multicomponent Feeds in the Minimum Number of Splits. *AIChE J.*, 33, pp. 694–696 (1987).

Muraki, M., and Hayakawa, T. Separation Process Synthesis for Multicomponent Products. *J. Chem. Eng. Jpn.*, 17, pp. 533–538 (1984).

Muraki, M., and Hayakawa, T. Multicomponent Products Separation Process Synthesis with Separation Sharpness. *J. Chem. Eng. Jpn.*, 20, pp. 195–198 (1987).

Muraki, M., and Hayakawa, T. Synthesis of a Multicomponent Multiproduct Separation Process with Nonsharp Separators. *Chem. Engg. Sci.*, 43, pp. 259–268 (1988).

Muraki, M., Kataoka, K., and Hayakawa, T. Evolutionary Synthesis of a Multicomponent Multiproduct Separation Process. *Chem. Engg. Sci.*, 41, pp. 1843–1851 (1986).

Nath, R. Studies in the Synthesis of Separation Processes. PhD Dissertation, University of Houston, Houston, TX (1977).

Nath, R., and Motard, R. L. Evolutionary Synthesis of Separation Processes.. *AIChE J.*, 27, pp. 578–587 (1981).

Paules, G. E., IV, and Floudas, C. A. Synthesis of Flexible Distillation Sequences for Multiperiod Operation. *Comput. Chem. Engg.*, 12, pp. 267–280 (1988).

Powers, G. J. Heuristic Synthesis in Process Development. *Chem. Engg. Prog.*, 68 (8), pp. 88–95 (1972).

Rodrigo, F. R., and Seader, J. D. Synthesis of Separation Sequences by Ordered Branch Search. *AIChE J.*, 21, pp. 885–894 (1975).

Seader, J. D., and Westerberg, A. W. A Combined Heuristic and Evolutionary Strategy for Synthesis of Simple Separation Sequences. *AIChE J.*, 23, pp. 951–954 (1977).

Siirola, J. J., Powers, G. J., and Rudd, D. F. Synthesis of System Designs: III. Toward a Process Concept Generator. *AIChE J.*, 17, pp. 677–682 (1971).

Siirola, J. J., and Rudd, D. F. Computer-Aided Synthesis of Chemical Process Designs from Reaction Path Data to the Process Task Network. *I & EC Fundamentals*, 10, pp. 353–362 (1971).

Souders, M. The Countercurrent Separation Process. *Chem. Engg. Prog.*, 60 (2), pp. 75–82 (1964).

Stephanopoulos, G., Johnston, J., Kriticos, T., Lakshmanan, R., Mavrovouniotis, M., and Siletti, C. DESIGN-KIT: An Object-Oriented Environment for Process Engineering, *Comput. Chem. Engg.*, 11, pp. 655–674 (1987).

Tedder, D. W. The Heuristic Synthesis and Topology of Optimal Distillation Networks. PhD Dissertation, University of Wisconsin, Madison, WI (1975).

Tedder, D. W., and Rudd, D. L. Parametric Studies in Industrial Distillation. Part II: Heuristic Optimization. *AIChE J.*, 24, pp. 316–322 (1978).

Westerberg, A. W. The Synthesis of Distillation-Based Separation Systems. *Comput. Chem. Engg.*, 9, pp. 421–429 (1985).

Index